SAS® Data Analytic Development

Wiley & SAS Business Series

The Wiley & SAS Business Series presents books that help senior-level managers with their critical management decisions.

Titles in the Wiley & SAS Business Series include:

Agile by Design: An Implementation Guide to Analytic Lifecycle Management by Rachel Alt-Simmons

Analytics in a Big Data World: The Essential Guide to Data Science and Its Applications by Bart Baesens

Bank Fraud: Using Technology to Combat Losses by Revathi Subramanian

Big Data, Big Innovation: Enabling Competitive Differentiation through Business Analytics by Evan Stubbs

Business Forecasting: Practical Problems and Solutions edited by Michael Gilliland, Len Tashman, and Udo Sglavo

Business Intelligence Applied: Implementing an Effective Information and Communications Technology Infrastructure by Michael Gendron

Business Intelligence and the Cloud: Strategic Implementation Guide by Michael S. Gendron

Business Transformation: A Roadmap for Maximizing Organizational Insights by Aiman Zeid

Data-Driven Healthcare: How Analytics and BI Are Transforming the Industry by Laura Madsen

Delivering Business Analytics: Practical Guidelines for Best Practice by Evan Stubbs

Demand-Driven Forecasting: A Structured Approach to Forecasting, Second Edition by Charles Chase

Demand-Driven Inventory Optimization and Replenishment: Creating a More Efficient Supply Chain by Robert A. Davis

Developing Human Capital: Using Analytics to Plan and Optimize Your Learning and Development Investments by Gene Pease, Barbara Beresford, and Lew Walker

Economic and Business Forecasting: Analyzing and Interpreting Econometric Results by John Silvia, Azhar Iqbal, Kaylyn Swankoski, Sarah Watt, and Sam Bullard

Financial Institution Advantage and the Optimization of Information Processing by Sean C. Keenan

Financial Risk Management: Applications in Market, Credit, Asset, and Liability Management and Firmwide Risk by Jimmy Skoglund and Wei Chen

Fraud Analytics Using Descriptive, Predictive, and Social Network Techniques: A Guide to Data Science for Fraud Detection by Bart Baesens, Veronique Van Vlasselaer, and Wouter Verbeke

Harness Oil and Gas Big Data with Analytics: Optimize Exploration and Production with Data Driven Models by Keith Holdaway

Health Analytics: Gaining the Insights to Transform Health Care by Jason Burke

Heuristics in Analytics: A Practical Perspective of What Influences Our Analytical World by Carlos Andre, Reis Pinheiro, and Fiona McNeill

Hotel Pricing in a Social World: Driving Value in the Digital Economy by Kelly McGuire

Implement, Improve and Expand Your Statewide Longitudinal Data System: Creating a Culture of Data in Education by Jamie McQuiggan and Armistead Sapp

Killer Analytics: Top 20 Metrics Missing from Your Balance Sheet by Mark Brown

Mobile Learning: A Handbook for Developers, Educators, and Learners by Scott McQuiggan, Lucy Kosturko, Jamie McQuiggan, and Jennifer Sabourin

The Patient Revolution: How Big Data and Analytics Are Transforming the Healthcare Experience by Krisa Tailor

Predictive Analytics for Human Resources by Jac Fitz-enz and John Mattox II

Predictive Business Analytics: Forward-Looking Capabilities to Improve Business Performance by Lawrence Maisel and Gary Cokins

Statistical Thinking: Improving Business Performance, Second Edition by Roger W. Hoerl and Ronald D. Snee

Too Big to Ignore: The Business Case for Big Data by Phil Simon

Trade-Based Money Laundering: The Next Frontier in International Money Laundering Enforcement by John Cassara

The Visual Organization: Data Visualization, Big Data, and the Quest for Better Decisions by Phil Simon

Understanding the Predictive Analytics Lifecycle by Al Cordoba

Unleashing Your Inner Leader: An Executive Coach Tells All by Vickie Bevenour

Using Big Data Analytics: Turning Big Data into Big Money by Jared Dean

Visual Six Sigma, Second Edition by Ian Cox, Marie Gaudard, and Mia Stephens.

For more information on any of the above titles, please visit www.wiley.com.

SAS® Data Analytic Development

Dimensions of Software Quality

Troy Martin Hughes

WILEY

Published by John Wiley & Sons, Inc., Hoboken, New Jersey.

Published simultaneously in Canada.

For general information on our other products and services or for technical support, please contact our Customer Care Department within the United States at (800) 762-2974, outside the United States at (317) 572-3993 or fax (317) 572-4002.

Wiley publishes in a variety of print and electronic formats and by print-on-demand. Some material included with standard print versions of this book may not be included in e-books or in print-on-demand. If this book refers to media such as a CD or DVD that is not included in the version you purchased, you may download this material at http://booksupport.wiley.com. For more information about Wiley products, visit www.wiley.com.

Library of Congress Cataloging-in-Publication Data:

Names: Hughes, Troy Martin, 1976– author.
Title: SAS data analytic development : dimensions of software quality / Troy Martin Hughes.
Description: Hoboken, New Jersey : John Wiley & Sons, 2016. | Includes index.
Identifiers: LCCN 2016021300 | ISBN 9781119240761 (cloth) | ISBN 9781119255918 (epub) | ISBN 9781119255703 (ePDF)
Subjects: LCSH : SAS (Computer file) | Quantitative research—Data processing.
Classification: LCC QA276.45.S27 H84 2016 | DDC 005.5/5—dc23
LC record available at https://lccn.loc.gov/2016021300

Cover Design: Wiley
Cover Image: © Emelyanov/iStockphoto

Printed in the United States of America

10 9 8 7 6 5 4 3 2 1

To Mom,
who dreamed of being a writer and,
through unceasing love, raised one,
and Dad,
who taught me to program
before I could even reach the keys.

Contents

Preface

Because SAS practitioners are software developers, too!

Within the body of SAS literature, an overwhelming focus on *data* quality eclipses *software* quality. Whether discussed in books, white papers, technical documentation, or even posted job descriptions, nearly all references to *quality* in relationship to SAS describe the quality of data or data products.

The focus on data quality and diversion from traditional software development priorities is not without reason. Data analytic development is software development but ultimate business value is delivered not through *software* products but rather through subsequent, derivative *data* products. In aligning quality only with data, however, data analytic development environments can place an overwhelming focus on software *functional* requirements to the detriment or exclusion of software *performance* requirements. When SAS literature does describe performance best practices, it typically demonstrates only how to make SAS software faster or more efficient while omitting other dimensions of software quality.

However, what about software reliability, scalability, security, maintainability, or modularity—or the host of other software quality characteristics? For all the SAS practitioners of the world—including developers, biostatisticians, econometricians, researchers, students, project managers, market analysts, data scientists, and others—this text demonstrates a model for software quality promulgated by the International Organization for Standardization (ISO) to facilitate the evaluation and pursuit of software quality.

Through hundreds of Base SAS software examples and more than 4,000 lines of code, SAS practitioners will learn how to define, prioritize, implement, and measure 15 dimensions of software quality. Moreover, nontechnical stakeholders, including project managers, functional managers, customers, sponsors, and business analysts, will learn to recognize the value of quality inclusion and the commensurate risk of quality exclusion. With this more comprehensive view of quality, SAS software quality is finally placed on par with SAS data quality.

Why this text and the relentless pursuit of SAS software quality? Because SAS practitioners, regardless of job title, are inherently software developers, too, and should benefit from industry standards and best practices. Software quality can and should be allowed to flourish in any environment.

OBJECTIVES

The primary goal is to describe and demonstrate SAS software development within the framework of the ISO software product quality model. The model defines characteristics of software quality codified within the *Systems and software Quality Requirements and Evaluation (SQuaRE)* series (ISO/IEC 25000:2014). Through the 15 intertwined dimensions of software quality presented in this text, readers will be equipped to understand, implement, evaluate, and, most importantly, value software quality.

A secondary goal is to demonstrate the role and importance of the software development life cycle (SDLC) in facilitating software quality. Thus, the dimensions of quality are presented as enduring principles that influence software planning, design, development, testing, validation, acceptance, deployment, operation, and maintenance. The SDLC is demonstrated in a requirements-based framework in which ultimate business need spawns technical requirements that drive the inclusion (or exclusion) of quality in software. Requirements initially provide the backbone of software design and ultimately the basis against which the quality of completed software is evaluated.

A tertiary goal is to demonstrate SAS software development within a risk management framework that identifies the threats of poor quality software to business value. Poor *data* quality is habitually highlighted in SAS literature as a threat to business value, but poor *code* quality can equally contribute to project failure. This text doesn't suggest that all dimensions of software quality should be incorporated in all software, but rather aims to formalize a structure through which threats and vulnerabilities can be identified and their ultimate risk to software calculated. Thus, performance requirements are most appropriately implemented when the benefits of their inclusion as well as the risks of their exclusion are understood.

AUDIENCE

Savvy SAS practitioners are the intended audience and represent the professionals who utilize the SAS application to write software in the Base SAS language. An advanced knowledge of Base SAS, including the SAS macro language, is recommended but not required.

Other stakeholders who will benefit from this text include project sponsors, customers, managers, Agile facilitators, ScrumMasters, software testers, and anyone with a desire to understand or improve software performance. Nontechnical stakeholders may have limited knowledge of the SAS language, or software development in general, yet nevertheless generate requirements that drive software projects. These stakeholders will benefit through the introduction of quality characteristics that should be used to define software requirements and evaluate software performance.

APPLICATION OF CONTENT

The ISO software product quality model is agnostic to industry, team size, organizational structure (e.g., functional, projectized, matrix), development methodology (e.g., Agile, Scrum, Lean, Extreme Programming, Waterfall), and developer role (e.g., developer, end-user developer). The student researcher working on a SAS client machine will gain as much insight from this text as a team of developers working in a highly structured environment with separate development, test, and production servers.

While the majority of Base SAS code demonstrated is portable between SAS interfaces and environments, some input/output (I/O) and other system functions, options, and parameters are OS- or interface-specific. Code examples in this text have been tested in the SAS Display Manager for Windows, SAS Enterprise Guide for Windows, and the SAS University Edition. Functional differences among these applications are highlighted throughout the text, and discussed in Chapter 10, "Portability."

While this text includes hundreds of examples of SAS code that demonstrate the successful implementation and evaluation of quality characteristics, it differs from other SAS literature in that it doesn't represent a compendia of SAS software best practices, but rather the application of SAS code to support the software product quality model within the SDLC. Therefore, code examples demonstrate software performance rather than functionality.

ORGANIZATION

Most software texts are organized around functionality—either a top-down approach in which a functional objective is stated and various methods to achieve that goal are demonstrated, or a bottom-up approach in which uses and caveats of a specific SAS function, procedure, or statement are explored. Because this text follows the ISO software product quality model and focuses on performance rather than functionality, it eschews the conventional organization of functionality-driven SAS literature. Instead, 15 chapters highlight a dynamic or static performance characteristic—a single dimension of software quality. Code examples often build incrementally throughout each chapter as quality objectives are identified and achieved, and related quality characteristics are highlighted for future reference and reading.

The text is divided into two parts comprising 18 total chapters:

Overview Three chapters introduce the concept of quality, the ISO software product quality model, the SDLC, risk management, Agile and Waterfall development methodologies, exception handling, and other information and terms central to the text. Even to the reader who is anxious to reach the more technically substantive performance chapters, Chapters 1, "Introduction," and 2, "Quality," should be skimmed to gleam the context of software quality within data analytic development environments.

Part I. Dynamic Performance These nine chapters introduce dynamic performance requirements—software quality attributes that are demonstrated, measured, and validated through software execution. For example, software efficiency can be demonstrated by running code and measuring run time and system resources such as CPU and memory usage. Chapters include "Reliability," "Recoverability," "Robustness," "Execution Efficiency," "Efficiency," "Scalability," "Portability," "Security," and "Automation."

Part II. Static Performance These six chapters introduce static performance requirements—software quality attributes that are assessed through code inspection rather than execution. For example, the extent to which software is modularized cannot be determined until the code is opened and inspected, either through manual review or automated test software. Chapters include "Maintainability," "Modularity," "Readability," "Testability," "Stability," and "Reusability."

Text formatting constructs are standardized to facilitate SAS code readability. Formatting is not intended to demonstrate best practices but rather standardization. All code samples are presented in lowercase, but the following conventions are used where code is referenced within the text:

- SAS libraries are capitalized, such as the *WORK library*, or the *PERM.Burrito data set within the PERM library*.
- SAS data sets appear in sentence case, such as the *Chimichanga data set* or the *WORK.Tacos_are_forever data set*.

- SAS reserved words—including statements, functions, and procedure names—are capitalized, such as the *UPCASE function* or the *MEANS procedure.*
- The DATA step is always capitalized, such as the *DATA step can be deleted if the SQL procedure is implemented.*
- Variables used within the DATA step or SAS procedures are capitalized, such as the *variable CHAR1 is missing.*
- SAS user-defined formats are capitalized, such as the *MONTHS format.*
- SAS macros are capitalized and preceded with a percent sign, such as the *%LOCKITDOWN macro prevents file access collisions.*
- SAS macro variables are capitalized, such as the *&DSN macro variable is commonly defined to represent the data set name.*
- SAS parameters that are passed to macros are capitalized, such as the *DSN parameter in the %GOBIG macro invocation.*

Acknowledgments

So many people, through contributions to my life as well as endurance and encouragement throughout this journey, have contributed directly and indirectly and made this project possible.

To the family and friends I ignored for four months while road-tripping through 24 states to write this, thank you for your love, patience, understanding, and couches.

To my teachers who instilled a love of writing, thank you for years of red ink and encouragement: Sister Mary Katherine Gallagher, Estelle McCarthy, Lorinne McKnight, Dolores Cummings, Millie Bizzini, Patty Ely, Jo Berry, Liana Hachiya, Audrey Musson, Dana Trevethan, Cheri Rowton, Annette Simmons, and Dr. Robyn Bell.

To the mentors whose words continue to guide me, thank you for your leadership and friendship: Dr. Cathy Schuman, Dr. Barton Palmer, Dr. Kiko Gladsjo, Dr. Mina Chang, Dean Kauffman, Rich Nagy, Jim Martin, and Jeff Stillman.

To my SAS spirit guides, thank you not only for challenging the limits of the semicolon but also for sharing your successes and failures with the world: Dr. Gerhard Svolba, Art Carpenter, Kirk Paul Lafler, Susan Slaughter, Lora Delwiche, Peter Eberhardt, Ron Cody, Charlie Shipp, and Thomas Billings.

To SAS, thank you for distributing the SAS University Edition and for providing additional software free of charge, without which this project would have been impossible.

Finally, thank you to John Wiley & Sons, Inc. for support and patience throughout this endeavor.

About the Author

Troy Martin Hughes has been a SAS practitioner for more than 15 years, has managed SAS projects in support of federal, state, and local government initiatives, and is a SAS Certified Advanced Programmer, SAS Certified Base Programmer, SAS Certified Clinical Trials Programmer, and SAS Professional V8. He has an MBA in information systems management and additional credentials, including: PMP, PMI-ACP, PMI-PBA, PMI-RMP, CISSP, CSSLP, CSM, CSD, CSPO, CSP, and ITIL v3 Foundation. He has been a frequent presenter and invited speaker at SAS user conferences, including SAS Global Forum, WUSS, MWSUG, SCSUG, SESUG, and PharmaSUG. Troy is a U.S. Navy veteran with two tours of duty in Afghanistan and, in his spare time, a volunteer firefighter and EMT.

CHAPTER 1

Introduction

DATA ANALYTIC DEVELOPMENT

Software development in which ultimate business value is delivered not through software products but rather through subsequent, derivative data products, including data sets, databases, analyses, reports, and data-driven decisions.

Data analytic development creates and implements software as a means to an end, but the software itself is never the end. Rather, the software is designed to automate data ingestion, cleaning, transformation, analysis, presentation, and other data-centric processes. Through the results generated, subsequent data products confer information and ultimately knowledge to stakeholders. Thus, a software product in and of itself may deliver no ultimate business value, although it is necessary to produce the golden egg—the valuable data product. As a data analytic development language, Base SAS is utilized to develop SAS software products (programs created by SAS practitioners) that are compiled and run on the SAS application (SAS editor and compiler) across various SAS interfaces (e.g., SAS Display Manager, SAS Enterprise Guide, SAS University Edition) purchased from or provided by SAS, Inc.

Data analytic software is often produced in a development environment known as end-user development, in which the developers of software themselves are also the users. Within end-user development environments, software is never transferred or sold to a third party but is used and maintained by the developer or development team. For example, a financial fraud analyst may be required to produce a weekly report that details suspicious credit card transactions to validate and calibrate fraud detection algorithms. The analyst is required to develop a repeatable SAS program that can generate results to meet customer needs. However, the analyst is an end-user developer because he is responsible for both writing the software and creating weekly reports based on the data output. Note that this example represents data analytic development within an end-user development environment.

Traditional, user-focused, or software applications development contrasts sharply with data analytic development because ultimate business value is conferred through the production and delivery of software itself. For example, when Microsoft developers build a product such as Microsoft Word or Excel, the working software product denotes business value because it is distributed to and purchased by third-party users. The software development life cycle (SDLC) continues after purchase, but only insofar as maintenance activities are performed by Microsoft, such as developing and disseminating software patches. In the following section, data analytic development is compared to and contrasted with end-user and traditional development environments.

DISTINGUISHING DATA ANALYTIC DEVELOPMENT

So why bother distinguishing data analytic development environments? Because it's important to understand the strengths and weaknesses of respective development environments and because the software development environment can influence the relative quality and performance of software.

To be clear, data analytic development environments, end-user development environments, and traditional software development environments are not mutually exclusive. Figure 1.1 demonstrates the entanglement between these environments, demonstrating the majority of data analytic development performed within end-user development environments.

The data-analytic-end-user hybrid represents the most common type of data analytic development for several reasons. Principally, data analytic software is created not from a need for software itself but rather from a

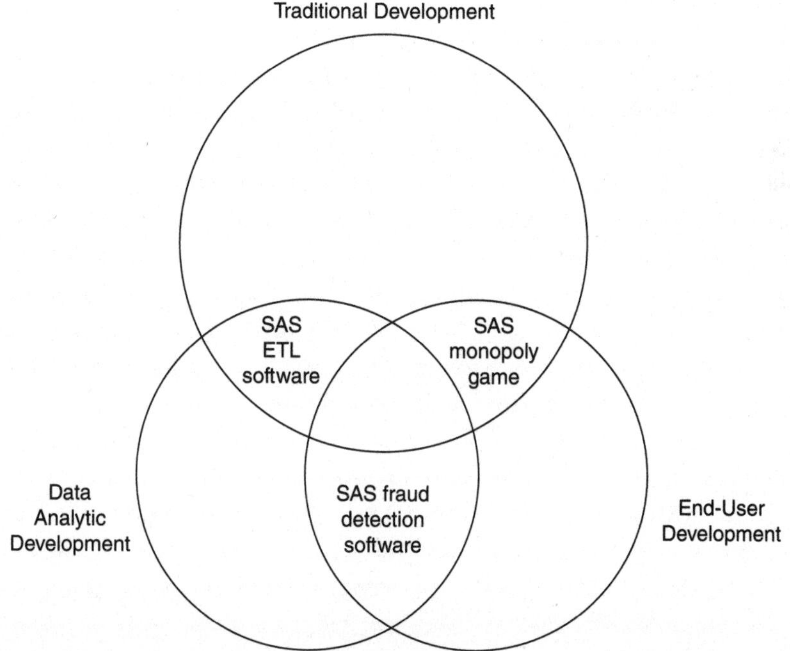

Figure 1.1 Software Development Environments

need to solve some problem, produce some output, or make some decision. For example, the financial analyst who needs to write a report about fraud levels and fraud detection accuracy in turn authors SAS software to automate and standardize this solution. SAS practitioners building data analytic software are often required to have extensive domain expertise in the data they're processing, analyzing, or otherwise utilizing to ensure they produce valid data products and decisions. Thus, first and foremost, the SAS practitioner is a financial analyst, although primary responsibilities can include software development, testing, operation, and maintenance.

Technical aspects and limitations of Base SAS software also encourage data analytic development to occur within end-user development environments. Because Base SAS software is compiled at execution, it remains as plain text not only in development and testing phases but also in production. This prevents the stabilization or hardening of software required for software encryption, which is necessary when software is developed for third-party users. For this reason, no market exists for companies that build and sell SAS software to third-party user bases because the underlying code would be able to be freely examined, replicated, and distributed. Moreover, without encryption, the accessibility of Base SAS code encourages SAS practitioners to explore and meddle with code, compromising its security and integrity.

The data-analytic-traditional hybrid is less common in SAS software but describes data analytic software in which the software does denote ultimate business value rather than a means to an end. This model is more common in environments in which separate development teams exist apart from analytic or software operational teams. For example, a team of SAS developers might write extract-transform-load (ETL) or analytic software that is provided to a separate team of analysts or other professionals who utilize the software and its resultant data products. The development team might maintain operations and maintenance (O&M) administrative activities for the software, including training, maintenance, and planning software end-of-life, but otherwise the team would not use or interact with the software it developed.

When data-analytic-traditional environments do exist, they typically produce software only for teams internal to their organization. Service level agreements (SLAs) sometimes exist between the development team and the team(s) they support, but the SAS software developed is typically neither sold nor purchased. Because SAS code is plain text and open to inspection, it's uncommon for a SAS development team to sell software beyond its organization. SAS consultants, rather, often operate within this niche, providing targeted developmental support to organizations.

The third and final hybrid environment, the end-user-traditional model, demonstrates software developed by and for SAS practitioners that is not data-focused. Rather than processing or analyzing variable data, the SAS software might operate as a stand-alone application, driven by user inputs. For example, if a rogue SAS practitioner spent a couple weeks of work encoding and bringing to life Parker Brothers' legendary board game *Monopoly* in Base SAS, the software itself would be the ultimate product. Of course, whether or not the analyst was able to retain his job thereafter would depend on whether his management perceived any business value in the venture!

Because of the tendency of data analytic development to occur within end-user development environments, traditional development is not discussed further in this text except as a comparison where strengths and weaknesses exist between traditional and other development environments. The pros and cons of end-user development are discussed in the next section.

End-User Development

Many end-user developers may not even consider themselves to be software developers. I first learned SAS working in a Veterans Administration (VA) psychiatric ward, and my teachers were psychologists, psychiatrists, statisticians, and other researchers. We saw patients, recorded and entered data, wrote and maintained our own software, analyzed clinical trials data, and conducted research and published papers on a variety of psychiatric topics. We didn't have a single "programmer" on our staff, although more than half of my coworkers were engaged in some form of data analysis in varying degrees of code complexity. However, because we were clinicians first and researchers second, the idea that we were also software developers would have seemed completely foreign to many of the staff.

In fact, this identity crisis is exactly why I use "SAS practitioners" to represent the breadth of professionals who develop software in the Base SAS language—because so many of us may feel that we are only moonlighting as software developers, despite the vast quantity of SAS software we may produce. This text represents a step toward acceptance of our roles—however great or small—as software developers.

The principal advantage of end-user development is the ability for domain experts—those who understand both the ultimate project intent and its data—to design software. The psychiatrists didn't need a go-between to help convey technical concepts to them, because they themselves were building the software. Neither was a business analyst required to convey the ultimate business need and intent of the software to the developers, because the developers were the psychiatrists—the domain experts. Because end-user developers possess both domain knowledge and technical savvy, they are poised to rapidly implement technical solutions that fulfill business needs without business analysts or other brokers.

To contrast, traditional software development environments often demonstrate a *domain knowledge divide* in which high-level project intent and requirements must be translated to software developers (who lack *domain* expertise) and where some technical aspects of the software must be translated to customers (who lack *technical* expertise in computer science or software development). Over time, stakeholders will tend to broaden their respective job roles and knowledge but, if left unmitigated, the domain knowledge divide can lead to communication breakdown, misinterpretation of software intent or requirements, and less functional or lower quality software. In these environments, business analysts and other brokers play a critical role in ensuring a smooth communication continuum among domain knowledge, project needs and objectives, and technical requirements.

Traditional software development environments do outperform end-user development in some aspects. Because developers in traditional environments operate principally as software developers, they're more likely to be educated and trained in software engineering, computer science, systems engineering, or other technically relevant fields. They may not have domain-specific certifications or accreditations to upkeep (like clinicians or psychiatrists) so they can more easily seek out training and education opportunities specific to software development. For example, in my work in the VA hospital, when we received training, it was related to patient care, psychiatry, privacy regulations, or some other medically focused discipline. We read and authored journal articles and other publications on psychiatric topics—but never software development.

Because of this greater focus on domain-specific education, training, and knowledge, end-user developers are less likely to implement (and, in some cases, may be unaware of) established best practices in software development such as reliance on the SDLC, Agile development methodologies, and performance requirements such as those described in the International Organization for Standardization (ISO) software product quality model. Thus, end-user developers can be disadvantaged relative to traditional software developers, both in software development best practices as well as best practices that describe the software development environment.

To overcome inherent weaknesses of end-user development, SAS practitioners operating in these environments should invest in software development learning and training opportunities commensurate with their software development responsibilities. While I survived my tenure in the VA psych ward and did produce much quality software, I would have improved my skills (and software) had I read fewer *Diagnostic and Statistical Manual of Mental Disorders* (DSM) case studies and more SAS white papers and computer science texts.

SOFTWARE DEVELOPMENT LIFE CYCLE (SDLC)

The SDLC describes discrete phases through which software passes from cradle to grave. In a more generic sense, the SDLC is also referenced as the *systems* development life cycle, which bears the same acronym. Numerous representations of the SDLC exist; Figure 1.2 shows a common depiction.

In many data analytic and end-user development environments, the SDLC is not in place, and software is produced using an undisciplined, laissez-faire method sometimes referred to as *cowboy coding*. Notwithstanding any weaknesses this may present, the ISO software product quality model benefits these relaxed development environments, regardless of whether the SDLC phases are formally recognized or implemented. Because the distinct phases of the SDLC are repeatedly referenced throughout this text, readers who lack experience in formalized development environments should learn the concepts associated

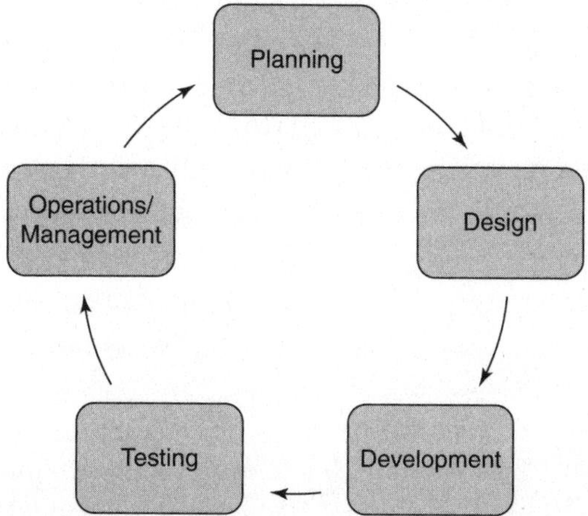

Figure 1.2 The Software Development Life Cycle (SDLC)

with each phase so they can apply them (contextually, if not in practice) to their specific environment while reading this text.

Planning Project needs are identified and high-level discussions occur, such as the "build-versus-buy" decision of whether to develop software, purchase a solution, or abandon the project. Subsequent discussion should define the functionality and performance of proposed software, thus specifying its intended quality.

Design Function and performance, as they relate to technical implementation, are discussed. Whereas planning is needs-focused, design and later phases are solutions- and software-focused. In relation to quality, specific, measurable performance requirements should be created and, if formalized software testing is implemented, a test plan with test cases should be created.

Development Software is built to meet project needs and requirements, including accompanying documentation and other artifacts.

Testing Software is tested (against a test plan using test cases and test data, if these artifacts exist) and modified until it meets requirements.

Acceptance Software is validated to meet requirements and formally accepted by stakeholders as meeting the intended functional and performance objectives.

Operation Software is used for some intended duration. Where software maintenance is required, this occurs simultaneously with operation although these discrete activities may be performed by different individuals or teams.

Maintenance While software is in operation, maintenance or modification may be required. Types of maintenance are discussed in Chapter 13, "Maintainability," and may be performed by users (in end-user development), by the original developers, or by a separate O&M team that supports software maintenance once development has concluded.

End of Life Software is phased out and replaced at some point; however, this should be an intentional decision by stakeholders rather than a retreat from software that, due to poor quality, no longer meets functional or performance requirements.

Although the SDLC is often depicted and conceptualized as containing discrete phases, significant interaction can occur between phases. For example, during the design phase, a developer may take a couple of days to do some development work to test a theory to determine whether it will present a viable solution for the software project. Or, during testing, when significant vulnerabilities or defects are discovered, developers may need to overhaul software, including redesign and redevelopment. Thus, while SDLC phases are intended to represent the focus and majority of the work occurring at that time, their intent is not to exclude other activities that would naturally occur.

SDLC Roles

Roles such as *customer, software developer, tester,* and *user* are uniquely described in software development literature. While some cross-functional development teams do delineate responsibilities by role, in other environments, roles and responsibilities are combined. An extreme example of role combination is common in end-user development environments in which developers write, test, and use their own software—bestowing them with developer, tester, user, and possibly customer credentials. SAS end-user developers often have primary responsibilities in their respective business domain as researchers, analysts, scientists, and other professionals, but develop software to further these endeavors.

A stakeholder represents the "individual or organization having a right, share, claim, or interest in a system or in its possession of characteristics that meet their needs and expectations."[1] While the following distinct stakeholders are referenced throughout the text, readers should interpret and translate these definitions to their specific environments, in which multiple roles may be coalesced into a single individual and in which some roles may be absent:

Sponsor "The individual or group that provides the financial resources, in cash or in kind, for the project."[2] Sponsors are rarely discussed in this text but, as software funders, often dictate software quality requirements.

Customer "The entity or entities for whom the requirements are to be satisfied in the system being defined and developed."[3] The customer can be the product owner (in Agile or Scrum environments), project manager, sponsor, or other authority figure delegating requirements. This contrasts with some software development literature, especially Agile-related, in which the term *customer* often represents the software end user.

SAS Practitioner/Developer These are the folks in the trenches writing SAS code. I use the terms *practitioner* and *developer* interchangeably, but intentionally chose *SAS practitioner* because it embodies the panoply of diverse professionals who use the SAS application to write SAS software to support their domain-specific work.

Software Tester Testers perform a quality assurance function to determine if software meets needs, requirements, and other technical specifications. A tester may be the developer who authored the code, a separate developer (as in software peer review), or an individual or quality assurance team whose sole responsibility is to test software.

User "The individual or organization that will use the project's product."[4] In end-user development environments, users constitute the SAS practitioners who wrote the software, while in other environments, users may be analysts or other stakeholders who operate SAS software but who are not responsible for software development, testing, or maintenance activities.

Waterfall Software Development

Waterfall software development methodologies employ a *stop-gate* or *phase-gate* approach to software development in which discrete phases are performed in sequence. For example, Figure 1.3 demonstrates that planning concludes before design commences, and all design concludes before development commences. This approach is commonly referred to as *big design up front*

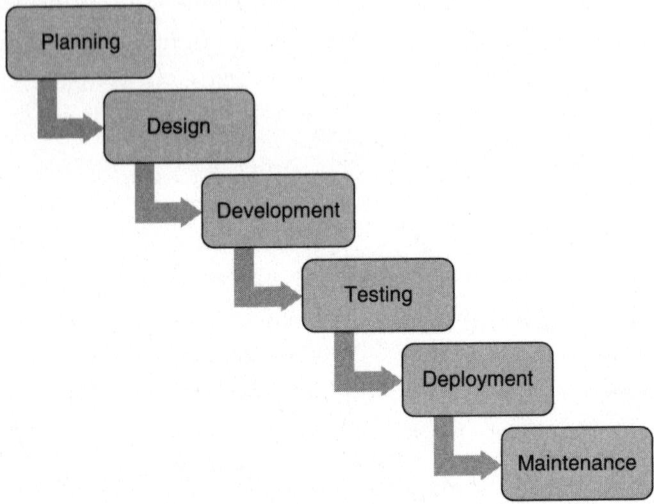

Figure 1.3 Waterfall Development Methodology

(BDUF), because the end-state of software is expected to be fully imagined and prescribed in the initial design documentation, with emphasis on rigid adherence to this design.

For years, Waterfall methodologies have been anecdotally referred to as "traditional" software development. Since the rise of Agile software development methodologies in the early 2000s, however, an entire generation of software developers now exists who (fortunately) have never had to experience rigid Waterfall development, so the "traditional" nomenclature is outmoded. Waterfall development methodologies are often criticized because they force customers to predict all business needs up front and eschew flexibility of these initial designs; software products may be delivered on time, but weeks or months after customer needs or objectives have shifted to follow new business opportunities. Thus, the software produced may meet the *original* needs and requirements, but often fails to meet all *current* needs and requirements.

Despite the predominant panning of Waterfall development methodologies within contemporary software development literature, a benefit of Waterfall is its clear focus on SDLC phases, even if they are rigidly enforced. For example, because development follows planning and design, software developers only write software after careful consideration of business needs and identification of a way ahead to achieve those objectives. Further, because all software is developed before testing, the testing phase comprehensively validates function and performance against requirements. Thus, despite its rigidity, the phase-gate approach encourages quality controls between discrete phases of the SDLC.

Agile Software Development

Agile software development methodologies contrast with Waterfall methodologies in that Agile methodologies emphasize responsible flexibility through rapid, incremental, iterative design and development. Agile methodologies follow the Manifesto for Agile Software Development (AKA the Agile Manifesto) and include Scrum, Lean, Extreme Programming (XP), Crystal, Scaled Agile Framework (SAFe), Kanban, and others.

The Agile Manifesto was cultivated by a group of 17 software development gurus who met in Snowbird, Utah, in 2001 to elicit and define a body of knowledge that prescribes best practices for software development.

Manifesto for Agile Software Development

We are uncovering better ways of developing software by doing it and helping others do it. Through this work we have come to value:

Individuals and interactions over processes and tools
Working software over comprehensive documentation
Customer collaboration over contract negotiation
Responding to change over following a plan

That is, while there is value in the items on the right, we value the items on the left more.[5]

In Agile development environments, software is produced through iterative development in which the entire SDLC occurs within a time-boxed iteration, typically from two to eight weeks. Within that iteration, software design, development, testing, validation, and production occur so that working software is released to the customer at the end of the iteration. At that point, customers prioritize additional functionality or performance to be included in future development iterations. Customers benefit because they can pursue new opportunities and business value during software development, rather than be forced to continue funding or leading software projects whose value decreases over an extended SDLC due to shifting business needs, opportunities, risks, and priorities. Figure 1.4 demonstrates Agile software development in which software is developed in a series of two-week iterations.

Agile is sometimes conceptualized as a series of miniature SDLC life cycles and, while this does describe the iterative nature of Agile development, it fails to fully capture Agile principles and processes. For example, because Agile development releases software iteratively, maintenance issues from previous iterations may bubble up to the surface during a current iteration, forcing developers (or their customers) to choose between performing necessary maintenance or releasing new functionality or performance as scheduled. Thus, a weakness ascribed to Agile is the inherent competition that exists between new development and maintenance activities, which is discussed in the "Maintenance in Agile Environments" section in Chapter 13,

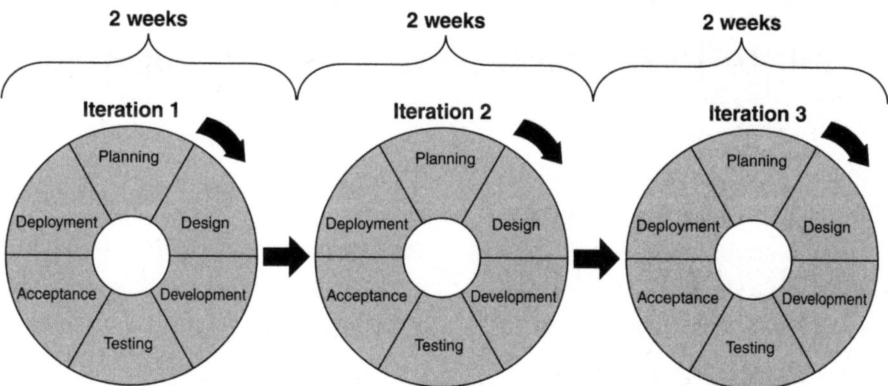

Figure 1.4 Agile Software Development

"Maintainability." This competition contrasts with Waterfall environments, in which software maintenance is performed primarily once software is in production and development tasks have largely concluded.

Despite this potential weakness, Agile has been lauded as a best practice in software development for more than a decade and has defined software development in the 21st century. Its prominence within traditional applications development environments, however, has not been mirrored within data analytic development environments. This is due in part to the predominance of end-user developers who support data analytic development and who are likely more focused on domain-specific best practices rather than software development methodologies and best practices.

Another weakness is found in the body of Agile literature itself, which often depicts an idealized "developer" archetype whose responsibilities seem focused narrowly on the development of releasable code rather than the creation of data products or participation in other activities that confer business value. In these software-centric Agile descriptions, common activities in data analytic development environments (such as data analysis or report writing) are often absent or only offhandedly referenced. Despite this myopia within Agile literature, Agile methodologies, principles, and techniques are wholly applicable to and advisable for data analytic development.

For those interested in exploring Agile methodologies, dozens of excellent resources exist, although these typically describe traditional software applications development. For an introduction to Agile methodologies to support data analytic development, I demonstrate the successful application of Agile to

SAS software development in a separate text: *When Software Development Is Your Means Not Your End: Abstracting Agile Methodologies for End-User Development and Analytic Application.*

RISK

From a software development perspective, basic risks include functional and performance failure in software. For example, a nonmalicious *threat* (like big data) exploits a software *vulnerability* (like an underlying error that limits efficient scaling when big data are encountered), causing *risk* (inefficient performance or functional failure) to business value. These terms are defined in the text box "Threat, Vulnerability, and Risk." While the *Project Management Body of Knowledge* (PMBOK) and other sources define *positive risk* as opportunity, only negative risks are discussed within this text.

THREAT, VULNERABILITY, AND RISK

Threat "A risk that would have a negative effect on one or more project objectives."[6] "A state of the system or system environment which can lead to adverse effect in one or more given risk dimensions."[7]

Vulnerability "Weakness in an information system, or cryptographic system, or components (e.g., system security procedures, hardware design, internal controls) that could be exploited."[8]

Risk "An uncertain event or condition that, if it occurs, has a positive or negative effect on one or more project objectives."[9] "The combination of the probability of an abnormal event or failure and the consequence(s) of that event or failure to a system's components, operators, users, or environment."[10]

Failure

Software failure is typically caused by threats that exploit vulnerabilities, but neither all threats nor all vulnerabilities will lead to failure. Errors (human mistakes) may lie dormant in code as vulnerabilities that may or may not be known to developers. Unknown vulnerabilities include coding mistakes (*defects*) that have not yet resulted in failure, while known vulnerabilities include coding mistakes (*latent defects*) that are identified yet unresolved. The "Paths to Failure" section in Chapter 4, "Reliability," further defines and distinguishes these terms.

For example, the following SAS code is often represented in literature as a method to determine the number of observations in a data set:

```
proc sql noprint;
   select count(*)
   into :obstot
   from temp;
quit;
```

The code is effective for data sets that have fewer than 100 million observations but, as this threshold is crossed, the &OBSTOT changes from numeric to scientific notation. For example, a data set having 10 million observations is represented as 10000000, while 100 million observations is represented as 1E8. To the SAS practitioner running this code to view the number of observations in the log, this discrepancy causes no problems. However, if a subsequent procedure attempts to evaluate or compare &OBSTOT, runtime errors can occur if the evaluated number is in scientific notation. This confusion is noted in the following output:

```
%let obstot=1E8;
%if &obstot<5000000 %then %put LESS THAN 5 MILLION;
%else %put GREATER THAN 5 MILLION;

LESS THAN 5 MILLION
```

Obviously 100 million is not less than 5 million but, because of two underlying errors, a vulnerability in the code exists. The vulnerability can be easily eliminated by correcting either of the two errors. The first error can be eliminated by changing the assignment of &OBSTOT to include a format that will accommodate larger numbers, as demonstrated with the FORMAT statement. The second error can be eliminated by enclosing the numeric comparison inside the %SYSEVALF macro function, which interprets 1E8 as a number rather than text. Both solutions are demonstrated and either correction in isolation eliminates the vulnerability and prevents the failure.

```
proc sql noprint;
   select count(*) format=15.0
   into :obstot
   from temp;
quit;

%if %sysevalf(&obstot<5000000) %then %put LESS THAN 5 MILLION;
%else %put GREATER THAN 5 MILLION;

GREATER THAN 5 MILLION
```

Because the failure occurs only as the number of observations increases, this can be described as a scalability error. The SAS practitioner failed to imagine (and test) what would occur if a large data set were encountered. But if the 100 million observation threshold is never crossed, the code will continue to execute without failure despite still containing errors. This error type is discussed further in the "SAS Application Thresholds" section in Chapter 9, "Scalability."

Developers often intentionally introduce vulnerabilities into software. For example, a developer familiar with the previous software vulnerability (exploited by the threat of big data) might choose to ignore the error in software designed to process data sets of 10,000 or fewer observations. Because the risk is negligible, it can be accepted, and the software can be released as is—with the vulnerability. In other cases, threats may pose higher risks, yet the risks are still accepted because the cost to eliminate or mitigate them outweighs the benefit.

Unexploited vulnerabilities don't diminish software reliability because no failure occurs. For example, the previous latent defect is never exploited because big data are never encountered. However, vulnerabilities do increase the risk of software failures; therefore, developers should be aware of specific risks to software. In this example, the risk posed is failure caused by the accidental processing of big data within the SQL procedure. When vulnerabilities are exploited and runtime errors or other failures occur, software reliability is diminished. The risk register, introduced in the next section, enables SAS practitioners to record known vulnerabilities, expected risks, and proposed solutions to best measure and manage risk level for software products.

Risk Register

A *risk register* is a "record of information about identified risks."[11] Risk is an inherent reality of all software applications, so risk registers (sometimes referred to as *defect databases*) document risks, threats, vulnerabilities, and related information throughout the SDLC. Developers and other stakeholders should decide which performance requirements to incorporate in software, but likely will not include all performance requirements in all software. While vulnerabilities will exist in software, it's important they be identified, investigated, and documented sufficiently to demonstrate the specific risks they pose to software operation.

A risk register qualitatively and quantitatively records known vulnerabilities and associated threats and risks to software function or performance, and can include the following elements:

- Description of vulnerability
- Location of vulnerability
- Threat(s) to vulnerability
- Risk if vulnerability is exploited
- Severity of risk
- Probability of risk
- Likelihood of discovery
- Cost to eliminate or mitigate risk
- Recommended resolution

Some risk registers, as demonstrated, are organized at the defect level while others are organized at the threat or risk level. Vulnerability-level risk registers are common in software development because while many threats lie outside the control of developers, programmatic solutions can often be implemented to eliminate or mitigate specific vulnerabilities. Moreover, general threats like "big data" can exploit numerous, unrelated vulnerabilities within a single software product.

Table 1.1 depicts a simplified risk register for two of the errors mentioned in the code. The risk severity, risk probability, likelihood of risk discovery, and cost to implement solution are demonstrated on a scale of 1 to 5, in which 5 is more severe, more likely to occur, less easy to discover, and more costly to repair.

The first and second risks describe separate vulnerabilities, each exploited by the threat of data sets containing 100 million or more observations. Despite the high severity (5) if the threat is encountered, the likelihood is low (1) because these file sizes have never been encountered in this environment.

Table 1.1 Sample Risk Register

Num	Vulnerability	Location	Risk	Risk Severity	Risk Probability	Risk Discovery	Risk Cost
1	%SYSEVALF should be used in evaluation	less-than operator	scientific notation won't be interpreted correctly	5	1	5	1
2	no format statement	SELECT statement of PROC SQL	scientific notation won't be interpreted correctly	5	1	5	1

If these two factors alone were considered, a development team might choose to accept the risk and release the software with the vulnerabilities, given their unlikelihood of occurrence. However, because the likelihood of discovery is low (5)—as no warning or runtime error would be produced if the threat were encountered—and because the cost to implement a remedy (modifying one line of code) is low (1), the development team might instead decide to modify the code, thus eliminating the risk rather than accepting it.

Not depicted, the recommended solution describes the path chosen to manage the risk—often distilled as *avoidance, transfer, acceptance,* or *mitigation,* and described in the following section. The recommended solution may contain a technical description of how the risk is being managed. For example, if a risk is being eliminated, the resolution might describe programmatically how the associated threat is being eliminated or controlled or how the associated vulnerability is being eliminated so it can't be exploited by the threat.

Risk Management

Risk management describes the "coordinated activities to direct and control an organization with regard to risk."[12] ISO further defines a *risk management framework* as a "set of components that provide the foundations and organizational arrangements for designing, implementing, monitoring, reviewing and improving risk management throughout the organization." The risk to operational software discussed throughout this text is software failure—functional- or performance-related; risk management accepts failure as an inevitable reality and strives to overcome it sufficiently to deliver an acceptable level of risk to stakeholders. Failure does not necessarily denote runtime errors but describes software that fails to meet functional or performance requirements.

The risk register typically includes not only theoretical vulnerabilities within software that have never been exploited but also risks that have materialized through actual software failure. These latter failures will also be demonstrated in the failure log (introduced in the "Failure Log" section in Chapter 4, "Reliability") if it is utilized. Risk resolution doesn't imply that risk is being eliminated; in many cases, once vulnerabilities are identified, a decision is made to accept the risk and to leave the code unchanged. Typical risk resolution patterns include:

Risk Avoidance "Informed decision not to be involved in, or to withdraw from, an activity in order not to be exposed to a particular risk."[13] For example, if your software is broken, stop using it.

Risk Sharing "Form of risk treatment involving the agreed distribution of risk with other parties."[14] For example, some SLAs state that if the power fails (causing production software to fail), the responsibility is transferred to the organization, and is not borne by the O&M team, developers, or users. *PMBOK and other sources use the alternative term "risk transfer."*[15]

Risk Retention "Acceptance of the potential benefit of gain, or burden of loss, from a particular risk."[16] Stakeholders retain or accept a risk when they evaluate and decide it's prudent not to resolve the risk. *PMBOK and other sources use the alternative term "risk acceptance."*[17]

Risk Mitigation "A course of action taken to reduce the probability of and potential loss from a risk factor."[18] For example, a programmatic solution may be implemented to lessen or eliminate a vulnerability in software, such as by removing a defect.

A risk register is important to production software because it provides a common framework for internal stakeholders to gauge the relative strength or weakness of software. In some cases, separate quantitative priorities are established in a single register; therefore, a developer might prioritize fixing one element of code while a customer might prioritize fixing a different element, allowing stakeholders effectively to vote for how to best improve software. By evaluating the aggregate risk of software, all stakeholders can gain a clearer understanding of its quality. And, because the risk register is meant to be a living, flexible document, if aggregate software risk becomes too high over time, risks that were once accepted can later be mitigated or eliminated to restore risk to an acceptable level.

WHAT'S NEXT?

In the next chapter, technical and nontechnical definitions of quality are discussed, including how personal perspective can influence the interpretation of quality. The ISO software product quality model is introduced and its benefit to SAS practitioners and development teams demonstrated. The roles of functional and performance (i.e., non-functional) requirements are demonstrated, and the importance of clear, measurable technical requirements is highlighted.

NOTES

1. ISO/IEC/IEEE 24765:2010. *Systems and software engineering—Vocabulary*. Geneva, Switzerland: International Organization for Standardization, International Electrotechnical Commission, and Institute of Electrical and Electronics Engineers.
2. Project Management Institute, 2013. *A guide to the project management body of knowledge (PMBOK guide)* (5th ed.). Newtown Square, PA: Project Management Institute.
3. IEEE Std 1233-1998. *IEEE guide for developing system requirements specifications*. Geneva, Switzerland: Institute of Electrical and Electronics Engineers.

4. Project Management Institute, 2013.
5. The Agile Alliance, *The Manifesto for Agile Software Development*. The Agile Alliance. Retrieved from www.agilealliance.org/the-alliance/the-agile-manifesto.
6. Project Management Institute, 2013.
7. ISO/IEC/IEEE 24765:2010.
8. IEEE Std C37.115-2003. *IEEE standard test method for use in the evaluation of message communications between intelligent electronic devices in an integrated substation protection, control and data acquisition system*. Geneva, Switzerland: Institute of Electrical and Electronics Engineers.
9. Project Management Institute, 2013.
10. IEEE Std 829-2008. *IEEE standard test method for use in the evaluation of message communications between intelligent electronic devices in an integrated substation protection, control and data acquisition system*. Geneva, Switzerland: Institute of Electrical and Electronics Engineers.
11. ISO/Guide 73:2009. *Risk management—Vocabulary*. Geneva, Switzerland: International Organization for Standardization.
12. Id.
13. Id.
14. Id.
15. Project Management Institute, 2013.
16. ISO/Guide 73:2009.
17. Project Management Institute, 2013.
18. ISO/IEC/IEEE 24765:2010.

CHAPTER 2
Quality

"It's baby alpaca. You like?"

"No me gusta." (I don't like it.)

"But it's very fine cloth. Very soft ... come feeeeeel!" There's nothing like having dusty alpaca thrust in your face, seconds after you've summarily rejected it. I sprang from my crouched position in the market stall, narrowly escaping the blow.

"No necessitoooo!" (No, I don't need it!)

"Que buscas, papi?" (What are you looking for?)

"Buscando por algo mas antigua." (I'm looking for something older ...)

Temuco, Chile. No, no one was trying to sell me a baby alpaca, but I was in yet another street-side *mercado* (market) perusing textiles and trinkets amid cloistered, canopied stalls. The vendors were Mapuche, an Andean people famed for, among other things, the intricate textiles they've meticulously produced for more than a millennium. With my broken Spanish and their broken English, we bartered through the afternoon as I was shown one *manta* (Andean Spanish for poncho) after another, each woven from alpaca, wool, or cotton. To the Mapuche, *baby alpaca* is code for *quality* but, to me, it means only *you're going to pay a lot for this manta!*

"Ay otras mas antiguas, con diseños o figuras?" (Anything older, with designs or figures?) As I disapprovingly rolled my eyes, thumbing over and pulling at the poor stitching of the last manta I'd been handed, Maria realized she wasn't dealing with the average tourist. You see, I'm the *manta* ringer ...

I'd been backpacking for months throughout Central and South America, touring textile museums in Cusco, La Paz, Santiago, Valparaiso, and elsewhere, and spending my days filming and interviewing indigenous artisans as they worked and wove, observing their techniques and learning how they value and distinguish textile quality.

"Baby alpaca?!"

I spun around, wincing at the tired phrase, but thankfully the attention of the *manta*-mongers had shifted to newly arrived gringos who immediately began exclaiming over the first poncho they were shown, "Oh, that's so soft!"

I audibly groaned. Suckers.

It was a machine-made poncho that definitely wasn't baby alpaca, would probably disintegrate when washed, and whose colors would definitely bleed—but they were satisfied and departed just as quickly as they came.

Maria returned to me with undivided attention and a small stack of antique *mantas* that had been recessed from view. Beautiful stitching, natural colors, intricate designs, historic figures: I'd finally found the high-quality textiles for which the Mapuche are famous and which had enticed me to Patagonia.

■ ■ ■

My pursuit of quality ponchos has remarkable similarities to the pursuit of software quality, excepting of course the distinction that I'm a *user* of ponchos but a *producer* of software. Several general tenets about quality can be inferred from the previous scenario:

Quality, like beauty, is in the eye of the beholder, in that one person's quality may not be another's. Maria believed (or at least was trying to convey) that "baby alpaca" denoted quality, while I valued stitching and classic design. Software quality also doesn't represent a tangible construct that can be unambiguously identified when separated from stakeholder perspective or intent.

Quality comprises many disparate characteristics, not all of which are valued by any one person. Characteristics can describe functionality, which first and foremost for a poncho is to keep you warm and dry, or can describe performance, such as durability over time or the ability to be washed without colors bleeding. Software quality models also include functional and performance dimensions that specify not only what the software should accomplish but also how well it should accomplish it. Just like a poorly stitched poncho, software can sometimes seem to meet functional needs, but unravel over time or under duress.

Quality (or the lack thereof) isn't always recognized and, without an organized understanding of the product or product type, it may be impossible to accurately determine (and prioritize) quality. I had studied textiles extensively and had objective criteria against which I could measure quality, but the tourists I encountered were not similarly knowledgeable. Dozens of characteristics that describe software quality exist in quality models, and the extent to which these dimensions and their interactions are understood can provide a framework and vocabulary that facilitate a defensible, standardized assessment of quality against internationally recognized criteria.

Quality standards should be measurable, in that while quality standards are beneficial, they are typically useless if they can't be measured and validated against a baseline. I can't distinguish baby alpaca fibers from adult alpaca fibers, so being told I'm paying a premium for something I can't even identify provides no value and is nonsensical. Software development similarly benefits from measurable requirements that not only guide development itself, but also enable demonstration of software quality to customers and stakeholders. Without measurable standards, how do SAS practitioners even know when they've completed their software or if it meets product intent? They don't!

Quality isn't always being pursued, because it has inherent trade-offs. As the tourists were shopping, I actually did interject and try to lead them to a "higher quality"—at least in my mind—poncho, to which they politely explained their rationale. The poncho was for a child so durability and performance were not valued—they just needed *something.* Moreover, they were running to catch a bus and had only a minute to shop, thus expedience outweighed product quality. Trade-offs always exist in software development, and the decision to choose one construct over another should be intentional, not made haplessly or out of ignorance. Function and performance define quality but inherently compete for a stake therein. Software quality, moreover, competes with project schedule and project cost, so many development projects may have goals that outpace quality, or embrace requirements that reference only function but not performance. The decision to exclude dimensions of quality is often justified but should be made judiciously.

The lack of quality incurs inherent risks, whether quality is negligently omitted or intentionally deprioritized. As the tourists were leaving, I hollered after them "Make sure you wash that thing in Woolite so it doesn't disintegrate!" I was alerting them to the risk posed by the threat of a rough washer that could exploit weaknesses (vulnerabilities) in their cheap poncho. Risks are inherent in all software development projects, but, especially where other constructs are prioritized over quality, developers should understand software vulnerabilities and the specific risks they pose. In this way, the benefits and business value of quality inclusion can be weighed against the risks of quality exclusion.

DEFINING QUALITY

In a general, nontechnical sense, *quality* can be defined as "how good or bad something is or a high level of value or excellence."[1] A tourist might exclaim "That's a quality poncho!" while having no knowledge of pre-Columbian textiles. In a technical sense, however, quality is always assessed with respect to needs or requirements. The Institute of Electrical and Electronics Engineers (IEEE) defines *quality* as "the degree to which a system, component, or process meets specified requirements."[2] The International Organization for Standardization (ISO) defines *software quality* as the "degree to which the software product satisfies stated and implied needs when used under specified conditions."[3] This technical distinction is critical to understanding the context in which *quality* is discussed throughout this text, since software quality cannot be evaluated without knowledge of the functional and performance intent of software. Quality cannot be judged in a vacuum.

An important aspect of software quality is its inclusion of both functional and performance requirements. Functional requirements specify the expected behaviors and objective of software, or *what* it does. Performance requirements (once termed *nonfunctional requirements*) specify characteristics or attributes of software operation, such as *how well* the expected behaviors are performed. For example, if SAS analytic software is designed to produce an HTML report, the content, accuracy, completeness, and formatting of that report would be specified as functional requirements, as would any data cleaning operations, transformations, or other processes. But the ability of the software to perform efficiently with big data (i.e., scalability) or the ability of the software to run across both Windows and UNIX environments (i.e., portability) instead demonstrate aspects of its performance.

To develop software effectively, SAS practitioners should understand the dimensions of quality that may be required, whether inferred through implied needs or prescribed through formal requirements. Quality begins during software planning and design, when needs are identified and technical requirements are specified. Without requirements, developers won't know whether they are developing a reliable, enduring product or an ephemeral solution intended to be run once by a single user for a noncritical system. Moreover, without an accepted software quality model that defines dimensions of quality and the ways in which they interact, it's difficult to demonstrate when software development has been completed and whether its ultimate intent was achieved.

Avoiding the Quality Trap

Quality gets bandied about so frequently in everyday conversation that it's important to differentiate its very specific definition and role within software development. In general, when you speak of *quality* in isolation such as *that's a quality alpaca*, you're expressing appreciation for product or service excellence. And, similarly, when you discover a *high-quality* Thai restaurant, you want to dine there often because the drunken noodle is delicious! In this sense, *quality* can represent an isolated assessment of the food, or it could represent a comparison of the *high-quality* Thai food to other Thai restaurants whose food was not exceptional. Thus, in assessing or describing *quality*, a comparison may intrinsically exist, but may be inferred rather than specified outright.

As a software *user*, I find this comparative use of *quality* is common. For example, I use Gmail because it offers greater functionality and performance than other email applications. In terms of functionality, I appreciate the threaded emails, overall layout, and ability to thoroughly customize the environment. In terms of performance, the high-availability service is extremely reliable and I trust the security of the Gmail servers and infrastructure. But in exclaiming that Gmail is high quality, I'm inherently making a comparison between this email application and email applications that I've previously used that offered less functionality or performance. Thus, a software *user* often determines quality based on his needs and his requirements and possibly a comparison of how well one software application meets those needs as compared to another.

As a software *developer*, however, I recognize that software quality is consistently defined by organizations like ISO and IEEE, which specify that quality must always be assessed against needs or requirements, rather than in isolation or comparatively to other software. These definitions are in line with the view espoused in product and project management literature: that quality represents "the degree to which a set of inherent characteristics fulfill requirements."[4] Without knowledge of the intent with which software was created, the software function and performance can be described, but it's impossible to determine software quality—this is the quality trap to which we fall victim, due to the chasm between generalized and industry-specific usage and definitions of quality.

So if the term *quality* carries so much baggage, how are you supposed to assess software quality or even discuss it? How do you tell a coworker that his software sucks? Software quality models are recognized as an industry standard because of the organized, standardized nomenclature that fully describes software performance. The primary objective of this text is to demonstrate the ISO software product quality model, its benefits, its limitations, and its successful implementation into SAS data analytic development and into the lexicon with which we discuss software and software requirements.

A New Quality Vocabulary

To begin a rudimentary assessment of software quality, imagine the following SAS data set exists:

```
data sample;
   length char1 $20 char2 $20;
   char1="I love SAS";
   char2="SAS loves me";
run;
```

You're now asked to assess the quality of the following SAS code, which simulates a much larger data transformation module within extract-transform-load (ETL) or other software:

```
data uppersample;
   set sample;
   char1=upcase(char1);
   char2=upcase(char2);
run;
```

So is the code high quality, low quality, or somewhere in the middle? In other words, would you rehire the SAS practitioner who wrote it, or fire him for being a screwup? Therein lies the quality trap. Given ISO and IEEE quality definitions, it's impossible to know whether this represents high- or low-quality software without first understanding the objective, needs, and requirements that spawned it.

You subsequently find an email from a manager requesting that the previous code be developed:

> Please transform the Sample data set to uppercase; I need this in 15 minutes for a presentation.

At this point, you can understand the ultimate project need and can assess that the code is high quality because it meets all requirements and was developed quickly. The only functional requirement given was capitalization, and the only performance requirement that can be inferred is speed of processing, since the manager needed the code written and executed quickly. It's not fancy, but it works and delivers the business value specified by the customer.

But now imagine that the manager instead had sent the following alternative email to request the software:

> Please create a module that can be used to capitalize all character variables within the Sample data set. The module should be able to be invoked as a macro from inside a DATA step and should specify the data set that is being capitalized as a parameter. The character variables should be dynamically identified and capitalized without manual reference to variable names.

This alternative requirements definition specifies software with identical functionality (at least when applied to the Sample data set), but is more dynamic and complex. The manager requested a modular solution, so a SAS macro probably should be created. Moreover, given that the macro requires that the data set and variable names be dynamic, the module should be able to be reused to perform the same function in other contexts and on future data sets. This would be of tremendous advantage if the same team later required hundreds of variables to be capitalized in a larger data set, which would take much longer to code manually.

Given the alternative more complex requirements, the original transformation code developed could be assessed to be of poor quality. It meets the ultimate functional objective (all variables are capitalized), but lacks performance requirements that specified it be modular, dynamic, and reusable. The SAS solution that does meet these additional performance requirements is demonstrated in the "Reusability Principles" section in Chapter 18, "Reusability." The increased code complexity also demonstrates the correlation between the inclusion of quality characteristics and increased complexity and development time, an inherent tradeoff in software development.

Dimensions of Quality

Given the specific use of *quality* within software development circles, the need still arises to describe software that exhibits a high (or low) degree of quality characteristics (i.e., performance requirements), irrespective of software needs or requirements. For example, back to the original question: If I'm unaware of the email that the manager sent the developer prescribing software requirements, how do I describe the quality of the code in a vacuum?

If backed into a corner, I might describe the code as *relatively low quality*, essentially making a comparison between *this* code and theoretical code that could have been written having the same functionality. More likely, I might state that the code is not *production software*, essentially evaluating the code against theoretical requirements that would likely exist if the code underpinned critical infrastructure, had dependencies, had multiple users, or was expected to have a long lifespan. However, a tremendous benefit of embracing a software quality model is that you can discuss the performance of software without dragging poor old *quality* into the fray. For example, I can instead say that the code lacks stability, reusability, and modularity—which concisely evaluates the software irrespective of needs and requirements.

The lexicon established through the adoption of software quality models benefits development environments because the derived vocabulary is essential to software performance evaluation. Only with unambiguous technical requirements that implement consistent nomenclature can quality be incorporated and later measured in software. It's always more precise and useful to state that software "lacked modularity and reusability" than to vaguely assert only that it "lacked quality." Even when stakeholders choose not to implement dimensions of software quality in specific products, they will have gained a recognized model and vocabulary about which to discuss quality and consider its inclusion.

Data Quality and Data Product Quality

Within SAS literature, when *quality* is referenced, it nearly always describes *data quality* and, to a lesser extent, the quality of data products such as analytic reports. In fact, in data analytic development, *quality* virtually never describes software or code quality. This contrasts sharply with traditional software development literature, which describes characteristics of software quality and the methods through which that quality can be designed and developed in software.

This shift in quality focus is somewhat expected, given that data analytic development environments usually consider a data product to be the ultimate product and source of business value, as contrasted with traditional software development environments in which software itself is the ultimate product. The misfortune is that by allowing data quality and data product quality to eclipse (or replace) code quality, data analytic development environments unnecessarily fail to incorporate performance requirements throughout the SDLC. Thus, code quality takes a backseat to data quality during requirements definitions, design, development, testing, acceptance, and operation.

If the software that undergirds data products lacks quality, though, how convincingly can those data products really be shown to demonstrate quality? To illustrate this all-too-common paradox, I'll borrow an analogy from a friend who, while working for the Department of Defense (DoD), encountered a three-star general who was adamant that some trendlines in a SAS-produced analytic report were "not purple enough." He was focused on the quality of the data product, to the total exclusion of interest in the quality of the underlying software. Was the software reliable? Was it robust, or could it be toppled by known vulnerabilities? Was it extensible in that it could be easily modified and repurposed if the general had a tactical need that required subtle variation of the analysis? No, none of these questions about code quality were asked, only "Can you make this more purple?!"

This text doesn't aim to diminish the importance of data quality and data product quality. Rather, it places code quality on par with data quality, with the understanding that any house built upon a poor foundation will fail.

SOFTWARE PRODUCT QUALITY MODEL

The software product quality model is a "defined set of characteristics, and of relationships between them, which provides a framework for specifying quality requirements and evaluating quality."[5] The ISO model, demonstrated in Figure 2.1, lists eight components and 38 subcomponents.

An important distinction of the ISO model is that functionality is viewed as equivalent to seven other quality components. Because functionality varies by each specific SAS software product, functionality is not a focus of this text despite lying at the core of every software project. Thus, unless explicitly stated, all SAS code examples are assumed to meet software functional requirements. Many of the quality model components and subcomponents of the software product quality model are described later in detail and represent individual chapters.

The software product quality model presented in Figure 2.1 differs slightly from but does not contradict the model inherently present in the organization of this text. The external quality dimensions, referenced as dynamic performance requirements, include reliability, recoverability, robustness, execution efficiency, efficiency, scalability, portability, security, and automation. The internal quality dimensions, referenced as static performance requirements, include maintainability, modularity, readability, testability,

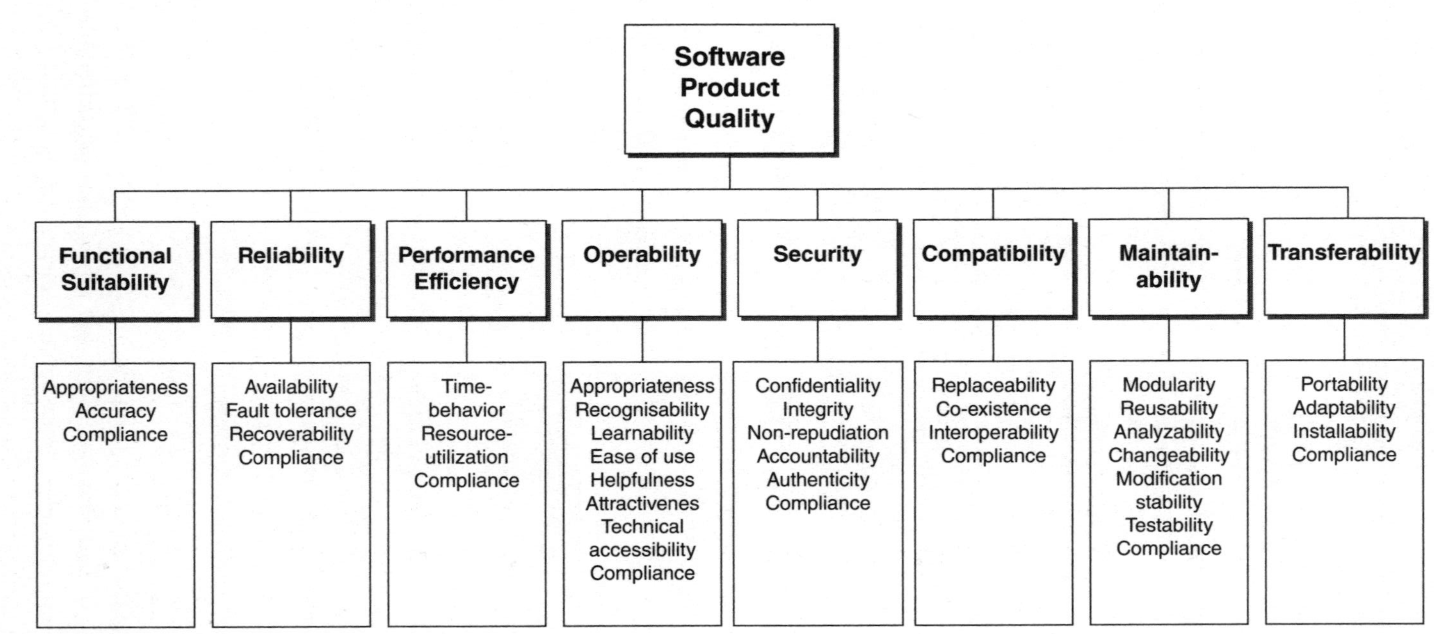

Figure 2.1 ISO Software Product Quality Model (reproduced from ISO/IEC 25000:2014).

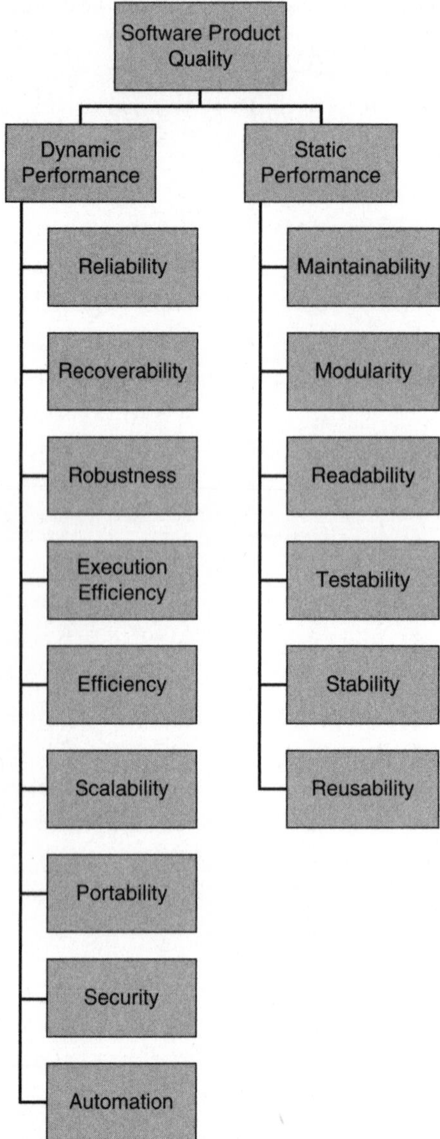

Figure 2.2 Software Quality Model Demonstrated in Chapter Organization

stability, and reusability. Figure 2.2 demonstrates the structure of this text as it fits within the ISO software product quality model.

Two quality dimensions—stability and automation—included in this text are not in the ISO model, so an explanation is warranted. Software *stability* describes software that resists maintenance and can be run without or with minimal modifications. In traditional software development environments, because software is produced for third-party users, developers have no access to their software once it has been tested, validated, and placed into operation. Thus, testing is valued because all defects and vulnerabilities should be eliminated before software is released. Failure to test software will necessitate that patches or updates are distributed to users, which can be costly and time-consuming and cause users to lose faith in software.

Base SAS software, unlike many software applications, can be run from an interactive mode in which the SAS application is opened manually, code is executed manually, and the log is reviewed manually. While sufficient for some purposes and environments, the interactive nature of SAS code encourages developers to continually modify code simply because they can, thus discouraging software stability. For this reason, an entire chapter distills the benefits of software stability, including its role as a prerequisite to code testing, reuse, extensibility, and automation.

Software *automation* describes software that can be run without (or with minimal) human intervention, and, in traditional software development, this is required because users expect to interact with an executable program rather than raw code they have to compile. The SAS interactive mode may be sufficient for executing some production software but, when SAS software must be reliably and regularly run, automation (and subsequent scheduling of batch jobs) can best achieve this objective. Automation of SAS software also includes spawning SAS sessions through software that improves performance through parallel processing.

External Software Quality

External software quality is "the degree to which a software product enables the behavior of a system to satisfy stated and implied needs when the system including the software is used under specified conditions."[6] External quality characteristics are those that can be observed by executing software. For example, software is shown to be reliable because it does not fail. Software is shown to be scalable because it is able to process big data effectively without failure or inefficiency.

External software quality is sometimes referred to as the *black-box* approach to quality, because the assessment of quality is made through examination of software execution alone, rather than inspection of the code. Thus, a black-box view of software efficiency can assess metrics such as execution time or resource utilization, but cannot determine whether SAS technical best practices that support efficient execution were implemented. Black-box testing is discussed in the "Functionality Testing" section in Chapter 16, "Testability."

In traditional software applications, users have no access to the underlying code and are only able to assess external software quality through a black-box approach. Thus, while internal software quality characteristics are important to building solid, enduring code that will be more easily maintained, only external software quality characteristics can be demonstrated to customers, users, and other stakeholders who may not have access to code. Because SAS software users often do have access to underlying code, the distinctions between black- and *white-box testing* (that is, testing internal software quality, which is further discussed in a later section) are less significant in SAS software development than in other languages.

Functionality

Functional suitability is "the degree to which the software product provides functions that meet stated and implied needs when the software is used under specified conditions."[7] Throughout this text, however, *functionality* is referenced rather than *functional suitability*, in keeping with prior ISO standards and the majority of literature.[8] Moreover, *functional requirements* describe technical rather than performance specifications within requirements documentation—for example, specifying the exact hue of purple on analytic reporting to satisfy the general.

The central characteristic of external software quality is functionality, without which software would have no purpose. Functionality is omitted from this and many other software development texts because it describes specific software intent while performance quality characteristics can be generalized. With functionality removed, the remaining external software characteristics comprise dynamic performance attributes, such as reliability or efficiency, but without function, there is nothing to make reliable or efficient.

Dynamic Performance Requirements

A *performance requirement* is "the measurable criterion that identifies a quality attribute of a function or how well a functional requirement must be accomplished."[9] Dynamic performance requirements are performance attributes that can be observed during software execution. In other words, they represent all characteristics of external software quality *except* functionality, as demonstrated

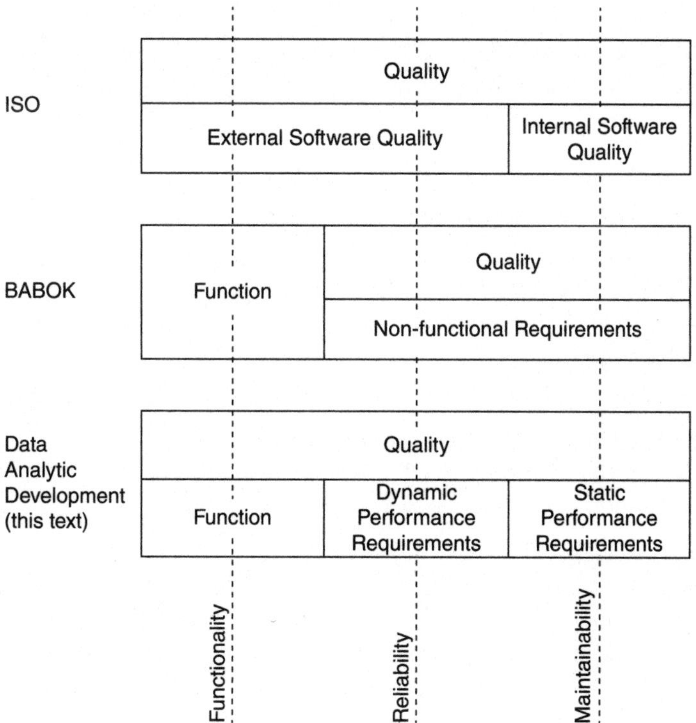

Figure 2.3 Interaction of Software Quality Constructs and Dimensions

in Figure 2.3. *Dynamic* refers to the fact that software must be executing (i.e., in motion) to be assessed, whereas static performance attributes must be assessed through code inspection when software is at rest.

Throughout this text, *performance requirements* are sometimes referred to as *performance attributes*, especially where the use is intended to demonstrate a quality characteristic that may not be required by specific software. For example, if a SAS practitioner is considering implementing scalability and security principles but decides to omit security from a software plan, security would be referenced as a *performance attribute*—not a *performance requirement*—to avoid confusion since the quality characteristic was not actually required by or implemented into the software. In general, however, the terms are interchangeable.

Because dynamic performance requirements can be observed by all stakeholders, they are traditionally more valued than static requirements, and are thus more readily prioritized into software design and requirements. For example, a customer can more easily comprehend the benefits of faster software because speed is measurable. However, unless the same customer has a background in software development, he may have a more difficult time comprehending the benefits of incorporating modularity or testability into software because these attributes cannot be directly measured or their effects observed during execution.

Internal Software Quality

Internal software quality is defined as the "degree to which a set of static attributes of a software product satisfy stated and implied needs when the software product is used under specified conditions."[10] These characteristics can only be observed through static examination of code and, sometimes in the case of readability, through additional software documentation that may exist. Testability sometimes can also be assessed through a formalized test plan and test cases, as discussed in Chapter 16, "Testability."

Internal software quality is sometimes referred to as the *white-box* (or *glass-box*) approach to quality, because the assessment of quality is made through code inspection rather than execution. A white-box view of software reusability assesses to what extent software can be reused based on reusability principles, but requires that code be reviewed either manually or through third-party software that parses code. Because users have no access to underlying code in traditional software applications that are encrypted, they have no way to assess internal software quality; if they lack technical experience, they may even have no awareness of its concepts.

Static Performance Requirements

Static performance requirements describe internal software quality, such as maintainability, modularity, or stability. Due to the inherent lack of visibility of internal software quality characteristics, in some organizations and for some software projects, it may be more difficult to encourage stakeholders to value and prioritize these characteristics into software as compared with the more observable dynamic performance requirements. For example, it's easy to demonstrate to a customer the benefits of increased speed, but to demonstrate the benefits of increased reusability requires a conversation not only about reusability but also software reuse.

In addition to being more difficult to observe, static performance requirements also provide less immediate gratification to stakeholders and instead represent an investment in the future of a software product. For example, when dynamic performance is improved to make software faster, the effect is not only *observable* but also *immediate*. When static performance is improved to make software more readable or modular, the changes are not observable (through software performance), and moreover, the improvement is only beneficial the next time the software needs to be inspected or modified. Modularity can facilitate improved dynamic performance, as discussed throughout Chapter 7, "Execution Efficiency," but, in general, static performance requirements are neither observable nor immediate. As the anticipated lifespan of software increases, however, static performance requirements become increasingly more valuable because they improve software maintainability, one of the most critical characteristics in promoting software longevity.

Commingled Quality

The organization of individual dimensions of software quality is straightforward; however, placement of quality dimensions within superordinate structures (such as quality models) can become complex due to nomenclature variations. For example, external software quality comprises both functional and performance requirements, a dichotomy commonly made in software literature. This is reflected in the ISO software product quality model, which subsumes functional suitability (i.e., functionality) under software product quality.[11]

Other quality models, however, include performance characteristics but omit functionality. For example, many definitions of the *iron triangle*—the nexus of project scope, schedule, and cost—define *scope* as having separate quality and functional components, thus intimating a model in which quality omits functionality. As one example of this somewhat divergent view of quality, the International Institute of Business Analysis in its *Guide to the Business Analysis Body of Knowledge*® (*BABOK Guide*®) distinguishes quality from functionality in its requirements definitions:

- "*Functional requirements*: describe the capabilities that a solution must have in terms of the behavior and information that the solution will manage."[12]
- "*Non-functional requirements or quality of service requirements*: do not relate directly to the behavior of functionality of the solution, but rather describe conditions under which a solution must remain effective or qualities that a solution must have."[13]

More quality model commingling occurs because software performance requirements comprise both dynamic and static performance requirements, with the former representing external software quality characteristics (with functionality omitted) and the latter representing internal software quality characteristics. Because this text focuses on software performance and excludes functionality (in assuming that functional requirements have been attained in all scenarios), the quality structure consistently referenced throughout this text reflects the dichotomy of dynamic versus static performance.

These commingled quality constructs are demonstrated in Figure 2.3, which highlights the roles of functionality, reliability, and maintainability through these various interpretations of quality. Whether functionality is a component of quality, or functionality and quality are both components of scope, one thing is clear—software functionality and software performance do compete for resources in a struggle to be prioritized into software product requirements. This and other tradeoffs of quality are discussed in the following section.

The Cost of Quality

As demonstrated in Figure 2.3, inherent trade-offs exist whether developing software or other products, often referred to as *constraints*. The *PMBOK® Guide* includes quality and five other constructs as the principal constraints to projects, but states that these should not be considered to be all-inclusive.[14] As a constraint, quality competes against other constructs for value and prioritization within software projects during planning and design. In fact, when developers omit performance from software, look at all the amazing things they can get instead!

Increased Scope Without the necessity to code performance requirements into software, developers instead can focus on additional functionality. For example, rather than making SAS software robust to failure, additional analytic reports could be developed.

Schedule The inclusion of performance requirements almost always makes code more lengthy and complex; as a result, designing, developing, and testing phases each will be lengthened. Without performance prioritized, the customer and other stakeholders can receive their software much faster.

Budget Because performance requirements take time to implement, and because time is money, project budgets typically increase with the increase in performance requirements. By omitting performance from software, a development team can either reduce budget or focus that money toward other priorities. Happy hour, anyone?

Resources Project resources include not only personnel but also maintaining the heat and power. By sacrificing performance requirements, fewer developers may be required to work on a project to achieve the same functionality so fewer resources will be consumed.

Risk Risk is not a benefit but rather the often-unintended consequence when performance is not prioritized. Because many performance requirements are designed to mitigate or eliminate vulnerabilities in software, risk typically increases as performance requirements are devalued or omitted.

While *constraint* implies a limiting or restrictive effect, constraints can also be viewed from the opportunity cost perspective. For example, developers can implement performance requirements into software, or, if they choose not to, they can produce quicker software, cheaper software, or software with increased functionality. To assess an opportunity, however, its value must be known. Thus, an objective of this text is to familiarize developers with the benefits of incorporating performance within software so that the opportunity cost of quality truly can be gauged. For example, if a stakeholder has decided to implement additional functionality (thus excluding additional performance), the stakeholder at least will have understood the value of the performance that could have been added, and thus will have made an informed decision based on the opportunity cost of that performance. Only when stakeholders understand the value of quality and the risks of the lack of quality can quality be measured against other project constraints such schedule, scope, or budget.

Establishing the value of quality can be difficult, especially where stakeholders disagree about the prioritization of constructs within a software development project. Sponsors who are funding software development will want it done inexpensively while users waiting for the software will want it developed quickly. Some developers may be more focused on the function of software and less concerned with performance attributes while other SAS practitioners may insist that the software be reliable, robust, maintainable, and modular. Thus, establishing consensus on what characteristics contribute to software quality and how to value that quality (against other project constructs) can be insurmountable. Nevertheless, stakeholders—including developers and non-developers alike—will be better positioned to discuss the inclusion of quality characteristics when they can utilize shared terminology demonstrated within the software product quality model.

QUALITY IN THE SDLC

The phases of the SDLC are introduced in the "Software Development Life Cycle (SDLC)" section in Chapter 1, "Introduction." Quality is best incorporated into software throughout the SDLC, including software planning before a single line of code has been written. In an oversimplified example, consider that you need to sort a data set to retain only unique observations. Functional needs such as this (but obviously more complex) often spawn software projects and, in many cases, only functional objectives are stated at the project outset.

But industry, organizational, and other regulations and standards, as well as overall software objectives, will necessitate not only what software must accomplish but also how well it must perform. Performance objectives include the manner in which software is intended to be run and, while they can change over the software lifespan, initial objectives and needs should be considered and discussed for inclusion as potential software performance requirements. Table 2.1 enumerates examples of performance-related questions that can be asked during software planning to spur performance inclusion.

Only through both function- and performance-related questions can stakeholders truly conceptualize software and all its complexities. Once the true needs of software are established in planning, other questions such as build-versus-buy can be discussed. For example, do you write (or modify) SAS software to sort the data set or purchase this functionality from a third-party vendor?

Another benefit of a formal design phase—even if for smaller software projects, this represents only an hour-long conversation—is the ability to

Table 2.1 Common Objectives and Questions

Performance Objective	Performance Question
Reliability, Longevity	Will the software be run only once, a few times, or is it intended to be enduring?
Reliability, Robustness	Will the software pose a risk if it fails?
Efficiency, Execution Efficiency	Does the software have any resource limitations or execution time constraints?
Scalability	How large are the input data sets and will they increase over time?
Portability	Should the software be able to run on different versions of the SAS or on different operating systems?
Reusability	Should components of the software be intended to be reused in the future?
Testability	Should the software be written in a testable fashion, for example, if a formalized test plan is going to be implemented?

redefine and prioritize needs of the project. In the initial scenario, the ultimate objective was stated to be the identification of unique observations that could be accomplished through sorting. However, a SAS index could also accomplish this objective without sorting any data, so a design phase would facilitate the discussion about the pros and cons of proposed technical solutions. The distinction between planning and design is sometimes made in that planning is needs-focused whereas design and development phases are solutions- and software-focused. Thus, the customer needs and objectives should be defined and understood during software planning, but the specific technical methods to achieve those objectives will materialize over time.

Requiring Quality

Everything about quality begins and ends with requirements. Requirements specify the technical objectives that software must achieve, guiding software through design and development. Whether those same requirements have been achieved is determined through testing that validates software completion and demonstrates intended quality.

In chapters that introduce dynamic performance requirements, a recurring *Requiring* section describes how technical requirements for the respective quality characteristic can be incorporated into requirements. The section is intended to demonstrate that, while the incorporation of performance is important and can occur throughout the SDLC, it's best implemented with planning and foresight and in response to specific needs and requirements.

As the number of stakeholders for a software project increases, the importance of formalized requirements commensurately increases. Where stakeholders represent developers who may be working together to build separate components of software, requirements help ensure they are developing toward identical specifications and a common goal. The extent to which requirements can unambiguously describe the body of development work enables developers to rely on information rather than interpretation. Stakeholders rarely enter project planning with identical conceptualizations of intended software function and performance; however, through codification and acceptance of requirements, stakeholders can inaugurate a software project with a shared view of software objectives and future functionality and performance.

Without collective software requirements, at software completion, stakeholders may disagree about whether software is high quality, low quality, or even complete. Without an accepted body of requirements, stakeholders have nothing to assess software against except their own needs, which may vary substantially by individual. Moreover, undocumented needs have a tendency

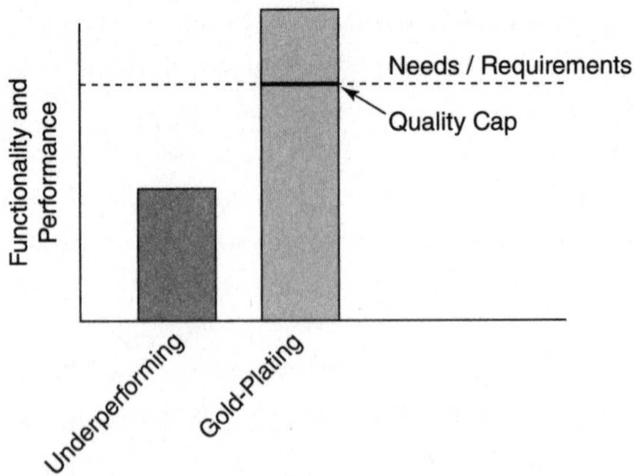

Figure 2.4 Underperformance and Gold-Plating

to morph throughout the SDLC through *scope creep*. Software requirements help avoid the misalignment that occurs when the software produced either exceeds or fails to meet customer needs, discussed in the following sections and demonstrated in Figure 2.4.

Note that while gold-plating may provide additional functionality or performance, this does not result in increased software quality because the customer did not need, had not required, or would not gain additional business value from this additional effort.

Avoid Underperformance

Underperforming doesn't make you an underachiever; at some point in your career, a customer will review your software and tell you it's not up to par. You failed to meet functional or performance objectives that were either implied or specified through technical requirements, so your software lacks anticipated quality. Underperformance represents a misalignment between the expected and delivered quality. However, when underperformance occurs in environments that rely primarily on performance needs that are implied rather than explicitly stated through technical requirements, this ambiguity can leave all stakeholders consternated. Developers may believe they delivered the intended software while customers and users feel otherwise, leaving a functional or performance gap.

A performance gap occurs when your manager requested efficient software and specified thresholds for run time, speed, and memory utilization, but the software you delivered fails to meet those objectives. The software is functionally sound, but its quality is diminished because performance is reduced. What if expected performance was only implied, not defined through requirements? Maybe the manager believed you understood that the software needed to complete in less than 15 minutes, but that information was never actually conveyed, or was confounded by other needs presented during software planning. The project manager wanted "fast SAS," and you thought you delivered "fast SAS," but because performance requirements were not defined or accepted (so they could be measured and validated), whose definition of "fast SAS" should be used to measure the quality of the software at completion?

Requirements that are stated and measurable can avoid underperformance, essentially by providing a checklist by which SAS practitioners can evaluate their code. Does it meet all functional requirements? Done. Does it meet all performance requirements? Done. Does it meet all test cases specified in a formalized test plan? Done. Establishing requirements ensures that quality is collectively defined during software planning and collectively measured and validated at software completion.

Avoid Gold-Plating

Continuing the development scenario in the "Quality in the SDLC" section, you're a researcher writing SAS software to sort data to select unique observations. You've asked yourself several performance-related questions during software planning and, in this end-user development scenario, business value is delivered almost exclusively through functionality, not performance. Thus, what matters is that the SAS software selects unique observations and produces an accurate data set. If the program fails, you as an end-user developer can restart it, so robustness is not that important. Because the data sets aren't very large, efficiency during execution also is not a priority because the processes will complete in seconds. And, because the data sets aren't expected to increase in size over time, data scalability should not be a priority in development.

This isn't to assert that quality doesn't matter in this scenario or in end-user development in general, but rather that if the incorporation of performance attributes pose no value and their absence poses a negligible (or at least acceptable) risk, then they shouldn't be included. Thus, in this scenario, if you spent hours empirically testing whether a SAS index, SQL procedure, or combined SORT procedure and DATA step would produce the fastest running code (or the most efficient use of system resources), that would have been a waste of time—at least for this project, because it would have delivered no additional business value while substantially delaying completion. Even if you did successfully achieve increased efficiency, the software quality by definition would not have increased because the additional performance fulfilled no performance requirements.

Gold-plating often occurs when software requirements lack clarity, leaving room for interpretation or assumption. A proud SAS practitioner might deliver an impressively robust and efficient program, expecting to be lauded by a manager, only to receive condemnation because the software didn't need to be robust. Rather, it needed to be completed *yesterday*, and the inclusion of the performance actually *decreased* software value because it delayed analysts' ability to use resultant data sets or data products. By creating technical requirements at project outset, a shared understanding of software value can be achieved that can guide SAS practitioners to focus on delivering only universally valued function and performance.

Gold-plating essentially answers the question: *Can you ever have too much quality?* Yes, you can. Delivering *unanticipated* performance can sometimes endear you to customers—your boss didn't ask you to make the software faster, but you did because you're just that good, and he heartily thanked you for the improvement. But delivering *unwanted* performance doesn't increase software quality and, as demonstrated, can decrease the value of software by causing schedule delays, budget overages, or the loss of valued function or performance that had to be eliminated because rogue developers gold-plated their software. Rogue is risky; don't do it.

Saying No to Quality

One technique to avoid gold-plating is not only to specify performance attributes that *should* be included in software, but also to specify quality characteristics that *should not* be included. Not all software is intended for production status or a long, happy life. Often in analytical environments, snippets of SAS code are produced for some tactical purpose and, after the subsequent results are analyzed, either discarded or empirically modified. This expedient software might not benefit from the incorporation of many performance characteristics. If its purpose shifts over time from tactically to strategically focused, however, or from supporting ancillary to critical infrastructure, commensurate performance should later be incorporated into the software.

This again illustrates the concept that some SAS software will be defined by functional requirements alone, to the exclusion of performance requirements. In cases in which the intent of software changes over time, reevaluation should determine whether the software requirements—including functional and performance—are still appropriate. But in cases where performance would provide no intrinsic value to software, performance characteristics should not be included and opportunity costs—such as additional functionality or reduced project cost—should instead be prioritized.

Saying Yes to Quality

Certain aspects of software intent or the development environment can warrant higher-performing software. In some circumstances, performance can be as highly valued and essential to software as functionality. This value may spur from additional performance being viewed as beneficial or, conversely, from a risk management perspective in which performance is prioritized to mitigate or eliminate specific risks. The following circumstances often dictate software projects requiring higher performance.

Industry Regulations Certain industries have specific software development regulations that must be followed; often, the software can or will be audited by government agencies. SAS software developed to support clinical trials research, for example, must typically follow the Federal Department of Agriculture (FDA) *General Principles of Software Validation* to ensure quality and performance standards are met.[15]

Organizational Guidelines Beyond industry regulations, organizations or teams often codify programmatic best practices that must be followed in software development and that often dictate the inclusion of quality characteristics. For example, some teams require that all SAS macros must include a return code that can demonstrate the success or failure of the macro to facilitate exception handling and robustness.

Software Underpinning Critical Infrastructure ETL or other SAS software that supports critical components of a data analytic infrastructure should demonstrate increased performance commensurate to the risk of the unavailability of the system, software, or resultant data products.

Scheduled Software SAS software that is automated and scheduled for recurring execution should demonstrate increased performance. Inherent in automating and scheduling SAS jobs is the understanding that SAS practitioners will not be manually reviewing SAS logs, so reliability and robustness should be high.

Software Having Dependent Processes or Users If SAS software is trusted enough to have dependent processes or users rely on it, its level of performance should reflect the risk of failure not only of the software itself, but also of all dependencies.

Software with Expected Longevity SAS software intended to be operational for a longer duration should demonstrate increased performance. Static performance requirements are especially valuable in enduring software products because they support software maintainability, an investment that increases in value with the intended software lifespan.

Implied Requirements

All this talk of requirements may have some readers concerned that quality is synonymous with endless documentation. Some software development environments do produce significant documentation out of necessity, especially when several or all of the characteristics in the "Saying Yes to Quality" section are present. However, from the ISO definition of quality, it's clear that software quality can address *implied* as well as *explicit* needs. Moreover, the second of four Agile software development values states that "[w]orking software [is valued] over comprehensive documentation."[16] Thus, at the organizational or team level, stakeholders will need to assess the quantity and type of requirements that should have actual documentation versus those that might be equally served through an implied understanding of performance needs and requirements.

To provide one example, SAS practitioners on a close-knit team hopefully recognize individual strengths and weaknesses, and know and trust each other's technical capabilities. If team members are highly qualified and have a keen awareness of quality characteristics that can facilitate high-performing software, a requirement that software be "built modularly through dynamic SAS macro code" would probably be gratuitous and possibly offensive if it landed in requirements documentation. Especially in teams that have performed together efficiently and effectively (and who have implemented a quality model against which to judge the inclusion of technical requirements in software definitions), many references to performance can be omitted from technical specifications without detriment.

Measuring Quality

Quality is a respected pursuit but even ambition cannot overcome an aimless journey toward an unknown or undefined destination. Backpacking through Central and South America, I rarely knew which chicken bus or taxi I was taking or where I would be sleeping, but I always knew the end-game: a flight out of Buenos Aires months later. Because the flight was defined, it was a clear metric that I would achieve, miss, or possibly modify to extend the journey. While backpacking, I could continually alter my route, travel modalities, and sightseeing to ensure I arrived in Buenos Aires on time.

The quest for software quality can be a similarly convoluted and tortuous pursuit but it should have a defined destination. Through performance requirements, SAS practitioners are made aware of the necessary quality characteristics that must be built into software. Even within Agile development environments that implement iterative, rapid development, the expected functionality and performance are defined and stable within each time-boxed iteration. Performance requirements not only guide and spur design and development, but also should be used to evaluate the attainment of quality at software completion and during operation.

Throughout the dynamic performance requirement chapters, the *Measuring* sections demonstrate ways that performance can be quantitatively and qualitatively measured. Dynamic performance requirements more readily lend themselves to quantitative measurement than static performance requirements. You can specify that software should complete in 15 minutes, be able to process 3 million observations per hour, only fail less than two times per month, or be operational within Windows but not UNIX. Dynamic performance requirements are not only measurable but also observable; thus, they are typically prioritized over static performance attributes in requirements documentation.

Static performance requirements are often more appropriately measured qualitatively or indirectly. For example, modularity might be approximated by counting lines per macro, or ensuring that all child processes generate return codes to signal success or failure to parent processes. However, modularity is more likely to be expressed as a desired trait of software and thus measured qualitatively. Some development environments similarly track how many times specific modules of code are reused and in what capacity or software, which can provide reuse metrics from which reusability can be inferred. But the best measurement of the inclusion (or exclusion) of static performance requirements is an astute SAS practitioner who can inspect the code and discern quality characteristics.

WHAT'S NEXT?

In Chapter 3, "Communication," techniques are demonstrated that support reliability, robustness, and modular software design by facilitating software communication. The robustness of software is facilitated through the identification and handling of return codes for individual SAS statements and automatic macro variables. User-generated return codes are demonstrated that facilitate communication between SAS parent and child modules, as are control tables that facilitate communication between different SAS sessions and batch jobs.

NOTES

1. "Quality." *Merriam-Webster.com*, 2016. http://www.merriam-webster.com (22 June 2016).
2. IEEE Std 829-2008. *IEEE standard test method for use in the evaluation of message communications between intelligent electronic devices in an integrated substation protection, control and data acquisition system*. Geneva, Switzerland: Institute of Electrical and Electronics Engineers.
3. ISO/IEC 25000:2014. *Systems and software engineering — Systems and software Quality Requirements and Evaluation (SQuaRE)*—Guide to SQuaRE. Geneva, Switzerland: International Organization for Standardization and Institute of Electrical and Electronics Engineers.
4. Project Management Institute, 2013. *A guide to the project management body of knowledge (PMBOK® guide)* (5th ed.). Newtown Square, PA: Project Management Institute.
5. ISO/IEC 25000:2014.
6. Id.
7. Id.
8. ISO/IEC 9126-1:2001. *Software engineering—Product quality—Part 1: Quality model*. Geneva, Switzerland: International Organization for Standardization and Institute of Electrical and Electronics Engineers.
9. IEEE Std 1220™-2005. *IEEE standard for the application and management of the systems engineering process*. Geneva, Switzerland: Institute of Electrical and Electronics Engineers.
10. ISO/IEC 25000:2014.
11. ISO/IEC 25010:2011. *Systems and software engineering — Systems and software Quality Requirements and Evaluation (SQuaRE)—System and software quality models*. Geneva, Switzerland: International Organization for Standardization and Institute of Electrical and Electronics Engineers.
12. International Institute of Business Analysis, 2015. *A guide to the business analysis body of knowledge® (BABOK® Guide v3)* (3rd ed.). Whitby, Ontario, Canada: International Institute of Business Analysis
13. Id.
14. Project Management Institute. 2013.
15. U.S. Food and Drug Administration (FDA). *General principles of software validation; Final guidance for industry and FDA staff*. Retrieved from www.fda.gov/MedicalDevices/DeviceRegulationandGuidance/GuidanceDocuments/ucm085281.htm.
16. The Agile Alliance, *The Manifesto for Agile Software Development*. Retrieved from www.agilealliance.org/the-alliance/the-agile-manifesto.

CHAPTER 3
Communication

From the Guatemalan treehouse perched high above the jungle that gave new meaning to the phrase "climbing into bed," to the Bolivian hotel built exclusively of salt blocks, to the Argentinian cattle ranch mooing into the night, I discovered one truth about lodging while backpacking abroad: the more remote, the seemingly more focus paid to securing communication for guests.

The days of traveling off the grid are still possible, but more so by ascetic choice than duress, as the Internet now inhabits even the most remote lodging and locales. Backpacking has become so interactive that in some ways you're more on the grid when abroad than when home. Backpackers are seemingly required to continuously Skype, Facebook, Twitter, Instagram, Snapchat, and email, littering the web with photo and video evidence that they're having an amazing time.

But some communication amounts to more than tourists getting their social media fix.

Having met in Guatemala, some Europeans and I discover we'll be in three other countries at similar times, exchange contact information, and finally meet up in Peru.

While checking email, I learn that the mountainous road to my Colombian hostel has washed out, so I rebook elsewhere, saving a trudge through the mud.

After checking Facebook and learning that a friend is having her first child, I buy a handwoven blanket for her in Ecuador.

When I'm stranded in a fog bank in Arequipa, Peru, and miss my flight to Easter Island, I not only reach my hotel to cancel my reservation, but also book travel onward to Chile.

And, after checking my bank account in Buenos Aires, I decide to fly home as scheduled rather than indefinitely extending my adventures.

■ ■ ■

Had I been unable to communicate while backpacking, I would have missed tremendous opportunities, been unnecessarily exposed to risks, and, in general, been oblivious to life beyond my surreal world. Some aspects of my trip would have been impossible without communication while in other cases the value of the trip was maximized through communication.

Communication can be enabling—an email to a Colombian hostel or 20 minutes booking a flight on the LAN website made the magic happen. Software often passes parameters to SAS macros and programs to facilitate functionality. Without these marching orders—plane tickets or parameters—the adventure can't even begin.

Other communication is bidirectional, often providing a validation function. The emails confirming my reservations demonstrated successful bookings and provided peace of mind. Return codes in software can demonstrate process success, notifying a parent process not only that its child started but also that it completed without exception or error.

In other cases, the content of communication may be untoward—informing you that a road has washed out or a fog bank isn't expected to lift for days—but invaluable nonetheless. Some risks can be avoided, mitigated, or eliminated through proactive communication patterns in both life and software. When failure—or the conditions that will lead to failure—is detected and communicated, software can sometimes still achieve full business value through dynamic exception handling routines.

But when failure is imminent, such as when you tearfully realize that you won't be playing with the Rapa Nui for a weekend, communication can find and prescribe alternatives. I missed seeing the *Moia* (monolithic human heads) in situ but serendipitously found one at the Corporación Museo de Arqueología e Historia Francisco Fonck in Valparaiso, Chile. Communication in software should also prescribe when a process or program should be terminated so further resources aren't wasted.

No backpacker is ever alone, as there's an endless stream of others carving strikingly similar paths through Central and South America—it's common to exchange information and rejoin weeks or months later for continued antics. Parallel processing similarly can involve individual processes taking divergent yet coordinated paths toward some collective goal, and communication between processes is often necessary to time, structure, and deconflict activities to facilitate success.

While examples thus far have described internal communication, external communication can be equally important. Banking activities, job-hunting activities, and blanket-buying activities (for new baby mamas) facilitated fluidity from the surreal to the real world and were necessary to ensure I didn't return to a life of chaos in the States. Software too doesn't just blather to itself—external communication is required to inform stakeholders of process successes, failures, and performance metrics.

■ ■ ■

RETURN CODES

Return codes are a preferred method to record and pass performance metrics within software. The Base SAS language generates return codes that reflect both normal and abnormal—or *exceptional*—functioning. When SAS code is syntactically correct but encounters an exception during execution, a note, warning, or runtime error is typically recorded in the log while one or more automatic return codes are automatically updated. For example, if the LIBNAME statement references an invalid logical location, a note is generated in the log and the SAS automatic macro variable &SYSLIBRC is set to a negative number.

When SAS code is executed with syntax errors, runtime errors are displayed in the log and one or more return codes are updated. If the LIBNAME statement instead contains a syntax error, such as an omission of quotations around the library reference, a runtime error is recorded in the log and the value of &SYSLIBRC is set to a positive number. In production software, syntax errors such as this one should have been eliminated during development and testing and thus will not be encountered during software operation. Thus, in production software, return codes are principally used to detect environmental and other exceptions that occur, as well as to validate software performance and completion. For example, the fail-safe path (discussed in the "Fail-Safe Path" section in Chapter 11, "Security") is facilitated through communication and handling of exceptions.

Just as SAS automatic macro variables operate as return codes for SAS procedures, functions, statements, and DATA steps, user-generated return codes perform commensurate communication and validation of SAS macros. By creating a global macro variable inside a macro, information about that macro's performance can be passed back to the parent process calling the macro. Validation that a macro or other SAS process has completed correctly can demonstrate its success, thus enabling subsequent processes to initiate. When used inside an exception handling framework, return codes facilitate dynamic, data-driven processing that flexibly responds to its environment during execution.

Faking It in SAS

For software developers more familiar with third-generation languages (3GLs), the previous explanation of return codes may seem insufficient or inaccurate. In fact, in many software languages, the purpose of a *return* statement is to signal the termination of a module and immediately return some value or values from a child process to its parent. System-generated return codes do perform this function in SAS; however, the %RETURN macro function "causes normal termination of the currently executing macro" but actually *returns* nothing.[1] No inherent methods exist within Base SAS to return values from a macro.

Given this limitation, true return codes cannot be generated in Base SAS and must instead be faked. By initializing a global macro variable inside a SAS macro, information can effectively be passed from the child back to the parent. While less efficient and less secure than other languages, it still gets the job done, and this limitation should not excuse SAS practitioners from failing to implement return codes in their software.

One disadvantage of having to fake return codes is the violation of loose coupling—the global macro variable, once created, is available for all subsequent processes, not just the parent process that called the macro. Thus, software must ensure that all global macro variables are defined uniquely to ensure that they don't interact unintentionally with or contaminate other processes. One method to achieve this security is to utilize the macro name as the base for the return code variable name. In this example, the %CHILD macro produces the return code &CHILDRC:

```
%macro child();
%let syscc=0;
%global childRC;
%let childRC=GENERAL FAILURE;
* do some process;
%if &syscc=0 %then %let childRC=;
%mend;
```

A second disadvantage of faking return codes is the tenacity of return codes to persist from previous processes or even previous programs run in the same SAS session. For example, with poor coding that fails to initialize a return code to a default value, it is possible for values from previous executions to persist into the current environment, thus contaminating the return code and results. As demonstrated, this vulnerability can be eliminated by assigning the return code to a general failure state at the start of the macro and by assigning a successful return code only once process success has been demonstrated. These vulnerabilities are discussed in the "Perseverating Macros" section in Chapter 11, "Security."

SYSTEM NUMERIC RETURN CODES

SAS system-generated automatic macro variables perform a variety of roles, including providing insight into the SAS environment and operating system (OS), controlling how SAS operates, and recording performance metrics. A subset of automatic macro variables includes return codes that describe software performance, such as the success or failure of specific SAS statements or functions. When warnings or runtime errors occur during execution, these often are automatically stored in return codes. One of the primary benefits of return codes is their ability to drive dynamic processing through automatic rather than manual identification of errors, allowing SAS practitioners to avoid the unhealthy practice of parsing log results to search for failure or to validate software success.

The majority of SAS automatic macro variables are described in the *SAS® 9.4 Macro Language: Reference*, 4th ed.[2] Additional SAS automatic macro variables are described in the *SAS® 9.4 SQL Procedure User's Guide*, 3rd ed.[3]

NUMERIC RETURN CODES

&SYSERR The System Error represents the error code of the most recent process, which resets at each boundary step.[4]

&SYSCC The System Current Code represents the highest error code produced throughout a SAS job and does not reset during a SAS session.[5]

&SQLRC The SQL Return Code represents whether the most recent SQL statement—that is *statement*, not *procedure*—executed correctly.[6]

&SYSFILRC The System File Return Code represents whether the FILENAME statement executed correctly.[7]

&SYSLIBRC The System Library Return Code represents whether the LIBNAME statement executed correctly.[8]

&SYSLCKRC The System Lock Return Code represents whether the LOCK statement executed correctly.[9]

&SYSRC The System Return Code records the status of statements that interact directly with the OS environment, such as X, SYSEXEC, SYSTASK, and WAITFOR.

SYSRC() While technically a function, not a return code, the System Return Code function returns the status of the most recently executed I/O function.[10]

&SYSERR

The &SYSERR macro variable is one of the most important return codes, as it demonstrates the success or failure of the last process to be executed. The following output demonstrates a successful DATA step in which the Final data set was created. Because no warning or error occurred, the value of &SYSERR is set to 0:

```
data final;
   set original; * data set exists;
run;

NOTE: There were 1 observations read from the data set WORK.ORIGINAL.
NOTE: The data set WORK.FINAL has 1 observations and 0 variables.
NOTE: DATA statement used (Total process time):
      real time           0.01 seconds
      cpu time            0.01 seconds

%put SYSERR: &syserr;
SYSERR: 0
```

When the Original data set does not exist, as demonstrated in the following output, &SYSERR is set equal to 1012, which represents the SAS general error condition:

```
data final;
   set original;
ERROR: File WORK.ORIGINAL.DATA does not exist.
run;

NOTE: The SAS System stopped processing this step because of errors.
WARNING: The data set WORK.FINAL may be incomplete.  When this
step was stopped there were 0 observations and 0 variables.
WARNING: Data set WORK.FINAL was not replaced because this step was
stopped.
NOTE: DATA statement used (Total process time):
      real time           0.01 seconds
      cpu time            0.01 seconds

%put SYSERR: &syserr;
SYSERR: 1012
```

The macro variable &SYSERR is read-only and cannot be manipulated directly by developers. Thus, &SYSERR will always contain the correct value, whereas through erroneous business logic, a read-write automatic macro variable can be manually reset to an invalid non-error condition. &SYSERR resets after each step boundary, including DATA steps and procedures. It's important to note that if two processes are executed in series, after which the value of &SYSERR is assessed, only the status of the second process will be evaluated. Thus, if testing and debugging serialized, monolithic SAS code, the value of &SYSERR may need to be repeatedly checked to ensure that faulty or failing processes are isolated and identified.

Many specific failure types, however, are only identified through specific return codes. For example, if the FILENAME statement fails, &SYSERR will still be 0 (representing error-free operation) because the exception is captured instead in &SYSFILRC. Especially when SAS code is wrapped in a macro framework, it's important to utilize all relevant return codes to capture all failures. For example, don't you just hate when a rogue *T* ends up in your software? In the following example, *5* is mistyped as *T*, causing failure even though the macro compiles:

```
%macro test;
%let syscc=0; * required to reset SYSCC any time it will be assessed;
data final;
%do i=1 %to T; * software defect in which T should have been 5;
   length char&i $10;
   %end;
run;
%mend;
```

When the code executes, the DATA step fails but produces only notes to the log while &SYSERR depicts error-free execution. Only &SYSCC accurately captures the syntax error that causes the failure:

```
%test;

NOTE: The data set WORK.FINAL has 1 observations and 0 variables.
NOTE: DATA statement used (Total process time):
      real time           26.44 seconds
      cpu time            0.90 seconds

ERROR: A character operand was found in the %EVAL function or %IF
condition where a numeric operand is required. The condition was: T
ERROR: The %TO value of the %DO T loop is invalid.
ERROR: The macro TEST will stop executing.

%put SYSERR: &syserr;
SYSERR: 0
%put SYSCC: &syscc;
SYSCC: 1012
```

The data set is still created, but because the LENGTH statements fail to execute, it has no variables. Notwithstanding this epic failure, &SYSCC must be used to capture the fault because &SYSERR will not demonstrate it.

&SYSCC

The automatic macro variable &SYSCC represents the highest value of any warning or runtime error encountered during a SAS session. For example, if runtime errors with values of 1012 and 1014 are encountered, &SYSCC will be set to 1014, thus masking the 1012 error. &SYSCC is cumulative as opposed to &SYSERR, which resets automatically when SAS boundary steps are crossed; therefore, &SYSCC will demonstrate that a warning or error has occurred at some point in the past even when &SYSERR has been overwritten with 0. &SYSCC does not identify all errors; thus, it should often be checked in combination with other return codes. For example, when the LIBNAME statement fails, it produces a note in the SAS log that cannot be programmatically detected through &SYSCC—only &SYSLIBRC.

When the following code is executed, the LIBNAME statement fails because the directory C:\neverland does not exist, producing a negative value for &SYSLIBRC that will persist until a subsequent LIBNAME statement resets it:

```
%let syscc=0;
libname lib 'c:\neverland\';
NOTE: Library LIB does not exist.

%put SYSCC: &syscc;
SYSCC: 0
%put SYSLIBRC: &syslibrc;
SYSLIBRC: -70008

data lib.mydata;
       length char1 $10;
run;

NOTE: Variable char1 is uninitialized.
ERROR: Library LIB does not exist.
NOTE: The SAS System stopped processing this step because of errors.
NOTE: DATA statement used (Total process time):
      real time           0.01 seconds
      cpu time            0.01 seconds
```

```
%put SYSCC: &syscc;
SYSCC: 1012
%put SYSLIBRC: &syslibrc;
SYSLIBRC: -70008
```

Because the LIBNAME statement failed, the LIB.Mydata data set cannot be located and the DATA step produces a runtime error. The value of &SYSCC is only set to an error code after this failure, despite being caused indirectly by a previous LIBNAME statement failure. &SYSCC should be used to detect and handle general or unknown errors that might not otherwise be captured, but should be used in conjunction with other error return codes. For example, had exception handling first assessed the value of &SYSLIBRC, the library assignment failure could have been detected and, through dynamic execution, the DATA step failure avoided.

Note that &SYSCC must be initialized to 0 to ensure its value does not persist from previous program executions. Because &SYSCC is cumulative, it does not reset automatically during a SAS session. For example, a SAS practitioner could run the previous code (and receive the 1012 error), go to lunch, run an unrelated program in the same SAS session, and the value of &SYSCC would still be 1012 despite successful execution of the second program. The following code simulates this subsequent, unrelated program execution in the same session:

```
data final;
   set original;
run;

NOTE: There were 1 observations read from the data set WORK.ORIGINAL.
NOTE: The data set WORK.FINAL has 1 observations and 0 variables.
NOTE: DATA statement used (Total process time):
      real time           0.01 seconds
      cpu time            0.00 seconds

%put SYSCC: &syscc;
SYSCC: 1012
```

The logic error occurs because &SYSCC is never automatically reset by the SAS application; thus an error code from one program can persist into and poison other programs that are open in the same session. To counter this persistence, and because &SYSCC is a read-write variable (unlike &SYSERR),

&SYSCC should always be reset to 0 before it is implemented or tested within code. The amended code now accurately reflects that no error has occurred in the DATA step:

```
%let SYSCC=0;
data final;
   set original;
run;

NOTE: There were 1 observations read from the data set WORK.ORIGINAL.
NOTE: The data set WORK.FINAL has 1 observations and 0 variables.
NOTE: DATA statement used (Total process time):
      real time           0.01 seconds
      cpu time            0.00 seconds

%put SYSCC: &syscc;
SYSCC: 0
```

A final caveat of &SYSCC is that because only the highest error value is retained, when multiple errors occur, errors having lower values will be masked. For example, when general errors occur, &SYSCC is set to 1012. However, if execution continues and the SAS application runs out of memory, &SYSCC will be overwritten and set to 1016. Further, if the 1016 error value occurs before 1012, &SYSCC will be set to 1016 but will never reflect 1012 because it is masked. This, too, demonstrates the necessity always to reset &SYSCC to 0 before later assessment.

One method to implement minimal exception handling is to reset &SYSCC to 0 at program start and test its value at program termination. In this simulated extract-transform-load (ETL) infrastructure, any &SYSCC value greater than 0 indicates that a warning or error has occurred, thus requiring scrutiny of the log:

```
%let SYSCC=0;

%macro extract;
%mend;

%macro transform;
%mend;

%macro load;
%mend;
```

```
%extract;
%transform;
%load;

%put SYSCC: &syscc;
```

In an actual ETL infrastructure, a non-zero value of &SYSCC at program termination might cause the program log file to be saved automatically so that developers could review it to analyze any warnings or runtime errors. If no error had occurred, the log file instead could be deleted automatically. A more robust implementation of &SYSCC might repeatedly reset and test its value within a single program. For example, in modularized code, once a macro is invoked, &SYSCC can be reset to 0, after which its value can be assessed before macro termination. This practice is identical to the previous example, except that it occurs repeatedly at the module level to better isolate where and why errors are occurring. Moreover, once an error is detected, program flow can be halted or dynamically rerouted so that it does not fail from cascading errors, as discussed in the "Cascading Failures" section in Chapter 6, "Robustness."

&SQLRC

The &SQLRC automatic macro variable functions similarly to &SYSERR but, rather than applying to SAS procedures and the DATA step, represents only the status of the most recently executed SQL statement within a SQL procedure. Note that &SQLRC resets only after SQL *statements*, not SQL *procedures*, each of which can contain several SQL statements. Unlike &SYSERR, &SQLRC is read-write and should be reset to 0 if assessed multiple times within the same SQL procedure. However, because &SQLRC is not cumulative beyond the SQL procedure boundary, it does not need to be reset to 0 before the first SQL statement within a SQL procedure.

The following output demonstrates an attempt to determine the number of observations in a data set, which fails because the Original data set does not exist. Note that because &SYSCC is both read-write and cumulative, it must be reset to 0; however, &SQLRC does not need to be similarly reset before its single use. Also note that the &SQLRC value of 8 represents a general error during SQL execution:

```
%let syscc=0;

proc sql noprint;
   select count(*)
   into :obstot
   from original;
quit;
```

```
ERROR: Table WORK.ORIGINAL doesn't have any columns. PROC SQL requires
each of its tables to have at least 1 column.
%put SYSERR: &syserr;
SYSERR: 1012
%put SYSCC: &SYSCC;
SYSCC: 1012
%put SQLRC: &SQLRC;
SQLRC: 8
```

One idiosyncrasy of &SQLRC is the fact that it resets after each SQL statement, even within a single SQL procedure. Because multiple SQL statements can be contained within a single SQL procedure, if &SQLRC is assessed at the termination of the SQL procedure, it will reflect the success or failure of only the most recent statement. In the following code, the first SQL statement fails because the Nodatanodatanodata data set does not exist, which sets &SQLRC briefly to 8:

```
data temp;
   length char1 $10;
run;

%let syscc=0;
proc sql noprint;
   select count(*)
   into :obstot1
   from nodatanodatanodata; * data set does not exist *;
%put SQLRC: &sqlrc;
%put SYSCC: &syscc;
   select count(*)
   into :obstot2
   from temp;
quit;
%put SQLRC: &sqlrc;
%put SYSCC: &syscc;
```

Because the second SQL statement succeeds, however, &SQLRC is reset from 8 to 0, thus masking the previous failure. Additionally, because &SYSCC is only set after a boundary step, the value of &SYSCC following the failure (but preceding the QUIT statement) misleadingly remains 0. Not until after QUIT does &SYSCC correctly reflect the error from the previous SQL statement:

```
%let syscc=0;
proc sql noprint;
   select count(*)
   into :obstot1
   from nodatanodatanodata;    * data set does not exist *;
```

```
ERROR: File WORK.NODATANODATANODATA.DATA does not exist.
%put SQLRC: &sqlrc;
SQLRC: 8
%put SYSCC: &syscc;
SYSCC: 0
   select count(*)
   into :obstot2
   from temp;
quit;
%put SQLRC: &sqlrc;
SQLRC: 0
%put SYSCC: &syscc;
SYSCC: 1012
```

As demonstrated, failure in single- or multi-statement SQL procedures can be detected through &SYSCC; however, &SYSCC fails to isolate where the warnings or runtime errors have occurred within the SQL procedure. Thus, to more specifically assess multi-statement SQL procedures, &SQLRC can be evaluated following each SQL statement.

&SYSFILRC

The automatic macro variable &SYSFILRC represents the success or failure of the most recent FILENAME statement and is 0 when FILENAME has executed correctly. One use of the FILENAME statement, as I demonstrate in a separate text, is to assess whether a SAS data set is locked by another user or process.[11] The following code determines whether a shared or exclusive lock is maintained on the WORK.Temp data set. If no lock exists, the data set is available and free for the taking:

```
data temp;
   length char $10;
run;

%let syscc=0;
%let lib=work;
%let tab=temp;
%let fil=%sysfunc(pathname(&lib))\&tab..sas7bdat;
filename myfile "&fil";
data _null_;
   excl=fopen('myfile','U');
```

```
   if excl=1 then call symput('LOCK','None');
   else do;
      shared=open("&lib..&tab");
      if shared=1 then call symput('LOCK','Shared');
      else call symput('LOCK','Exclusive');
      end;
run;
%put LOCK: &lock;
%put SYSFILRC: &sysfilrc;
%put SYSCC: &syscc;
```

When executed, the FILENAME statement succeeds (&SYSFILRC is 0) and the data set is not locked (&LOCK is None). If the Temp data set does not exist, however, no runtime error occurs, but &SYSFILRC is set to 1 to reflect an exception in the FILENAME statement. The output further demonstrates that the value of &SYSCC is not affected by &SYSFILRC or the missing file:

```
%put LOCK: &lock;
LOCK: None
%put SYSFILRC: &sysfilrc;
SYSFILRC: 1
%put SYSCC: &syscc;
SYSCC: 0
```

Without examining &SYSFILRC, the output mistakenly indicates that no lock is available for the Temp data set; but the data set is not locked—it is missing. When the FILENAME statement produces an exception, the FOPEN function creates the file Temp.sas7bdat, which in turn is then shown (by the OPEN function) not to be locked. This demonstrates a significant idiosyncrasy of &SYSFILRC that occurs when FILENAME references a missing file, which generates the return code 1; however, the file reference produced is valid and, when referenced by FOPEN, creates the missing file. This circular logic is necessary, however, because to create the file from scratch, FOPEN requires a valid file reference, generated by the FILENAME statement. Necessity of this logical fallacy is further demonstrated later in the "Mutexes and Semaphores" section.

An assessment of &SYSFILRC immediately after the FILENAME statement would have prevented FOPEN from executing and producing this failure. Moreover, because the file created (Temp.sas7bdat) is not an actual SAS data set, although it now appears in the WORK library, it can neither be opened nor deleted in SAS and must instead be deleted through the OS.

Another common use of the FILENAME statement is to create piped commands to the OS, although here the &SYSFILRC macro variable operates differently. In the following code, the %GET_FILELIST macro accepts a parameterized directory and creates a space-delimited macro variable (&FILELIST) that contains all files in that directory. It's a very useful tool; however, as the output indicates, when a nonexistent directory (C:\neverland) is specified in the parameter, this exception is indicated in the SAS log but not in &SYSFILRC, which indicates 0 for success:

```
%macro get_filelist(dir=);
%let syscc=0;
%local i;
%local fil;
%global filelist;
%let filelist=GENERAL ERROR;
filename f pipe "dir /b &dir*.sas";
data _null_;
   length filelist $10000;
   infile f truncover;
   input filename $100.;
   filelist=catx('',filelist,filename);
   call symput('filelist',filelist);
   retain filelist;
run;
%mend;

%get_filelist(dir=c:\neverland\);

NOTE: The infile F is:
      Unnamed Pipe Access Device,
      PROCESS=dir /b c:\neverland\*.sas,RECFM=V,
      LRECL=32767

Stderr output:
The system cannot find the file specified.
NOTE: 0 records were read from the infile F.
NOTE: DATA statement used (Total process time):
      real time           0.05 seconds
      cpu time            0.01 seconds

264  %put SYSCC: &syscc;
SYSCC: 0
```

```
265  %put SYSERR: &syserr;
SYSERR: 0
266  %put SYSFILRC: &sysfilrc;
SYSFILRC: 0
```

To overcome this limitation in &SYSFILRC, it might first be necessary to validate the existence of the directory with a first FILENAME statement that omits the PIPE command and, only after that directory has been confirmed to exist, to issue a second FILENAME statement that includes the PIPE command.

&SYSLCKRC

The automatic macro variable &SYSLCKRC demonstrates the success of the most recent LOCK statement, which was historically used to demonstrate that a data set was not in use and could be opened exclusively by a process. Because the LOCK statement was shown to lack integrity in Base SAS software and often produced invalid results, the LOCK statement and &SYSLCKRC should only be used within a SAS/SHARE environment, but are demonstrated solely for comprehension of legacy SAS code. A methodology that replaces LOCK functionality is demonstrated later in the "Mutexes and Semaphores" section.

The following output demonstrates legacy use of the LOCK statement to gain an exclusive lock on the PERM.Final data set so that no other process or user (in theory) can access the data set:

```
data perm.final;
   length char $10;
run;

NOTE: The data set PERM.FINAL has 1 observations and 1 variables.
NOTE: DATA statement used (Total process time):
      real time           0.01 seconds
      cpu time            0.01 seconds

lock perm.final;
NOTE: PERM.FINAL.DATA is now locked for exclusive access by you.

%macro test;
%if &SYSLCKRC=0 %then %do;
   proc sort data=perm.final;
      by char;
   run;
   %end;
```

```
lock PERM.final clear;
%mend;

%test;

NOTE: There were 1 observations read from the data set PERM.FINAL.
NOTE: The data set PERM.FINAL has 1 observations and 1 variables.
NOTE: PROCEDURE SORT used (Total process time):
      real time           0.14 seconds
      cpu time            0.01 seconds

NOTE: PERM.FINAL.DATA is no longer locked by you.
```

In many cases, the LOCK statement does successfully lock the referenced data set, preventing other users or SAS sessions from simultaneously accessing the data set. However, in some cases, the LOCK statement can produce a &SYSLCKRC value of 0 even though the data set is exclusively locked by another SAS session. Given this inconsistent performance, use of the LOCK statement is not recommended.

&SYSLIBRC

The automatic macro variable &SYSLIBRC demonstrates the success or failure of the most recent LIBNAME statement, and reflects 0 if LIBNAME assigns a valid library. One dynamic use of LIBNAME is to assign a library to the logical location of the named SAS program that is executing. For example, if the SAS program C:\SAS\temp\test.sas is executing in SAS Display Manager, a developer might want to assign the SAS library CURRENT to C:\SAS\temp:

```
* must be saved to a named SAS program to execute;
%let syscc=0;
%let path=%sysget(SAS_EXECFILEPATH);
%let path=%substr(&path,1,%length(&path)-%length(%scan(&path,-1,\)));
libname current "&path";
%put SYSLIBRC: &SYSLIBRC;
%put SYSCC: &syscc;
```

The following code initializes the CURRENT library and, because the LIBNAME statement executes correctly, the value of &SYSLIBRC is set to 0:

```
libname current "&path";
NOTE: Libref CURRENT was successfully assigned as follows:
      Engine:        V9
      Physical Name: C:\SAS\temp
```

```
%put SYSLIBRC: &SYSLIBRC;
SYSLIBRC: 0
%put SYSCC: &syscc;
SYSCC: 0
```

If the LIBNAME statement erroneously references &PATCH instead of &PATH, however, SAS generates a warning because it cannot resolve the macro variable &PATCH. The warning is detected and demonstrated in the macro variable &SYSCC, which is assigned the warning code 4. This error also causes the LIBNAME statement to fail, which is reflected in the negative &SYSLIBRC value.

```
libname current "&patch";
WARNING: Apparent symbolic reference PATCH not resolved.
NOTE: Library CURRENT does not exist.
%put SYSLIBRC: &SYSLIBRC;
SYSLIBRC: -70008
%put SYSCC: &syscc;
SYSCC: 4
```

Several methods could detect these failures programmatically. For example, the macro variable &PATH could be assessed with %SYMEXIST to determine if it's missing, or %LENGTH could assess whether &PATH has a zero character length. Following the LIBNAME statement, the value of &SYSCC could also be validated to prevent subsequent code from executing if a warning or runtime error was detected. The most reliable method, however, is to assess &SYSLIBRC after the LIBNAME statement and, if the value it not 0, to terminate the program or skip subsequent processes that require the CURRENT library.

For example, when the following code is executed and the directory C:\neverland does not exist, the negative value for &SYSLIBRC reflects this exception, but &SYSCC still indicates 0 because an exception rather than a runtime error has occurred:

```
%let syscc=0;
libname current "c:\neverland";
WARNING: Library CURRENT does not exist.
NOTE: Libref CURRENT was successfully assigned as follows:
      Engine:        V9
      Physical Name: c:\neverland
%put SYSCC: &syscc;
SYSCC: 0
%put SYSLIBRC: &syslibrc;
SYSLIBRC: -70008
```

In robust design, the value of &SYSLIBRC should be assessed immediately following the LIBNAME statement, and additional exception handling should monitor &SYSCC and other relevant automatic macro variables throughout execution.

&SYSRC

The &SYSRC automatic macro variable records the status of statements that interact directly with the OS environment, such as X, SYSEXEC, SYSTASK, and WAITFOR. Although not described in detail in this text, &SYSRC will have a 0 value after successful completion of these statements. Because &SYSRC is not reset until a subsequent X, SYSEXEC, SYSTASK, or similar statement, and because the macro variable is read-write, its value should be reset manually before later assessment, similar to &SYSCC initialization.

The following output demonstrates an initial success of the X statement in a Windows environment to pipe the contents of the current folder into the text file Dir.txt, followed by a subsequent failure when the DOS command DIR is misspelled DIRT:

```
%let sysrc=0;
%let syscc=0;
data _null_;
   x "dir > dir.txt";
run;

NOTE: DATA statement used (Total process time):
      real time           1.66 seconds
      cpu time            0.07 seconds

%put SYSRC: &sysrc;
SYSRC: 0
%put SYSCC: &syscc;
SYSCC: 0
data _null_;
   x "dirt > dir.txt"
!                        ;
run;

NOTE: DATA statement used (Total process time):
      real time           0.77 seconds
      cpu time            0.07 seconds
```

```
%put SYSRC: &sysrc;
SYSRC: 9009
%put SYSCC: &syscc;
SYSCC: 0
```

The &SYSCC return code is not affected by the value of &SYSRC; thus, exceptions in these statements must be handled by assessing &SYSRC. The &SYSRC macro variable is demonstrated in the "Batch Exception Handling" section in Chapter 12, "Automation," where it is used in an exception handling framework to demonstrate the success of SYSTASK statements.

SYSRC()

The SYSRC function, not to be confused with the &SYSRC automatic macro variable, returns the completion status of the most recently executed input/output (I/O) function. Because I/O functions are often implemented within the %SYSFUNC macro function, the value of SYSRC() can be saved as a macro variable and used to drive program flow through exception handling.

The following code and output display the number of variables in the Final data set:

```
%let dsid=%sysfunc(open(perm.final, i));
%let vars=%sysfunc(attrn(&dsid, nvars));
%let close=%sysfunc(close(&dsid));

%put VARS: &vars;
VARS: 1
```

Because the OPEN function requires a shared lock on the PERM.Final data set, however, the code will fail if a separate user or SAS session has an exclusive lock on the data set:

```
%let dsid=%sysfunc(open(perm.final, i));
%let vars=%sysfunc(attrn(&dsid, nvars));
WARNING: Argument 1 to function ATTRN referenced by the %SYSFUNC
or %QSYSFUNC macro function is out of range.
NOTE: Mathematical operations could not be performed during %SYSFUNC
function execution. The result of the operations have been set to
a missing value.
%let close=%sysfunc(close(&dsid));

%put VARS: &vars;
VARS: .
```

When OPEN fails, &DSID is not assigned; thus, the ATTRN function subsequently fails. The remedy is to always test the return code of I/O functions (like OPEN) with the SYSRC function before performing subsequent, dependent actions, such as the ATTRN or CLOSE functions. Because a value of 0 is returned under normal I/O functioning, the following code performs ATTRN only if OPEN is successful:

```
%macro test;
%let dsid=%sysfunc(open(perm.final, i));
%if %sysfunc(sysrc())=0 %then %do;
   %let vars=%sysfunc(attrn(&dsid, nvars));
   %let close=%sysfunc(close(&dsid));
   %end;
%else %let vars=FAILURE;
%put VARS: &vars;
%mend;

%test;
```

Another method to achieve the same program control is to test &DSID immediately after the OPEN function and, if the value is 0 (indicating a failure to open the data set), to abort the process.

SYSTEM ALPHANUMERIC RETURN CODES

SAS alphanumeric return codes are gratuitous in that they only mirror the numeric value of &SYSERR. They are introduced here, but due to several idiosyncrasies, their implementation is never recommended.

ALPHANUMERIC RETURN CODES

&SYSWARNINGTEXT The System Warning Text contains the last warning message generated.[12]

&SYSERRORTEXT The System Error Text contains the last error message generated.[13]

&SYSWARNINGTEXT

The automatic macro variable &SYSWARNINGTEXT contains the text of the last warning message produced in the current SAS session or printed to the SAS log. A warning message does not imply that an actual warning occurred.

For example, printing "WARNING: FAKE" with a %PUT statement will cause &SYSWARNINGTEXT to change to FAKE, as demonstrated in the following output:

```
%put BEFORE: &syswarningtext    LENGTH: %length(&syswarningtext);
BEFORE:     LENGTH: 0
%put WARNING: FAKE;
WARNING: FAKE
%put AFTER: &syswarningtext     LENGTH: %length(&syswarningtext);
AFTER: FAKE    LENGTH: 4
```

The following output demonstrates an actual warning that occurs when an uninitialized macro variable &NOTAREALMACRO is referenced:

```
%put BEFORE: &syswarningtext    LENGTH: %length(&syswarningtext);
BEFORE:     LENGTH: 0
%put &notarealmacro;
WARNING: Apparent symbolic reference NOTAREALMACRO not resolved.
&notarealmacro
%put AFTER: &syswarningtext     LENGTH: %length(&syswarningtext);
AFTER: Apparent symbolic reference NOTAREALMACRO not resolved.
      LENGTH: 55
```

Because &SYSWARNINGTEXT is read-only and resets only when replaced by a new warning, it cannot be manually reset to missing. Despite only producing a warning (as opposed to a runtime error) in the SAS log, the code execution still represents a failure because the desired variable was not displayed. For this reason, while displayed as a "WARNING" in the SAS log, in this instance, the missing macro variable constitutes an error, defined as *a human mistake*.

Because &SYSWARNINGTEXT cannot be reset, the macro variable will persist into subsequent programs until the SAS session is terminated. For example, after the previous code is executed, the following valid DATA step is subsequently run in a separate program albeit in the same SAS session, which produces very perplexing output:

```
%let syscc=0;
data final;
   length char $10;
run;

NOTE: The data set WORK.FINAL has 1 observations and 1 variables.
NOTE: DATA statement used (Total process time):
      real time           0.03 seconds
      cpu time            0.03 seconds
```

```
%put SYSCC: &syscc;
SYSCC: 0
%put SYSWARNINGTEXT: &syswarningtext;
SYSWARNINGTEXT: Apparent symbolic reference NOTAREALMACRO not resolved.
```

Despite resetting the &SYSCC value to 0 and not encountering any warnings or runtime errors, the code confusingly displays a &SYSWARNINGTEXT value that perseverates from the previous, unrelated program.

To overcome this limitation and ensure that &SYSWARNINGTEXT is assessed only when relevant, code should only reference &SYSWARNINGTEXT when &SYSCC is equal to 4 and when &SYSCC was reset immediately before the process. However, as demonstrated in the following "&SYSERRORTEXT" section, because a &SYSCC value greater than 4 (indicating an error) will mask a warning code, in some circumstances, there is no programmatic way to demonstrate that &SYSWARNINGTEXT is current and relevant. For this reason as well as the other caveats previously demonstrated, use of &SYSWARNINGTEXT is never recommended.

&SYSERRORTEXT

The &SYSERRORTEXT automatic macro variable is idiosyncratic and is introduced for no other reason than to discourage its use. It contains the text of the last error message produced in the current SAS session or printed to the SAS log. An error message does not imply that an actual runtime error occurred. For example, printing "ERROR: FAKE" with a %PUT statement will cause &SYSERRORTEXT to change to FAKE, as demonstrated in the following output:

```
%put BEFORE: &syserrortext   LENGTH: %length(&syserrortext);
BEFORE:     LENGTH: 0
%put ERROR: FAKE;
ERROR: FAKE
%put AFTER: &syserrortext    LENGTH: %length(&syserrortext);
AFTER: FAKE     LENGTH: 4
```

The following output demonstrates an actual runtime error that occurs when an extra dollar sign in the LENGTH statement creates a syntax error. Because the macro variable &SYSERRORTEXT is read-only and resets only when replaced by a new error, it cannot be manually reset to missing:

```
%put BEFORE: &syserrortext   LENGTH: %length(&syserrortext);
BEFORE:     LENGTH: 0
```

```
data final;
length char $$10;
                      -
                      391
                      76
ERROR 391-185: Expecting a variable length specification.
ERROR 76-322: Syntax error, statement will be ignored.
run;

NOTE: The SAS System stopped processing this step because of errors.
WARNING: The data set WORK.FINAL may be incomplete.  When this
step was stopped there were 0 observations and 0 variables.
NOTE: DATA statement used (Total process time):
      real time           0.08 seconds
      cpu time            0.06 seconds

%put AFTER: &syserrortext    LENGTH: %length(&syserrortext);
AFTER: 76-322: Syntax error, statement will be ignored.    LENGTH: 48
```

Another caveat of &SYSERRORTEXT is that it only captures the final runtime error while some defects will produce several consecutive runtime errors. In this example, the extra dollar sign produces two consecutive errors (391-185 and 76-322); thus when &SYSERRORTEXT is assessed immediately after the DATA step, only the second error "Syntax error, statement will be ignored" is retained. This is unfortunate, because the first error statement was slightly more informative: "Expecting a variable length specification."

&SYSERRORTEXT also can mask use of &SYSWARNINGTEXT. As described in the prior "&SYSWARNINGTEXT" section, &SYSWARNINGTEXT should only be assessed when &SYSCC was reset before a process and when it equals 4 after the process. However, in this example, because an error—in addition to a warning—is generated, &SYSCC is assigned the higher error value rather than the warning value of 4. In this example, there is no way to determine programmatically that the value of &SYSWARNINGTEXT is current and relevant because &SYSCC is set to an error code, not a warning code.

Similar to &SYSWARNINGTEXT, the value of &SYSERRORTEXT persists throughout a SAS session, so an irrelevant value of &SYSERRORTEXT will infiltrate subsequent, unrelated programs. For example, after executing the previous code, a separate program executes a DATA step and successfully creates the FINAL2 data set without any runtime errors, as demonstrated by the 0 values of both &SYSERR and &SYSCC. However, the value of &SYSERRORTEXT from the original program persists in the output because the read-only macro variable cannot be reset:

```
%let syscc=0;
data final2;
   length num 8;
run;

NOTE: The data set WORK.FINAL2 has 1 observations and 1 variables.
NOTE: DATA statement used (Total process time):
      real time           0.03 seconds
      cpu time            0.03 seconds

%put SYSERR: &syserr;
SYSERR: 0
%put SYSCC: &syscc;
SYSCC: 0
%put SYSWARNINGTEXT: &syswarningtext;
SYSWARNINGTEXT: Data set WORK.FINAL was not replaced because this step
was stopped.
%put SYSERRORTEXT: &syserrortext;
SYSERRORTEXT: 76-322: Syntax error, statement will be ignored.
```

To overcome this final weakness and ensure that &SYSERRORTEXT is assessed only when relevant, code should only reference &SYSERRORTEXT when &SYSCC is greater than 4 and when &SYSCC was reset immediately before the process. Notwithstanding, due to the numerous caveats mentioned and its limited utility, &SYSERRORTEXT is never recommended.

USER-GENERATED RETURN CODES

Just as SAS system return codes demonstrate the success or failure of SAS procedures, DATA steps, functions, and statements, developers should validate their own macros, processes, and programs with user-generated return codes. During software design, vulnerabilities that can lead to software failure should be identified, after which routines that detect these vulnerabilities can be implemented into an exception handling framework. Chapter 6, "Robustness," introduces defensive programming and provides examples of vulnerabilities that can exist for even seemingly straightforward SAS routines like reading a data set.

One software best practice is to require that at least one return code be generated for every macro that is developed. Assigning &SYSCC to 0 in the first line of the macro and checking the value of &SYSCC at macro termination can provide some defense against unexpected failure, not necessarily in the child process being evaluated, but in the parent process calling the macro and depending on the child's success. For example, before a DATA step, you may want to verify that the referenced data set is not empty.

The macro %GETOBS determines the number of observations in a data set. If this fails for any reason, the return code &GETOBS_RC will demonstrate the failure:

```
data temp;
run;

* determines the number of observations in a data set;
%macro getobs (dsn= /* data set name is LIB.DSN or DSN format */);
%let syscc=0;
%global obstot;
%let obstot=;
%global getobs_rc;
%let getobs_rc=GENERAL FAILURE;
proc sql noprint;
   select count(*)
   into :obstot
   from &dsn;
quit;
%if &syscc>0 %then %let getobs_rc=something aint right!;
%else %let getobs_rc=; * empty shows success;
%mend;

%getobs(dsn=temp);
%put RC: getobs_rc;
RC: something aint right!
```

Even if the developer doesn't realize that the SQL statement will fail when the data set has no variables (as is the case with Temp), this potentially unforeseen error is captured in &SYSCC. A runtime error will still have occurred, but the parent process calling %GETOBS will be able to dynamically reroute or terminate gracefully rather than succumbing to cascading failures. Note that the return code was uniquely named to ensure it doesn't conflict with return codes from other processes, as discussed in "Perseverating Macro Variables" in

Chapter 11, "Security." User-generated return codes are typically operationalized through two methods: in-band and out-of-band signaling, discussed in the next two sections.

In-Band Signaling

In-band signaling refers to passing metadata over an existing channel through which data are primarily intended to be transferred. In respect to software development, in-band return codes transfer performance information—*metadata*—in the same macro variable that *data* are otherwise transferred. A benefit of in-band return codes is simplification, since a separate global macro variable does not need to be created. The example in the "User-Generated Return Codes" section demonstrates an out-of-band return code (&GETOBS_RC) that is separate from the &OBSTOT macro variable. An in-band solution instead could have assigned the value of &OBSTOT to "something ain't right!" if an exception or error occurred, thus enabling &OBSTOT to contain either the number of observations or an error message in the event that the observation count could not be obtained.

The biggest caveat of in-band return codes is that they must be able to be unambiguously distinguished from valid data that otherwise would appear in the macro variable. For example, a macro variable assigned to represent the number of variables in a data set will always contain numeric data, lending itself to convey additional return code values through alphanumeric data. The alphanumeric return codes would not be confused with valid numeric data.

One example of a user-generated return code is partially demonstrated in the previous "SYSRC()" section. When the %TEST macro fails, the local macro variable &VARS is set to FAILURE. The %GLOBAL macro statement creates the return code and the %LET statement initializes the value. This initialization ensures that if the ATTRN function fails and &VARS is not set to the number of variables in the data set, the failure will be apparent:

```
%macro test;
%let syscc=0;
%global vars;
%let vars=GENERAL FAILURE;
%let dsid=%sysfunc(open(perm.final, i));
%if %sysfunc(sysrc())=0 %then %do;
   %let vars=%sysfunc(attrn(&dsid, nvars));
   %let close=%sysfunc(close(&dsid));
   %end;
```

```
%else %let vars=FAILURE;
%mend;

%test;
%put VARS: &vars;
```

In this example, the in-band return code &VARS has three possible outcomes. If &VARS is a number, it represents the number of variables present in the PERM.Final data set. If &VARS is FAILURE, it represents that OPEN encountered an exception such as a missing or locked data set. And, if &VARS is GENERAL FAILURE, it represents that some other exception or error occurred that prevents &VARS from being assigned.

In-band return codes are far less common than out-of-band, because in-band signaling requires that the metadata be unambiguously distinguished from the data. Thus, if a macro variable is designed to include alphanumeric text, it becomes more difficult to distinguish a valid value from performance metadata that are passed when a valid value cannot be generated.

Out-of-Band Signaling

Out-of-band return codes are much more common and useful than in-band return codes, because no ambiguity exists about whether a value constitutes valid data or performance metadata reflecting an error or failure. Also, because the return code still represents a global macro variable that must be initialized and reset, an empty return code can demonstrate a process that completed without warning or runtime error. The only potential downside of out-of-band signaling is the additional macro variable that must be created to pass the return code.

The in-band example from the "In-Band Signaling" section is modified to demonstrate out-of-band signaling. In this example, &VARS now only represents the number of variables in the data set, and no longer conveys performance failure. Thus, &VARS remains missing (rather than being set to FAILURE) if an exception occurs:

```
%macro test;
%global vars;
%let vars=;
%global testrc;
%let testrc=GENERAL FAILURE;
%let dsid=%sysfunc(open(perm.final, i));
%if %sysfunc(sysrc())=0 %then %do;
   %let vars=%sysfunc(attrn(&dsid, nvars));
```

```
   %let close=%sysfunc(close(&dsid));
   %let testrc=;
   %end;
%else %let testrc=FAILURE;
%mend;

%test;
%put VARS: &vars;
%put TESTRC: &testrc;
```

Another benefit of out-of-band return codes is that when an empty macro variable is defined to represent process success, the %LENGTH macro function can be used to assess whether an exception occurred, since the length of the return code will be 0 under normal functioning.

The following modification demonstrates the %RUNTEST macro, the parent routine that calls the child %TEST macro. If %TEST is successful, %RUNTEST will print the number of variables in PERM.Final; if it is unsuccessful, %TEST will print an error to the log. Of course, in actual production software, rather than printing messages to the log, an exception handling framework would dynamically alter the program flow:

```
%macro runtest;
%test;
%if %length(&testrc)=0 %then %put VARS: &vars;
%else %put An exception caused a failure;
%mend;

%runtest;
```

The logic creating an out-of-band return code must ensure than the return code remains synchronized with the macro variables of processes that it represents. For example, in the previous code, &TESTRC represents performance of the %TEST macro function, which produces &VARS; thus &TESTRC should be perfectly correlated with &VARS. If &VARS contains a value, &TESTRC should be missing; if &VARS is missing, &TESTRC should be set to FAILURE or GENERAL FAILURE.

In the previous %TEST macro, however, if an exception or error occurs in the ATTRN function, the value of &VARS will not be set, so it will be empty. However, &TESTRC will also be empty because exception handling only tests for failure of the OPEN function and does not test for failure of ATTRN. While this failure is unlikely to occur, this logic error could cause a failure in which both macro variables are empty. To close this loophole, either the success of the ATTRN function must additionally be tested or the value of &VARS must be tested to validate that it was successfully assigned. The successful use of return codes is further demonstrated in the "Exception Handling" section in Chapter 6, "Robustness."

PARALLEL PROCESSING COMMUNICATION

Return codes are critical to software performance, but they are only viable within a single SAS session. Once the session is terminated, return codes and all other macro variables are eliminated. In complex software designs that use the SYSTASK command to spawn batch jobs, or in environments in which SAS sessions run in parallel and must communicate with each other, return codes are of no use. Mutexes, semaphores, control tables, and the SYSPARM parameter instead are required to facilitate cross-session communication. Additional communication complexities that can exist in parallel processing are described in the "Complexities of Concurrency" section in Chapter 11, "Security."

Mutexes and Semaphores

A *semaphore* is "a shared variable used to synchronize concurrent processes by indicating whether an action has been completed or an event has occurred."[14] Semaphores are flags that represent events or conditions in software, and are typically represented by variables. In SAS software, however, because macro variables cannot be read across SAS sessions, files or data sets must be used as semaphores to signal between sessions, similar to flag semaphores that are used to signal between passing naval vessels. One of the most common uses of semaphores is to signal whether an object (such as a data set) is available or whether it is in use.

In some cases, an object can only be used by one process at a time, for example, when an exclusive lock is required to access and modify a SAS data set. A *mutex semaphore* (or *mutex*) represents a class of semaphore that ensures mutually exclusive access to some object, and is defined as "a mechanism for implementing mutual exclusion."[15] A mutex ensures that not only is current data set use flagged, but also that other processes cannot interrupt this use when the flag is detected. Because of the inconsistency demonstrated in the LOCK statement, which was described previously in the "&SYSLCKRC" section, a mutex is required to exclusively lock a data set to facilitate communication between sessions in parallel processing.

In this example, the mutex is operationalized as an empty text file that is created and similarly named to a data set. If the mutex (i.e., text file) exists and is not exclusively locked by any SAS session, the current SAS session locks the mutex and can subsequently access the associated data set securely; when data set access is no longer needed, the mutex lock is released. If, however, a SAS session encounters a mutex that is locked, exception handling prevents access to the associated data set. This logic enables two or more concurrent sessions of SAS to take turns accessing a data set without fear of file access collisions and runtime errors.

To demonstrate use of mutexes, the following code can be run in one session of SAS:

```
libname perm 'c:\perm';

data perm.test;
   length char1 $10;
run;

%macro mutex(dsn= /* in LIB.DSN format */,
   cnt= /* number of iterations to run */);
%local lib;
%local tab;
%local closed;
%local dsid;
%let lib=%scan(&dsn,1);
%let tab=%scan(&dsn,2);
%let loc=%sysfunc(pathname(&lib))\&tab..txt.lck;
filename myfil "&loc";
%do i=1 %to &cnt;
   %let dsid=%sysfunc(fopen(myfil,u));
   %if &dsid>0 %then %do;
      data &dsn;
         set &dsn;
      run;
      %let closed=%sysfunc(fclose(&dsid));
      %end;
   %end;
%mend;

%mutex(dsn=perm.test, cnt=1000);
```

In a second session of SAS, immediately execute the LIBNAME statement and %MUTEX macro (but not the DATA step) so that both sessions run in parallel. The program will create the mutex (i.e., text file C:\perm\test.txt.lck), which both sessions will continuously test for access to the PERM.Test data set. Only when a session can exclusively lock the mutex will that session execute the subsequent DATA step, thus preventing runtime errors that would otherwise occur from collisions.

Control Tables

A *control table* represents data stored externally to program code used to facilitate data-driven processing, often facilitating bidirectional communication in which software both writes to and reads from the control table. In SAS literature, control tables are sometimes referred to as *control data sets*, reflecting SAS parlance that abandons the relational database terminology *table* in favor of *data set*. In addition to containing data that dynamically drive program flow, some control tables additionally collect performance metrics from processes, sometimes in turn using those metrics to drive further processing. Thus, while some control tables contain just enough information to prescribe program flow, others can additionally contain a chronology of past performance data. Sample control tables are demonstrated in Table 7.1 in Chapter 7, "Execution Efficiency," and in Table 9.1 in Chapter 9, "Scalability."

Both control tables and configuration files dynamically drive program flow, facilitating flexibility that decreases hardcoding while increasing usability. Configuration files are distinguished, however, because communication flows only from the configuration file to the software. Configuration files also are typically only read during software initialization, whereas control tables can be read throughout software execution. Configuration files tend to be updated manually so that varied results can be produced the next time software is executed, whereas control tables more commonly are created and modified by software itself while it executes. Configuration files are demonstrated in the "Custom Configuration Files" section in Chapter 17, "Stability."

Because macro variables cannot be passed across independent SAS sessions running in parallel, control tables can facilitate cross-session communication, allowing software in each session to take turns leaving and retrieving information in the same control table. Because an exclusive lock is required to modify a data set, and because file access collisions will occur if two or more users or processes attempt to exclusively access the same SAS data set simultaneously, a mutex must be implemented to prevent collisions and runtime errors in parallel processing. Mutexes are introduced previously in the "Mutexes and Semaphores" section. The use of control tables to support parallel processing is demonstrated in the "Parallel Processing" section in Chapter 7, "Execution Efficiency."

Another use of control tables is to return data and metadata from child to parent processes. When the SYSTASK statement is used to create a batch process, the child process is spawned in a new session of SAS apart from the parent. While the parent can pass information to the child via the SYSPARM parameter (discussed in the next section), the child has no inherent method to pass data or metadata across session boundaries back to the parent. For this reason, similar to the way in which control tables are operationalized to facilitate parallel processing, child batch processes can dump data and metadata into a control table that a parent process can later retrieve to validate success of the batch process and to retrieve other data or output.

When a batch process is spawned with SYSTASK and warnings or runtime errors occur in the child process, the parent is informed via the &SYSRC macro variable. The &SYSRC macro variable is introduced in the "&SYSRC" section, and is demonstrated in the "Euthanizing Programs" section in Chapter 11, "Security." Because the &SYSRC value following a SYSTASK statement will have limited numeric values that convey little useful information, however, custom exception handling frameworks must typically be engineered to detect and respond to warnings and runtime errors that occur in batch processes.

SYSPARM Parameter and Macro Variable

SYSPARM is a command-line option that enables information to be passed during invocation of a SAS batch job, either through a SAS program via the SYSTASK statement or directly from the OS. Because macro variables cannot be passed from one SAS session to another, a parent process spawning a child process can pass information only to that child as quoted text within the SYSPARM parameter. The child process in turn is able to access the text through the automatic macro variable &SYSPARM.

For example, in the simplest invocation, the following SYSTASK statement in a parent process spawns a new SAS session, runs the program

C:\perm\prog.sas, and passes a SAS title (&TITLE) to be used in a child process via the SYSPARM parameter:

```
systask command """%sysget(SASROOT)\SAS.exe"" -noterminal -nosplash
   -sysparm ""&title"" -sysin ""c:\perm\prog.sas""";
```

In the child process (not shown), the value of &SYSPARM would automatically be set to the value of &TITLE from the parent process when the batch job is run. SYSPARM is discussed throughout Chapter 12, "Automation," with the section "Passing Parameters with SYSPARM" demonstrating how to pass multiple parameters through tokenization and parsing of SYSPARM. Stress testing demonstrated in the "Stress Testing" section in Chapter 16, "Testability," shows that due to length limitations of SYSTASK, SYSPARM values should be no more than 8,000 characters.

WHAT'S NEXT?

In the next nine chapters, dynamic aspects of software performance are discussed, describing those dimensions of quality that are directly observable through software execution. Reliability, described in the next chapter, is often considered paramount to other software performance requirements that are subsumed under it. With software reliability, stakeholders can be assured that SAS software will function correctly, consistently, and for the intended lifespan of the software.

NOTES

1. "%RETURN Statement." *SAS® 9.4 Macro Language: Reference, Fourth Edition*. Retrieved from http://support.sas.com/documentation/cdl/en/mcrolref/67912/HTML/default/viewer.htm#p0qyygqt5a69xnn1rhfju3kjs8al.htm.
2. "Automatic Macro Variables." *SAS® 9.4 Macro Language: Reference, Fourth Edition*. Retrieved from http://support.sas.com/documentation/cdl/en/mcrolref/67912/HTML/default/viewer.htm#n18mk1d0g1j31in1q6chazvfseel.htm.
3. "Interfaces with the SQL Procedure." *SAS® 9.4 Macro Language: Reference, Fourth Edition*. Retrieved from http://support.sas.com/documentation/cdl/en/mcrolref/67912/HTML/default/viewer.htm#p1mw9xt2ti8f2xn1ozbct4305aqa.htm.
4. "SYSERR Automatic Macro Variable." *SAS® 9.4 Macro Language: Reference, Fourth Edition*. Retrieved from http://support.sas.com/documentation/cdl/en/mcrolref/67912/HTML/default/viewer.htm#n1wrevo4roqsnxn1fbd9yezxvv9k.htm.
5. "SYSCC Automatic Macro Variable." *SAS® 9.4 Macro Language: Reference, Fourth Edition*. Retrieved from http://support.sas.com/documentation/cdl/en/mcrolref/67912/HTML/default/viewer.htm#p11nt7mv7k9hl4n1x9zwkralgq1b.htm.

6. "Using the PROC SQL Automatic Macro Variables." *SAS® 9.4 SQL Procedure User's Guide, Third Edition*. Retrieved from http://support.sas.com/documentation/cdl/en/sqlproc/69049/HTML/default/viewer.htm#p0xlnvl46zgqffn17piej7tewe7p.htm.
7. "SYSFILRC Automatic Macro Variable." *SAS® 9.4 Macro Language: Reference, Fourth Edition*. Retrieved from http://support.sas.com/documentation/cdl/en/mcrolref/67912/HTML/default/viewer.htm#p1dziosxz7yamln0zvp6r4sr64c5.htm.
8. "SYSLIBRC Automatic Macro Variable." *SAS® 9.4 Macro Language: Reference, Fourth Edition*. Retrieved from http://support.sas.com/documentation/cdl/en/mcrolref/67912/HTML/default/viewer.htm#p0vnvc6k3fo349n1aszgd2ls6oxk.htm.
9. "SYSLCKRC Automatic Macro Variable." *SAS® 9.4 Macro Language: Reference, Fourth Edition*. Retrieved from http://support.sas.com/documentation/cdl/en/mcrolref/67912/HTML/default/viewer.htm#p1rkc7ffzq9jnnn12gulu6k81g6y.htm.
10. "SYSRC Function." *SAS® 9.4 Functions and CALL Routines: Reference, Fourth Edition*. Retrieved from https://support.sas.com/documentation/cdl/en/lefunctionsref/67960/HTML/default/viewer.htm#n1xxj8ih9sjbznn1r17pgxuf8kpr.htm.
11. Troy Martin Hughes, "From a One-Horse to a One-Stoplight Town: A Base SAS Solution to Preventing Data Access Collisions through the Detection and Deployment of Shared and Exclusive File Locks." Presentation to Western Users of SAS Software (WUSS), 2014.
12. "SYSWARNINGTEXT Automatic Macro Variable." *SAS® 9.4 Macro Language: Reference*, 4th ed. Retrieved from http://support.sas.com/documentation/cdl/en/mcrolref/67912/HTML/default/viewer.htm#p1ikib7e81mbaon1iygs8ebnvhx8.htm.
13. "SYSERRORTEXT Automatic Macro Variable." *SAS® 9.4 Macro Language: Reference*, 4th ed. Retrieved from http://support.sas.com/documentation/cdl/en/mcrolref/67912/HTML/default/viewer.htm#n0f6lmit5jr1xen1dl1owpkufvfj.htm.
14. ISO/IEC/IEEE 24765:2010. *Systems and software engineering—Vocabulary*. Geneva, Switzerland: International Organization for Standardization, International Electrotechnical Commission, and Institute of Electrical and Electronics Engineers.
15. ISO/IEC/IEEE 21451-1:2010. *Information technology—Smart transducer interface for sensors and actuators—Part 1: Network Capable Application Processor (NCAP) information model*. Geneva, Switzerland: International Organization for Standardization, International Electrotechnical Commission, and Institute of Electrical and Electronics Engineers.

PART I

Static Performance

CHAPTER 4

Reliability

Chicken buses.

For the uninitiated, they're the decommissioned big, yellow school buses, shipped, driven, or towed south from the United States to Guatemala for a new life and revitalization. But to be reborn as a chicken bus, you need four things—chickens, color, bling, and exhaust.

With large segments of the Maya people living in fairly remote, inaccessible regions, public buses are largely favored over personal vehicles in highland pueblos. As an agrarian people, the Maya transport not only their produce but also their livestock to large communal markets twice or thrice weekly. I've encountered live chickens on chicken buses several dozen times—but also goats, sheep, and one "dog" that was so ugly I still believe a family was bussing their chupacabra to market to sell.

The Guatemalan highlands are the most colorful landscape and people I've ever encountered, boasting the largest concentrations of ethnic Maya in the world. The orchidaceous colors woven for centuries into their textiles are vibrantly reincarnated on the buses that crisscross the landscape's mountainous roads. Vibrant paintings of saints, the Virgin Mother, and biblical quotations are common chicken bus adornments, as are tie-dye and Christmas themes.

Every bus is also chromed to the max—its grill, bumper, and wheels, naturally, but also vintage air horns, Mercedes and Rolls Royce emblems, and spurious smoke stacks, ladders, railings, and anything else that can be polished to a mirrored finish. That is to say that despite the cobblestone and dirt roads that chicken buses must daily navigate, their exteriors are impeccably maintained.

As for the exhaust—well, Guatemala boasts neither the emission standards nor maintenance requirements of the United States, so plumes of black smoke trail nearly all the buses, while passengers are left to subtly choke and question the integrity and reliability of the service.

I was en route from Antigua to Panajachel, the largest pueblo bordering the pristine Lake Atilán. We'd endured the groaning engine, straining gears, and tumultuous shifting for nearly two hours—not to mention the stench of livestock—when finally the chicken bus slowed on an uphill climb and drifted off the road to a halt. After 20 minutes of commotion, we were ordered to disembark. No, it wasn't the *bandidos* (bandits) this time, just another engine gone awry.

Now I never heard the original instructions from the *conductor* (driver), over the braying sheep and brash cry of infants, but as I strode around the roadside with my 100 companions, it was clear that we were supposed to be searching for something. But what?

A young student, amused with my consternation, finally translated that we'd been ordered to scour the countryside for—now wait for it—pieces of metal and twine.

I watched with a mixture of respect and fear as the *conductor* tinkered in the engine and "repaired" our chicken bus with roadside twine and other sundry scraps that had been gathered by passengers. An hour later, we were again slowly making our way to the lake, everyone seemingly undaunted by the day's events. They'd undoubtedly seen this scene play out enough times that it neither raised eyebrows nor diminished spirits. We'd reach Pana when we reached Pana.

■ ■ ■

To many Guatemalans, chicken bus reliability is not a priority in life. They do value bus security (i.e., no *bandidos*), but if a bus picks them up hours late or breaks down en route, they don't get too ruffled. In fact, they're happy to pitch in and search the brush for metal, twine, and other scraps and to lend a helping hand to unexpected roadside maintenance.

In many end-user development environments, software reliability is not valued as highly as other quality characteristics. Especially in data analytic development environments, emphasis is placed on getting the right data and analytic solution, but not necessarily on producing software that can run reliably without assistance or regular maintenance.

If software encounters unpredictable exceptions or otherwise fails, SAS practitioners often can fix it on the spot without additional assistance. They may have been the original developers of the software but, even if not, as software developers, SAS practitioners are able to get their software back on the road.

Thus, in some ways, end-user software development mimics chicken buses in that the *conductor* is expected not only to be a formidable driver, but also to maintain the bus. The benefits are great, because a skilled *conductor* who can moonlight as a mechanic is more likely to discern when a bus is running poorly, struggling, and about to fail. And, with this mechanical knowledge of the bus, he's often able to get the bus—for better or for worse—running again without a tow.

Because data analytic developers understand not only their industry and business through domain expertise, but also software development, they're more likely to detect when software is running poorly or producing invalid output or data products. Following detection of runtime errors, SAS practitioners are able to gather the necessary scraps to get their code functioning again.

A downside of data analytic development, however, is its reliance not on the integrity of software but on the end-user developers themselves who maintain it. Build software for a third-party and separate the developers from the users so the two never shall meet, and users must trust that *software* will be reliable, consistent, and provide accurate results for its intended lifespan. But imbed users with developers (including end-users who are developers), and software reliability typically plummets because reliability is instead placed in the *people*, not the *software*.

This chapter introduces software reliability which requires a shift from trusting in the personnel who maintain software to trusting in the software itself to be robust to failure, to function consistently, and to perform for its intended lifespan without excessive maintenance or modification. A hallmark of reliable SAS software is that it can be executed if necessary by non-developers with minimal Base SAS knowledge.

DEFINING RELIABILITY

Reliability is defined in the International Organization for Standardization (ISO) software product quality model as "the degree to which the software product can maintain a specified level of performance when used under specified conditions."[1] The Institute of Electrical and Electronics Engineers (IEEE) additionally includes the concept of software endurance or longevity, defining *reliability* as "the ability of a system or component to perform its required functions under stated conditions for a specified period of time."[2] Historical definitions typically incorporated *failure* into reliability—for example, "the probability that software will not cause the failure of a system for a specified time under specified conditions."[3]

A more general definition of reliability is a *measure of performance against failure*, often defined mathematically as the probability of error-free operation. A reliable product or service can be depended upon to deliver expected functionality and performance in a consistent, repeatable manner and for some expected duration. Because an unreliable product is often considered worthless, reliability is typically considered paramount to other performance requirements. Within the ISO software product quality model, the reliability dimension additionally includes subordinate characteristics of availability (described in this chapter), fault-tolerance (described in Chapter 5), and recoverability (described in Chapter 6).

The ACL triad that includes availability, consistency, and longevity is introduced, demonstrating qualities that reliable software must espouse. Software failure and the risk thereof are explored through the contributing effects of errors, exceptions, defects, faults, threats, and vulnerabilities. Through reliable software, SAS practitioners demonstrate their technical prowess and earn the trust and respect of stakeholders who learn to trust the software—not only the maintenance skills—of their developers.

PATHS TO FAILURE

If failure distinguishes reliability from unreliability, it's important first to define software failure and distinguish this from other related, often commingled terms, including *defect, fault, bug,* and *error*. Software reliability is ultimately judged from the perspective of the customer or sponsor, the source of software requirements. Failures are identified as customers detect software deviations from stated requirements, and as users detect deviations from software marketing, instruction, and other documentation. Failure can also be interpreted and defined differently based on stakeholder perspective, as demonstrated in the "Redefining Failure" section in Chapter 5, "Recoverability."

Because reliability can be thwarted by functional or performance failures and because numerous pathways lead to software failure, some differentiation between these terms is beneficial. The accompanying textbox defines failure-related software development terms. In this text, *defect* is used to represent incorrect code that causes a failure, in keeping with IEEE definitions. *Errors* are distinguished from defects, with the former causing the latter, described in the following section.

RELIABILITY TERMS

Reliability "The ability of a system or component to perform its required functions under stated conditions for a specified period of time."[4]

Failure "Termination of the ability of a product to perform a required function or its inability to perform within previously specified limits."[5]

Defect "A generic term that can refer to either a fault (cause) or a failure (effect)."[6]

Exception "An event that causes suspension of normal program execution. Types include addressing exception, data exception, operation exception, overflow exception, protection exception, underflow exception."[7] Specific exception types are not distinguished in this text.

Fault "A manifestation of an error in software."[8] *Faults* represent defects or errors that are discovered during execution, but are not further distinguished in this text.

Error "A human action that produces an incorrect result."[9]

Bug "An error or fault."[10] Because of the ambiguity with which *bug* has traditionally been used and defined, the term is not further distinguished or used in this text.

Runtime Error Throughout this text, *runtime errors* are distinguished as failures that occur during SAS software execution that cause syntax and other errors to be printed to the SAS log. This distinction is important because it is common in SAS to have errors (human mistakes) that produce defects in code which result in incorrect output or results (i.e., failure) but which do not produce runtime errors.

Errors, Exceptions, and Defects

In most software literature, errors are described as some variation of *human-derived mistakes*. Errors are not limited to code and in fact often point to derelict, incomplete, or misconstrued software requirements. Errors can also occur when incomplete or otherwise invalid data or inputs are received. The entry of invalid data into software can signal errors in data entry or data cleaning processes but if the software was intended to be reliable, the inability of SAS software to detect and handle invalid data appropriately represents another error—the lack of quality controls.

Errors printed to the SAS log are referenced in this text as *runtime errors* to distinguish them from the IEEE definition of *error*, which reflects a human mistake. Runtime errors can occur when SAS practitioners make errors in code (syntax errors) or for various reasons such as out-of-memory conditions. When SAS runs out of memory (and produces a runtime error), the underlying error (human mistake) could represent inefficiently coded SAS syntax, the failure of requirements documentation to state memory usage thresholds, or the failure of a SAS administrator to procure necessary memory. Thus, the source of runtime errors may not be programmatic in nature despite causing programmatic failure.

The SAS Language Reference lists several types of errors that can occur in software, including syntax, semantic, execution-time, data, and macro-related.[11] Syntax errors, for example, occur when invalid or imperfectly formed code is executed, a common example being the infamously missing semicolon:

```
data final
   set original;
run;
```

For the purpose of this text, syntax and semantic errors are not distinguished because they represent invalid code that should not exist in production software and which likely will have been discovered during software testing. In this example, the error—the missing semicolon—can also be considered the defect, because the code is corrected by adding the semicolon, which prevents further software failure.

Exceptions occur when software encounters an unexpected, invalid, or otherwise problematic environmental state, data, or other element or action. For example, if data standards specify variable value ranges, receiving invalid data beyond those ranges represents an exception. Or if SAS software is designed and developed to run on Windows, an exception occurs when SAS encounters a UNIX environment. Exceptions are described in the section "Exception Handling" in Chapter 6, "Robustness," which demonstrates methods to detect and handle exceptions to facilitate robust, reliable execution.

Some software exceptions are automatically handled by Base SAS native exception handling routines, such as division-by-zero detection, which prints a note to the SAS log that a zero denominator was detected but does not produce a runtime error. In fact, one of the primary goals of exception handling is to detect some abnormal state (like a zero denominator) and prevent a runtime error from occurring. Base SAS has very few native exception-handling capabilities, so in most cases developers will need to build an exception-handling framework that forestalls exceptional events and other vulnerabilities that can cause software failure.

A division-by-zero exception is detected and handled natively by Base SAS, which converts the attempted fraction to a missing value, as demonstrated in the following output:

```
data temp;
   num=5;
   denom=0;
   div=num/denom; * div set to missing;
run;
```

```
NOTE: Division by zero detected at line 1786 column 12.
num=5 denom=0 div=. _ERROR_=1 _N_=1
NOTE: Mathematical operations could not be performed at the
following places. The results of the operations have been set
to missing values.
```

Because SAS contains built-in exception handling to detect division by zero, a note rather than error is printed to the log, and both the automatic macro values for &SYSCC and &SYSERR remain 0, representing error-free execution.

If this is the intended outcome of division by zero, it's acceptable to allow SAS native exception handling to handle division by zero. Thus, in the previous example, no additional user-defined exception handling is required because the missing value is desirable.

In other cases, however, software requirements might specify that native exception handling is not sufficient. For example, rather than setting a value to missing, perhaps the entire observation should have been deleted. The zero value might indicate that more substantial data errors exist elsewhere, requiring termination of the process or program. In these cases, division by zero could represent not only an exception, but also a failure caused by software errors.

In some cases, division by zero can be caused by faulty or missing exception handling that fails to detect the zero denominator before division. In the following example, exception handling attempts to prevent division by zero but, due to a typographical error, instead eliminates values of one. Thus, while the exception—the division by zero—occurs in line four, line three contains the typographical error and thus the defect that must be corrected:

```
data final;
   set original;
   if denom^=1 then do;    * typographical error and defect occur here;
      div=num/denom;    * exception occurs here;
      end;
   else delete;
run;
```

Another version of the division-by-zero exception occurs when exception handling to detect a zero denominator was omitted because no quality control check was specified in software requirements. Thus, while the exception occurs in line three of the code, the real error is a failure of creativity in software requirements: SAS practitioners overlooked the possibility that zero values could be encountered. During software design, protocols for exceptional data should be discussed so that software can respond in a predictable manner if they are encountered:

```
data final;
   set original;
   div=num/denom;    * execution error occurs here;
run;
```

A final version of the division-by-zero exception occurs when quality control (to prevent zero denominators) was incorporated into requirements but the requirements were not implemented. For example, perhaps the requirements stated that the first DATA step (which creates Original) should delete all observations for which the value of Denom was 0 or missing. However, because this quality control was omitted, an exception occurs in the second module when invalid data are passed that include zero Denom values. This error ultimately represents a failure in communication or interpretation of technical requirements:

```
* MODULE 1;
data original;
   set temp;    * error (omission of quality control) occurs here;
run;

* MODULE 2;
data final;
   set original;
   div=num/denom;        * exception occurs here;
run;
```

In this scenario, software requirements prescribe that data quality controls should have been placed in the first module, so the overarching error occurs here, as well as the defect that should be corrected. However, this error goes undetected in the first DATA step because the exception—the division by zero—does not occur until the second module. To remedy the exception, the defect in the first DATA step can be corrected by implementing the required exception handling there, or the requirements can be modified so that the exception handling can be implemented in the second DATA step. Either solution will correct the failure, but this demonstrates that exception- and error-causing code can often be defect-free while pointing to defects and errors in separate code or modules.

While division by zero represents a handled exception, many runtime errors are caused by unhandled exceptions. For example, the DELETE procedure was introduced in SAS 9.4, so use of the procedure in previous SAS versions will cause runtime errors and software failure. For SAS practitioners who implement the DELETE procedure in code, exception handling would be required to make the software portable to previous versions of SAS, possibly by instead deleting data sets with the DATASETS procedure. In this way, while the exception—encountering SAS 9.1—still occurs, it is detected and handled, preventing runtime errors and software failure. This practice is discussed further in the "SAS Version Portability" section in Chapter 10, "Portability."

Failure

Failure occurs when software does not function as needed, as intended, or as specified in technical requirements. A functional failure occurs when software does not provide the specified service, result, or product. In data analytic development, SAS software is often developed to produce an analytic product that may incorporate separate data ingestion, cleaning, transformation, and analysis steps. If the ultimate analysis is incorrect due to errors in any of these steps, this represents a functional failure because the intent of the software—the analytic product—is invalid.

In the scenario in the "Errors, Exceptions, and Defects" section, the failure to delete observations that contained a division by zero is a functional failure because the software creates an invalid data product. On the other hand, when the analysis is correct but is produced more slowly than specified by technical requirements, this instead represents a performance failure. Because this text focuses on the performance side of quality, unless otherwise specified, all scenarios and code examples depict fully functional solutions that deliver full functionality and valid data products; all failures represent performance failures.

In addition to producing incorrect behavior, a software failure by most definitions must also be observable—not necessarily the cause, but at least the effect. That a failure *can be* observed, however, does not indicate that it *has been* observed; as a result, some failures can go unnoticed for days, weeks, or even the entire software lifespan. An extreme case of functional failure occurs when SAS software produces invalid results every time it is ever run due to erroneous formulas having been applied to transform data. An extreme case of performance failure is extract-transform-load (ETL) software that produces valid data products, but which never executes rapidly enough to meet deadlines, possibly due to inefficient code. In both cases, business value is substantially impaired if not eliminated.

In the case of failures that produce SAS runtime errors, clues often exist that can assist developers in diagnosis and defect resolution. For example, in the "Errors, Exceptions, and Defects" section, the division-by-zero exception prints a note in the SAS log to inform users. A benefit of SAS runtime errors, regardless of type, is that they are easily discoverable via the SAS log. This doesn't mean, however, that defects are easily resolved or necessarily pinpointed, but it does provide a starting point for debugging efforts. As previously demonstrated, an error in one module can point to underlying software defects in other modules. In other cases, a runtime error actually points to a failure in the definition, interpretation, or implementation of technical requirements.

Many failures, however, produce no runtime errors. Functional failures can unknowingly produce invalid results and worthless data products while demonstrating a clean SAS log, especially where incorrect yet syntactically valid business logic exists. One hallmark of unreliable software is that it produces the correct result *some* of the time but not *all* of the time, which can especially infuriate developers. The following code is intended to concatenate a series of values into a space-delimited global macro variable and, when executed the first time, prints the correct result:

```
%macro test;
%global var;
%do i=1 %to 5;
   %let var=&var &i;
   %end;
%mend;

%test;
%put VAR: &var;
VAR: 1 2 3 4 5
```

When executed a second time from the same SAS session, however, while still producing no runtime errors, the code fails functionally by producing invalid results:

```
%test;
%put VAR: &var;

VAR: 1 2 3 4 5 1 2 3 4 5
```

In this example, the SAS practitioner mistakenly assumed that the %GLOBAL macro statement both creates and initializes a macro variable, resetting it to a blank value. As stated in the *SAS Macro Language: Reference*, however, "If a global macro variable already exists and you specify that variable in a %GLOBAL statement, the existing value remains unchanged."[12] To prevent this failure, the &VAR macro variable must be initialized with one additional line of code:

```
%macro test;
%global var;
%let var=;        * corrects the defect;
%do i=1 %to 5;
   %let var=&var &i;
   %end;
%mend;
```

This failure reduces the reliability of the code because its results are inconsistent—sometimes accurate and sometimes not. In this example, because the ultimate function of the code is also compromised, the uncorrected version demonstrates a lack of accuracy that contributes to its unreliability.

Just as the previous failure did not produce any runtime errors, performance failures can also occur while demonstrating a clean SAS log. When a software requirement specifies that an ETL process must complete in two hours, yet the process occasionally completes in three, this represents a performance failure because on-time completion is slow and unreliable. The underlying error could be a programmatic one, in which inefficient coding caused the occasional delay. A related programmatic error could be the implementation of SAS processes that lacked scalability, in that larger data sets caused processes to slow beyond the acceptable time threshold.

Nonprogrammatic errors could also contribute to the unreliability and slow execution. For example, attempting to run the ETL process on antiquated SAS software, a slow network, or faulty hardware could also contribute to the unsatisfactory three-hour run time. It's also possible that both the code and hardware were of sufficient quality while the technical requirements themselves were unrealistic. Thus, faced with this failure, SAS practitioners must decide whether to implement a programmatic solution or a nonprogrammatic solution, or whether they should modify the requirements (to a three-hour execution time) so that all performance requirements can be reliably met.

Sources of Failure

Throughout this text, only programmatic sources of failure and their respective programmatic remedies are discussed. In some environments, SAS practitioners are lucky enough to concern themselves only with the software they write and have no other information technology (IT) administrative responsibilities. This paradigm exists when a development team is responsible for authoring SAS software while a separate operations and maintenance (O&M) team installs, updates, maintains, optimizes, and troubleshoots the SAS installation. In some team structures, one or more developers may be designated as SAS platform administrators and bear the brunt of O&M responsibilities, freeing others to focus on development.

This division of labor can be very beneficial to SAS practitioners because it allows them to focus solely on software development and other responsibilities, avoiding the weeds of often enigmatic IT resources and infrastructure.

It benefits the biostatistician who, although adept at writing SAS programs that support analysis and research, might be uncomfortable administrating a SAS server, purchasing and optimizing hardware, and configuring a high-availability environment.

The reality of many SAS teams, especially those in end-user development environments, is that IT administrative responsibilities are shared. Even if one member is primarily responsible for managing the SAS infrastructure and other administrative tasks, these responsibilities may be secondary to SAS development and non-IT activities. Due to this lower prioritization of SAS administrative tasks, in these environments, the SAS server may never truly be customized or optimized for performance.

It benefits SAS practitioners to understand their infrastructure, including hardware, network, related software, and SAS administrative functions. A researcher might be faced with SAS code that fails to complete in the required time period or terminates abnormally due to nonprogrammatic issues. Without either a designated O&M team or SAS administrator, the reliability and overall performance of SAS software may be severely diminished. Some nonprogrammatic sources of failure include:

- *SAS system options*—While some SAS system options can be modified programmatically, others must be incorporated into the SAS configuration file or command line when run in batch mode. If the SORTSIZE system option is too small, for example, sorting can fail or perform inefficiently as memory is consumed.
- *SAS infrastructure*—Improper configuration of SAS servers can cause failure. For example, too many SAS sessions running on too few CPUs can spell disaster.
- *Third-party software*—Whether the Java installation was recently updated or a Windows patch was installed, these changes can cause the SAS application to fail due to interoperability challenges.
- *Hardware*—Processors can run out of memory and disk drives can run out of space. Even more pernicious hardware failures include the permanent destruction of SAS code or data due to server or drive failure.
- *Network infrastructure*—If the network pipe is slow enough, it can cause performance failure. If the network fails entirely, SAS practitioners can lose access to their software and/or data.
- *User error*—Software requirements and documentation, in prescribing how software should function and perform, either explicitly or implicitly

denote the manner in which users must interact with software. When users run software outside of technical specifications—either accidentally or maliciously—failure can occur.

With the exception of SAS system options that can be modified programmatically, these sources of failure are not discussed further in this text. Nevertheless, SAS practitioners can substantially increase the value and performance of their software by gaining a fuller understanding of these nonprogrammatic factors that influence software performance and failure.

Failure Log

Whereas a *risk register* contains all known software vulnerabilities—theoretical and exploited, as well as related threats and risks—a *failure log* includes entries for each software failure. Failure does not necessarily denote that the software terminated with a runtime error or provided an invalid solution, but rather denotes that one or more functional or performance requirements were not achieved. Risk registers are introduced in the "Risk Register" section in Chapter 1, "Introduction."

Within end-user development environments, operational failures—once software is in production—are typically identified by developers themselves. In more traditional software development environments, software users more commonly encounter failures and report them to an O&M team, who pass this information to developers. However and by whomever failures are recorded, it's critical that developers receive as much information as possible about failures so they can recreate, investigate, and remedy all errors.

In some cases a failure will occur that was never first identified as a risk, representing an error that lingered in code or in requirements. This is never preferred, as it demonstrates a failure of developers to identify and prevent threats or vulnerabilities. Thus, even when failure is inevitable, it is always better for developers to have identified a weakness and predicted its consequences rather than to have been surprised by a failure. Developers who face repeated failures of various errors that were not identified, documented, or discussed lose the confidence of stakeholders. A failure log is demonstrated in the "Measuring Reliability" section and, by demonstrating that actualized failures were already recorded as recognized risks within a risk register, SAS practitioners can show their awareness of software vulnerabilities and can validate or modify risk management strategies.

Qualitative Measures of Failure

Quantitative measures of reliability are beneficial to production software because they not only track performance over time but also can be contractually required to ensure software meets or exceeds performance requirements. As the criticality, frequency, or intended longevity of software increases, the value of reliability measures also increases due to the increased risk that software failure poses. Characteristics that often prescribe relatively higher quality in software are described in the "Saying Yes to Quality" section in Chapter 2, "Quality."

Whether or not quantitative reliability metrics are required or recorded, qualitative measures are often assessed after a failure. This can occur through a formalized process in which the qualitative assessment is incorporated into the failure log, or can be an informal discussion between customers, developers, and other stakeholders to collect information about the failure. Some of the questions that a customer certainly will want answered include:

- What caused the failure?
- Was the cause internal or external? In other words, is there anything SAS practitioners could have done to prevent the failure?
- Was this a functional failure or a performance failure?
- Was the failure discovered through automated quality control processes or manually?
- Was the source of the failure recorded in the risk register, or did the failure uncover new software threats or vulnerabilities?
- Has this failure occurred in the past? If so, why was it not corrected or why did the correction fail to resolve the vulnerability?
- Will this failure occur again in the future?

Whether explicitly recorded in a failure log or discussed casually after software failures, these qualitative questions can form quantitative trends over time. For example, assessing the cause of a single failure—say, a locked data set—is qualitative but, over time, if a failure log demonstrates that 67 percent of failures occurred for this reason, this represents an alarming, quantitative trend. In this way, when higher reliability is sought during software operation, failure logs can be utilized to mitigate or eliminate risks that are causing the most failures for specific software products.

ACL: THE RELIABILITY TRIAD

Reliable software must guard against failure, but how is this accomplished? The reliability principles of availability, consistency, and longevity (ACL) each must be present; otherwise, reliability may suffer. *Available* software is operational when it needs to be; in data analytic development, this often implies that resultant data products are also available for use. *Consistency* delivers the correct solution with unwavering function and performance. *Longevity* (or endurance) demonstrates consistency over the expected lifespan of the software.

Availability

Because this text focuses squarely on software quality rather than system quality, reliability and availability describe only programmatic endeavors to achieve that end. System components such as the network or infrastructure, however, contribute greatly to software availability. As discussed in the upcoming "Requiring Reliability" section, *high availability* is a reliability objective commonly specified in service level agreements (SLAs) that govern critical infrastructures, often defined as the percent of time that software or their data products must be available.

In a simplistic system environment, an end-user developer might utilize a single instance of SAS Display Manager on a laptop. Students and researchers taking advantage of SAS University Edition also enjoy the benefits of a system devoid of servers, networks, and other working parts. When I boot up SAS Enterprise Guide on my laptop, for example, I can immediately run SAS software without any network connection or prerequisites. Fewer working parts typically signify less chance of failure.

While availability sometimes is measured as the length of time without a software failure, a more common metric describes the average amount of time that software is functional, thus incorporating failure duration. This second metric incorporates recoverability and the recovery period—the amount of time that it takes to bring software or resultant data products back online to a functional state. Thus, the downside of a simple infrastructure like my laptop is that when I fry it by bringing it to the pool (thinking this was a good location to write a book), I have to buy another computer, reinstall the OS, call SAS to modify my license agreement to reflect the new machine, and reinstall the SAS application before I can start running SAS software again. Simple infrastructures are only highly available until you have to recover quickly, at which point the benefits of more complex systems become salient.

At this other end of the complexity spectrum lie robust SAS systems with metadata servers, web servers, grids, clusters, redundant array of independent disks (RAID) storage, and client machines, all of which lie on a network that must be maintained. Complex SAS systems are often complex not only because they can support larger user bases with phenomenally higher processing power and storage, but also because they can support high availability, such as hardware systems that can fail over immediately to redundant equipment when primary equipment fails.

Another benefit of SAS servers over single-user SAS clients is their ability to run for extended periods of time with minimal interruption. In many professional environments, users are required to logout from workstations at the end of the day. Without SAS servers, this practice effectively limits the ability of SAS practitioners to run long jobs. I've worked in an organization that forced mandatory nightly logouts but also required developers to use single-user SAS clients. We'd run SAS software during the day and, if a job was still executing when we left work, we locked our workstations, knowing that the job would be killed sometime around midnight when we were forcefully logged off from the system. SAS servers solve this conundrum by allowing users to log off their workstations while their jobs continue to run unfettered on the server, thus supporting significantly higher availability.

The key point is that although programmatic best practices can facilitate higher availability through code that fails less often, true high availability will be achieved only through a robust SAS system and infrastructure in which every chain in the system is equally reliable. In the "Outperforming Your Hardware" section in Chapter 7, "Execution Efficiency," the perils of writing software that is *too* good for your system are discussed. Because high availability can only be achieved through both programmatic and nonprogrammatic endeavors, it's also possible to outperform your hardware when chasing the reliability prize. For example, if the infrastructure or network that undergirds your SAS environment is so buggy that it crashes unexpectedly every week, programmatic *recoverability*—not *reliability*—principles might provide more value to your team because you're constantly having to recover and restart software.

Consistency

Variability is said to be the antithesis of reliability; inconsistent behavior can compromise or eliminate product value. One aspect of the desirability of chain restaurants is that patrons can go to a known establishment in any part of the country (or world) and purchase a product of known taste, texture, quantity, and quality. For example, while Chipotle prides itself on using diverse product sources through its Local Grower Support Initiative, patrons nevertheless can expect a consistent, quality product whether dining in California or Virginia.[13]

Consistency in software is no less important; it describes the similar function, performance, output, and general experience that users should expect each time they execute software. Most importantly, software should maintain functional consistency in that the correct output or solution is produced each time. In the prior "Failure" section, one example demonstrates functional inconsistency in which a SAS macro generates the correct result (1 2 3 4 5) only the first time it's executed, and incorrect results thereafter from within the same SAS session. Any time a functional failure occurs, such as incorrect results or a program terminating with runtime errors, reliability is inherently reduced.

In data analytic development, one of the primary challenges to software reliability is the variability of data injects that software must be able to accommodate. Especially where third-party data outside the command and control of SAS practitioners must be relied upon, data set structure, format, quantity, completeness, accuracy, and validity each can vary from one data inject to another, so software must predict and respond to these and other sources of variability. Because variable data—when that variability is within expected norms—will inherently produce variable results or output, another challenge is the validation of software results and output to demonstrate success and consistency.

Because an error in software should not denote end of product life, software reliability is measured as mean time *between* failures (MTBF) as opposed to mean time *to* failure (MTTF), signifying that the software is not scrapped because an error is encountered. All errors do not cause software termination, but when failure occurs or when software must be terminated for corrective maintenance, these periods of unavailability are defined as *recovery periods*. The recovery period and recoverability are discussed in Chapter 5, "Recoverability."

Consistency is revisited repeatedly in subsequent chapters. Portability, for example, seeks to deliver consistent functionality and performance across disparate OSs and environments. Scalability aims to deliver consistently efficient performance as data or demand increases. In some cases, the consistency of performance will be diminished due to data, environmental, or other variability but, through implementation of an exception handling framework that detects and dynamically responds to this variability, the consistency of functionality can be maintained.

Longevity

The IEEE definition of *reliability* includes the aspect of duration in that software should perform "for a specified time."[14] This duration can reflect that the software should be executed only once, run for two weeks, or run for two years, but some objective for the software lifespan should be commonly understood among all stakeholders before development commences. The intent of software endurance moreover should reflect whether or to what extent software is expected to be modified over its lifespan. For example, if software becomes irrelevant, faulty, or begins to perform poorly due to data or environmental changes, this performance shift could demonstrate a lack of endurance when measured against software's intended lifespan. However, if this variability is predictable, maintenance may reflect a necessary component of software endurance and the only way it can achieve product longevity.

Software is unique because failure doesn't denote end of product lifespan. For example, if a SAS server crashes, the production jobs that were executing will suddenly halt. Yet the jobs should not be damaged and should be able to be restarted once server availability is restored. Software differs substantially from other products in which failure does demonstrate end of product life. For example, drop your cell phone and trample it on the treadmill, and you may be purchasing a new one in the morning. For many physical products, *durability* is a key reliability principle, and products that are not durable will succumb to poor craftsmanship or the environment and be rendered useless more quickly.

In theory, software should just keep going and going like the Energizer Bunny; however, in practice this is seldom the case. Software relies on underlying hardware, OS, third-party software such as Java, and often a network infrastructure, each of which may be slowly evolving. The Base SAS language itself gradually changes over time, even though effort is made to ensure that newer versions are backward compatible with prior versions. Thus, while not as significant, software, too, exhibits a decay continuum similar to (yet not as pronounced as) physical products.

In the case of cell phones, they often aren't destroyed in a single, tumultuous moment—sometimes one key stops working after six months, the screen gets a small crack after nine months, and around 18 months you lose the ability to turn off the ringer and finally decide it needs to be replaced. Software doesn't actually wizen or weaken with age; however, because SAS data analytic environments are vibrant and dynamic, software can become outmoded as data injects or other environmental elements change over time. For example, even if you're fortunate enough to work in an uncommonly stable environment in which the hardware, OS, and other infrastructure that support SAS software do not change, reliance on third-party data sources can require software maintenance to facilitate continued reliability.

Another factor contributing to the decay of data analytic software performance is not variability of input or the environment, but the accretionary nature of data processed. As a system initially conceptualized to process data streams of 1 million observations per hour grows to 100 million observations per hour, system resource limitations may cause diminishing performance and lead to more frequent functional failure of software. This gradual decline in software performance, whether due to changes in the environment, data structure, or quantity, is represented in the software decay continuum in Figure 4.1.

Another contributing factor toward the lack of software endurance is not a change in performance, but rather the changing needs and requirements of the customer. When software is planned and designed, its expected lifespan should be considered and established and, while code stability is beneficial to that lifespan, many data analytic projects are too dynamic to pin down indefinitely. For example, while building a complex ETL infrastructure to ingest data from international sources, I was already aware that the format of data streams would be changing in the upcoming months. But, rather than delay development, we released the initial version of the software on time so that the customer could receive business value sooner, with the understanding that an overhaul of the ingestion module would occur when third-party data structures were modified. Thus, the lifespan of the software was not diminished, but it did explicitly account for necessary maintenance.

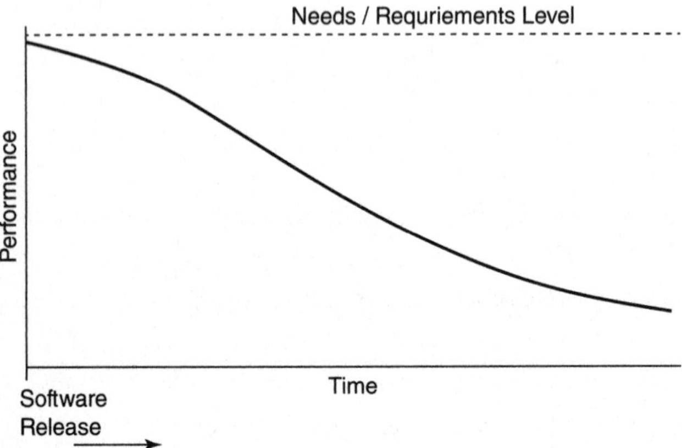

Figure 4.1 Software Decay Continuum

In many cases, however, customers may not be able to predict shifts in needs or requirements. In another example, a separate ETL infrastructure was designed, developed, and released for production that featured once-daily data ingestion and reporting functionality. The customer naturally loved the SAS software so much that after a couple months of reliable operation, he decided that he wanted hourly—rather than daily—ingestion and reporting. A couple weeks later, while stakeholders were still discussing the feasibility of redesigning software to be faster and more efficient to meet this significantly higher threshold, a separate customer prioritized the addition of a separate data stream to the ETL software. Two more weeks passed, and the first customer then made a third request that the software reporting functionality should be updated.

In this classic example of software scope creep, because of these (and several other) change requests that were submitted by various customers, we decided to pool all new requirements over a period of three months, after which we designed, developed, tested, and released an overhauled version of the ETL infrastructure. During the requirements gathering and subsequent SDLC phases prior to release of the new version, however, the *relative* performance of the software continued to decline (as compared to the changing and increasing requirements that were being levied), despite the actual performance remaining stable and reliable. When the software was rereleased, software quality was again on par, as all functional and performance requirements were met.

Figure 4.2 demonstrates the performance of the initial software release, which decreases very slightly due to the expected decay continuum, but which is influenced more significantly by the continuous requirements being levied against the product. While this pattern continued after the first software

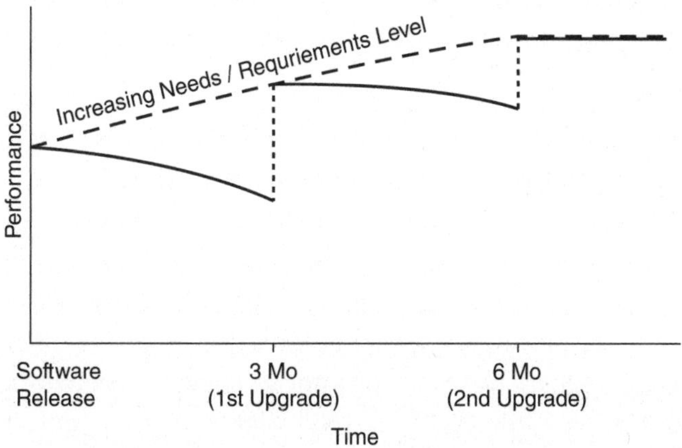

Figure 4.2 Increasing Requirements in Operational Phase

upgrade, after the second software upgrade at six months, the software requirements were stabilized and performance was again acceptable.

Software maintenance plays a pivotal role in software longevity in that software that is more maintainable is more readily modified and thus more likely to be maintained rather than scrapped. In the second example, we were able to modify the ETL software and release a new version (rather than entirely new software) because we had built software with maintainability principles such as flexibility and extensibility in mind. SAS software that lacks maintainability is more likely to more quickly fail to meet requirements as its performance and relative performance decrease in response to a shifting environment or shifting customer needs. This aspect is discussed in the "Stability Promoting Longevity" section in Chapter 17, "Stability," as well as in Chapter 18, "Reusability."

RELIABILITY IN THE SDLC

As the most commonly included performance attribute within requirements documentation, software reliability generally makes an appearance wherever technical requirements exist. Even where requirements are informal or implied, however, a general impression of the reliability that will be necessary for a software product should be established during software planning and design. Developers need to understand whether they're building a Mercedes or a Miata—in other words, will this software become an enduring cornerstone of SAS architecture, or a clunker that just needs to scrape by for a month or two? Both of these extremes—and everything in between—have their place in SAS development, but SAS practitioners must understand where on the reliability spectrum a software product lies before development commences.

Requiring Reliability

Because reliability is often mathematically defined within organizations and contractually defined in SLAs, reliability metrics are typically prioritized over other performance metrics. The type of reliability metrics defined will depend on software intent, heavily influenced by the expected software lifespan as well as the intended frequency of execution. Thus, the reliability objective will be very different for single-use SAS software as opposed to enduring software. Software developed within end-user development environments also may not require higher availability or reliability because SAS practitioners can perform maintenance or restart the software when necessary.

A SAS analytic program may only need to be run once to achieve a solution, after which the software is either empirically modified or archived for posterity and possible future auditing. Empirically developed software common in data analytic development environments is often produced and evolves through rapid, incremental development that chases some analytic thread toward a solution. Reliability in single-use software may be represented exclusively or predominantly by function rather than performance, for example, by demonstration that the correct solution and data products were generated. Neither longevity nor consistency is prioritized because software is not intended to be reused or rerun; availability is valued only insofar as it reflects a correct, available data product.

Other SAS software may specify a longer lifespan such as one year, but may not require frequent execution. For example, an ETL process may ingest data once a day or once per week, executing for some duration and then lying dormant until its next scheduled run. In these cases, reliability may be evaluated largely as a ratio of successful executions to total executions. For example, a process required to run once daily, seven days a week for a month would have 30 opportunities for success in September. If the software failed to complete or produced invalid results two of those days, this performance represents 28/30 or 93.3 percent reliability. Requirements can be this straightforward and simply state the reliability threshold that must be maintained by software.

Especially where software is consistently executed during its lifespan on some recurring schedule, another valuable metric is MTBF. If a SAS program runs once daily and the MTBF is four weeks, you can expect the software to choke about once a month. One consideration when establishing MTBF requirements is how to evaluate failures that can't be rectified before a next scheduled run. For example, if a daily SAS process fails two days in a row because it cannot be fixed immediately following the first failure, should the second failure be calculated into the MTBF calculation, even though it represents a continuation of the original unresolved error or defect? Thus, although fairly straightforward, the definition of MTBF in this example must resolve whether to include the second failure as a new event or a continuation of the first.

Still other SAS software may exhibit much higher execution frequency, such as systems that are constantly churning. Systems that offer continuous services, such as web servers, databases, or email providers, will often instead define reliability through availability metrics. For example, Amazon Web Services (AWS) guarantees "99.999% availability" for one of its server types, effectively stating that the AWS server will be down for no more than

5.5 minutes per year.[15] Availability is measured not by the number or ratio of successful software executions but rather by the percentage of total time that a service or product is functional.

External Sources of Failure

Your SAS software fails, but it's not because of faulty code; an El Niño storm front took out the power to your block, bringing down not only the SAS server but the entire network. Your production SAS jobs that are scheduled to run every hour are obviously a lost cause for the day. But, while sitting in your dark cube twiddling your thumbs, a coworker mentions, "This is really going to kill our reliability stats for the month!" Should an act of God or other external factors discount the reliability of otherwise admirably performing SAS software? Moreover, what about that scheduled maintenance for next month to upgrade from SAS 9.3 to 9.4? How will that affect reliability performance required by your contract and collected through metrics?

One thing is clear—having this discussion literally in the dark, after a failure has occurred and brought down your system, is too late. Better late than never, though. External sources of failure, whether unplanned (such as power outages) or planned (such as required software upgrades), represent threats to reliability that should be discussed even if the decision is made to accept the risks they pose. These discussions and decisions typically occur at the team or organizational level rather than at the project level, and are thus conceptualized to occur outside (or before) the SDLC.

In some environments, the risk of failure due to external cause is essentially transferred from the developers or O&M team to the customer, in that the customer acknowledges and accepts that power outages, server outages, network outages, and other external sources of failure do occur and should not discount software reliability statistics. SAS practitioners should develop software that recovers from these failures using recoverability principles, but the practitioners aren't liable for securing backup generators, servers, or other hardware or infrastructure to prevent or mitigate external failures. For example, if a university's network connection drops, preventing researchers from accessing their SAS application, software, or data products, this likely would not be recorded as a software failure because it was beyond the control of the SAS practitioners.

Other environments are less permissive and require higher or high availability of at least the SAS server. SAS practitioners might be responsible for maintaining a duplicate server and data storage to support rapid system failover, but still would not be required to have backup power sources because that responsibility (and risk) would be borne at the organizational level. In this example, if the SAS server crashed and caused production software to fail, reliability metrics would reflect that failures had occurred because the SAS system failed to meet high availability requirements. After a power failure, however, reliability metrics would be undiminished because that risk would have been transferred to the customer or organization.

Reliability Artifacts

Reliability, more so than other quality characteristics, is benefited not only from quality code but also from external software artifacts, including a failure log and risk register. The risk register should be utilized throughout software development and testing; therefore, SAS practitioners and other stakeholders can use it to predict future software performance by assessing the cumulative nature, severity, and frequency of risks that exist. Thus, when software is released, developers should understand its strengths and weaknesses and effectively how reliably it should perform in production even before any failure metrics have been collected.

In software that will demand high reliability, requiring a risk register is a good first step to implement during software design to ensure that vulnerabilities are detected, documented, and discussed to calculate the ultimate risk that will be accepted when the software is released. Because software maintenance generally continues once software is operational, failures that may occur should be documented in a failure log but also should be included in the risk register if they exposed previously unknown or undocumented threats or vulnerabilities. Risk registers are described in the "Risk Register" section in Chapter 1, "Introduction."

The second reliability artifact, the *failure log,* is a necessity when the collection and measurement of reliability metrics are prescribed. Unlike the risk register, which only calculates theoretical risk of failure to software, the failure log records actual deviations from functional and performance requirements. It's often necessary to categorize failure by type or order them by severity. For example, if a SAS process is required to complete in less than 30 minutes but completes in 33 minutes one morning because of a slow network, this performance failure is worth recording, but might be significantly less detrimental than a functional failure in which the software abruptly terminates and produces no usable output. Reviewing, organizing, and analyzing failure patterns—rather than merely recording failures in a log—is critical to understanding and improving software reliability over its operational lifespan.

Reliability Growth Model

Imagine software that demonstrates a 10 percent failure rate—that is, one out of every ten times the software is executed, it incurs a functional or performance failure. If your Gmail crashed one out of ten times it was opened, or failed to send one out of ten emails, these functional failures would probably cause you to abandon the service in favor of an alternative email provider. Because reliability is so paramount to software operated by third-party users such as Gmail subscribers, companies such as Google attempt to deliver increasing performance over the lifespan of software, following the *reliability growth model*. For example, when a glitch occurs in Gmail and the box pops up asking if you'd like to submit data regarding the error, this is to facilitate Google's understanding of and ability to eliminate the exception or error in the future.

Reliability growth is "the improvement in reliability that results from correction of faults."[16] The reliability growth model describes software whose reliability increases over time, irrespective of other functional or performance changes that may occur. When a product is released, exceptions may be discovered that were not conceptualized during planning and design, or software defects may be discovered that were unknown during software testing. Thus, if software is maintained throughout its operation, and as defects are identified and resolved, reliability tends to increase asymptotically toward perfect availability over the software lifespan as depicted in Figure 4.3. Also note that because reliability and availability were prioritized into software design, the initial reliability at production was already 98 percent.

In Figure 4.3, reliability continues to increase because stakeholders have prioritized its value within software requirements. Due to required

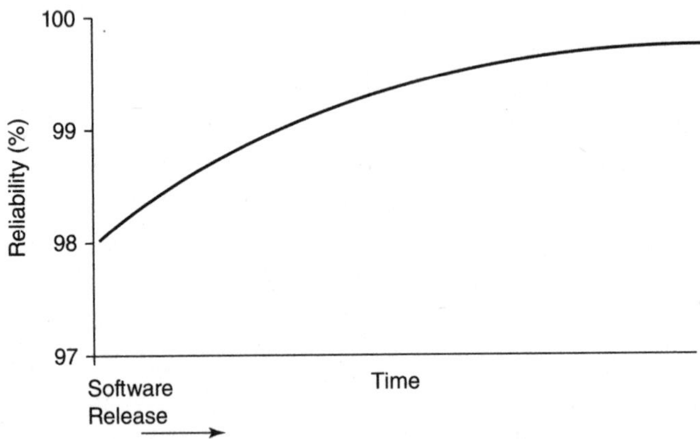

Figure 4.3 Reliability Growth Curve with Decreasing Failure Rate

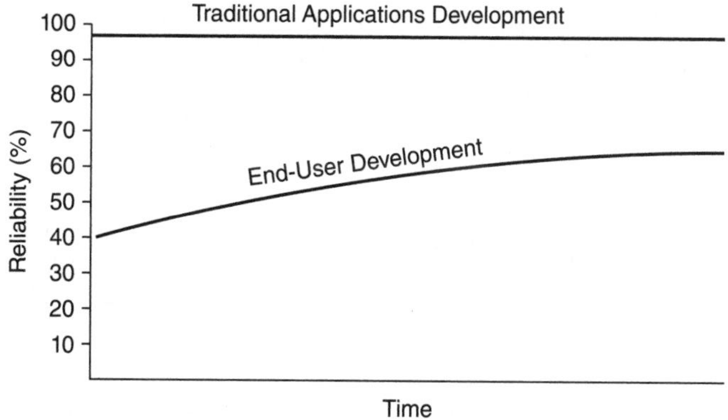

Figure 4.4 Traditional and End-User Development Reliability Growth Curves

maintenance also being prioritized to further higher levels of reliability, both the failure rates and MTBF will decrease over time. Thus, another way of conceptualizing the reliability growth curve is to imagine that the distance between failures tends to increase as sources of failure are mitigated or eliminated. If stakeholders decide to deprioritize maintenance, reliability will typically tend to stabilize, then diminish, following the decay continuum described in the earlier "Longevity" section.

In end-user development environments, reliability is often deprioritized in favor of greater functionality or in favor of a decreased software development period. Especially in data analytic development environments, SAS practitioners may be anxious to view software output so they can get to what we call "playtime"—turning data into information and knowledge. Moreover, because many SAS practitioners are end-user developers who are savvy software developers, when functional or performance failures occur, they can often immediately rerun, refactor, or revise the SAS software on the spot, avoiding painful change management processes that are necessary in some more traditional development environments. Thus, while end-user development will also typically follow the reliability growth curve, the initial reliability may be exceedingly less than exhibited by software applications intended for third-party users while the ultimate reliability achieved will also be significantly lower. Figure 4.4 demonstrates a typical end-user development reliability growth curve overlaying the traditional software growth curve observed in Figure 4.3.

Reliability growth models and decay continuums are useful in software planning and design because they can be used to model stakeholder expectations for software lifespan. If a customer wants an exaggeratedly high level of reliability or availability on day one of software release, that stakeholder must recognize the incredible amount of testing that will be required, not to mention beta releases to test the software with actual users. Even Gmail had beta releases while functionality and performance were still being tweaked. More appropriately, stakeholders can describe lower levels of reliability that might be acceptable in the first couple weeks or months of software operation, during which time reliability should advance toward higher reliability objectives as software errors and defects are eliminated through corrective maintenance.

Measuring Reliability

Reliability and availability metrics can have vastly different uses based on software intent, as well as how the risks of internal and external sources of failure are perceived. SAS practitioners must first understand which failure types contribute detrimentally toward reliability metrics and which (if any) represent accepted organizational risks such as power outages, network failures, or necessary software upgrades.

Table 4.1 demonstrates an extremely abbreviated failure log collected for SAS software from March through April. Reliability metrics are calculated from the failure log, as demonstrated in the remaining subsections of this chapter. This table is reprised in the "Measuring Recoverability" section in Chapter 5, "Recoverability," to demonstrate associated recoverability metrics.

Reliability can be complicated when it describes software that not only has failed but also must recover. For example, when the power fails, it must first be restored, after which the network and SAS server are restored, and finally SAS software is recovered. Software recovery is often captured separately in

Table 4.1 Sample Failure Log for SAS Software Run Daily for Two Months

Error Number	Failure Date	Recovery Date	External/ Internal	Severity	Description	Resolution
1	3-16 3 PM	3-17 9 AM	Ext	1	power outage	waited
2	3-18 2 PM	3-19 10 AM	Int	1	failure related to power outage	made more restorable
3	3-25 10 AM	3-25 11 AM	Ext	1	network outage	waited
4	4-5 3 PM	4-5 4 PM	Int	1	out of memory SAS error	restarted software
5	4-22 9 AM	4-22 1 PM	Int	2	minor data product HTML errors	modified HTML output

recovery metrics but the recovery period contributes to overall reliability when availability metrics are utilized. Chapter 5, "Recoverability," describes the interaction between these two intertwined dimensions of software quality.

RELIABILITY FORMULAS

Availability = functional time / total time in service

Failure Rate = number of failures / total number of executions

Inclusive MTBF = summation of the time between successive failures / number of failures

Exclusive MTBF = summation of the time between recovery and the next failure / number of failures

Availability

As a measurement of reliability, *availability* is defined as "the degree to which a system or component is operational and accessible when required for use."[17] ISO goes on to clarify that it is "often expressed as a probability. Availability is usually expressed as a ratio of the time that the service is actually available for use by the business to the agreed service hours." Availability is arguably the reliability metric most commonly utilized in software requirements documentation and SLAs.

Failure rate and MTBF each fail to describe reliability completely, because they typically are not weighted by the amount of time that software is failing and thus unavailable. In other words, did a subtle performance failure cause a momentary shudder in execution that was rectified in minutes, or did SAS practitioners have to rebuild the software over a grueling week of development during which time the software could not be run? Availability captures the percent of time that a system or software is functioning as compared to existing in a failed state. In doing so, availability incorporates not only reliability, but also the subcomponent of recoverability.

Availability metrics are most commonly implemented in software that requires higher or high availability. For example, if SAS analytic software is only run once per week, it only needs to be available for that one hour, so capturing availability could be difficult or pointless. But many SAS servers are required to be available 24×7, so even in environments for which no SAS software benefits from recording availability metrics, availability is the most common metric used to capture server reliability.

Where software is run frequently and consistently, availability can complement failure rate and MTBF because it incorporates the recovery period during which software is not operational. It primarily differs from the latter two metrics in that availability weights each failure by duration, while failure rate and MTBF do not. In essence, if a SAS job should execute once per hour but fails one Monday morning and cannot be restored until four hours later, this represents only one failure, even though four of the eight expected software executions for that day either failed or were skipped while subsequent maintenance was occurring. The failure rate would depict only one failure, while the availability would depict all four failures.

In Table 4.1, five software failures of external and internal cause are recorded in the failure log. If stakeholders have required that the software warrants redundant power, network, or other systems, then the team is liable for the risks posed by both internal and external vulnerabilities, and all five failures should be calculated into the availability metrics. In this case, if software is intended to be executed once per hour 24×7, the first entry results in 18 failures (because 18 executions were missed), the second entry represents 20 failures, and so forth. With 31 days in March, 30 days in April, and 24 hours per day, the software should have executed 1,464 times but, because it failed or was in a failed state 44 times, the software completed only 1,432 times. Thus, the availability is 1,420 / 1,464 or 96.99 percent.

Using the same failure log, consider that instead of running a 24×7 operation the SAS practitioners are responsible for lower availability software that operates 10×7 (8 AM to 6 PM, seven days a week). Thus, the first entry represents only four software failures (three on March 16 and one on March 17) rather than 18. Moreover, the total number of expected software executions drops to 610 or (61 days × 10 hours). Given this new paradigm, 16 failures occurred out of 610 software executions, so the availability is 594 / 610, or 97.38 percent. This underscores the importance of accurately determining when the software must be operational and when it can be allowed to be turned off or is undergoing maintenance.

As a final example, imagine a third team demonstrating the same software failure log, operating software from 8 AM to 6 PM seven days a week, but whose customer has stated that the SAS practitioners are not required to maintain the network, hardware, or other infrastructure. This is a common operational model in which the risk of failure of infrastructure components is effectively transferred to the customer or other IT teams. Where a power or network outage occurs, it's understood that the SAS practitioners are not liable, so these failures, while possibly recorded as "External" in the failure log, will not be included in reliability calculations.

When external causes of failure are removed from the reliability metrics calculations, only 11 software executions failed or could not execute because they were in a failed state. Thus, availability increases to 599/610 or 98.2 percent. This spread of availability rates in these three examples underscores the importance of adopting a risk management strategy before software release, as well as defining the scope of when and how often software is expected to execute.

Failure Rate

Failure rate is "the ratio of the number of failures of a given category to a given unit of measure."[18] The failure rate is a straightforward method of quantifying reliability, tracking the number of failures over time and within specified time periods. When failure rate is tracked longitudinally, the acceleration or deceleration rate of reliability moreover can demonstrate whether software performance is increasing or decreasing over time. For example, assessing the number of failures per week over a six-month period might demonstrate that while software is performing within specified reliability limits, its increasing failure rate signals that maintenance should be performed before the reliability threshold is exceeded.

Because failure rate can be calculated irrespective of the frequency of software execution or time interval between executions, it's extremely valuable when software is intended to be enduring but may not be run consistently or frequently. For example, if analytic software is only run the first week of the quarter but then lies dormant for three months, failure rate would be a preferred method to calculate its reliability because the metric does not take into account the dormant periods that complicate calculation of MTBF.

A limitation of failure rate is that it fails to account for the length of time that software is inoperable following a failure, the recovery period. Thus, to gain a more complete understanding of software's performance, it's often beneficial to calculate and track the recovery period as well. For example, when SAS software fails to meet performance requirements and completes in three hours rather than one, the performance failure should be documented, but no recovery period exists (because the software did not terminate with runtime errors or produce invalid results) so the software can be run again immediately. In other failure patterns, however, the software might be unavailable for an extended period for corrective maintenance following a failure.

Especially in these cases, incorporation of recovery metrics can provide more context through which to interpret failure rates and severity.

In environments in which software is run consistently, such as software that is operated 24×7 in March and April, the failure rate will represent the unweighted availability. Thus, when the power outage caused the software to fail for 18 hours between March 16 and 17, this outage is treated as a single failure rather than 18 consecutive failures because it points to a single threat origin. Failure rate is easy to calculate, in this case five failures in two months or 2.5 failures per month. Because of the ease with which failure rate can be calculated, it is often tracked longitudinally to demonstrate acceleration or deceleration of failures, for instance three failures in January and two in February.

Because failure rate is simplistic, many teams choose to track failure rate by other qualitative characteristics in the failure log—often distinguishing liability such as internal versus external sources of failure. This could enable SAS practitioners and other stakeholders to demonstrate, for example, that *software* reliability might be increasing despite decreasing reliability in the SAS server or external infrastructure. Or, it might demonstrate that more significant functional failures are decreasing but performance failures are increasing. These metrics can enable stakeholders to prioritize maintenance to improve software performance. Thus, while simplistic and limited in some ways, failure rates gain power when combined with other qualitative metrics collected during and after software failure.

MTBF

Mean time between failures (MTBF) is "the expected or observed time between consecutive failures in a system or component."[19] It is measured by averaging the number of minutes, hours, or days between failures over a specified period. MTBF typically includes the failure period (i.e., recovery period) in which software is in a failed state and thus is measured from the incidence of one failure to the next. For example, if SAS software fails on a Friday, is restored to functionality on Tuesday, but fails again the next Friday, the MTBF is still seven days, representing the Friday-to-Friday span. Some definitions of MTBF conversely count from the point of recovery to the next failure, so it's important to define this metric. These two definitions are represented in Figure 4.5.

MTBF is more sensitive to software execution frequency as well as the interval between software runs. For example, if SAS software is executed the first day of every month, and fails on March 1, the earliest it would again be

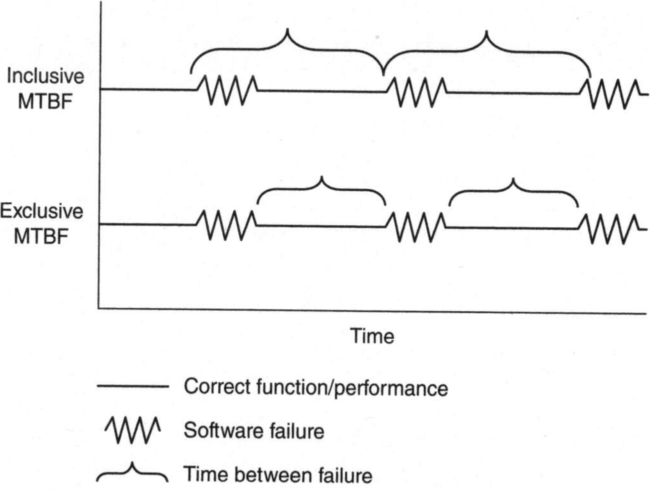

Figure 4.5 MTBF Competing Definitions, Inclusive and Exclusive of Recovery

executed (for production) is the April 1 execution, so the MTBF will always be at least one month. This one-month MTBF metric also fails to capture whether it took one day or three weeks to restore the software functionality, given that it was not required to be run again in production until April 1. For this reason, MTBF is best applied to software that runs frequently and consistently, but can also be incorporated with availability or recoverability metrics, both of which also take into account the duration of software failure in addition to its frequency.

Recall the five software failures recorded in the failure log in Table 4.1. The software warrants redundant power, network, and other systems, so the team is liable for the risks posed by both internal and external threats, and all five failures should be calculated into the MTBF calculations. In this case, if software is intended to be executed once per hour 24×7, the first failure occurs after 15 days and 15 hours (counting hours from 12:01 AM March 1). The second failure occurs 47 hours after the first; thus the first two metrics contributing to MTBF are 15.63 and 1.96 days and, continuing with this logic, the remaining time periods between failures are 6.83, 11.21, and 16.75 days, respectively. The final period from April 22 through May 1 is 8.63 days but isn't included because May 1 represents the end of the evaluation period rather than a failure. Thus, the MTBF for March and April is 52.38/5 or 10.48 days.

The inclusive MTBF definition will roughly approximate the total measurement period divided by the number of failures, in this case 61/5 or 12.2 days.

Table 4.2 Inclusive versus Exclusive MTBF Calculations

Error Number	Failure Date	Restore Date	Inclusive MTBF	Exclusive MTBF
1	3-16 3 PM	3-17 9 AM	15.63	15.63
2	3-18 2 PM	3-19 10 AM	1.96	1.21
3	3-25 10 AM	3-25 11 AM	6.83	6
4	4-5 3 PM	4-5 4 PM	11.21	11.67
5	4-22 9 AM	4-22 1 PM	16.75	16.71
	MTBF		**10.48**	**10.14**

In development environments that don't demand specificity, this can be an acceptable alternative to MTBF. A slightly more complex definition of MTBF excludes the recovery period from calculation. For example, if the power fails and brings down your SAS server, and cannot be revived for two days, reliability is not penalized for the duration that the system or software remains down. The calculations for inclusive and exclusive MTBF are compared in Table 4.2.

While the difference in the calculated metrics is subtle and can seem like splitting hairs, the exclusive MTBF eliminates time spent restoring software. The calculation for exclusive MTBF is more complex because both failure and recovery points must be recorded. A benefit of this method, however, is it more accurately demonstrates time that the software is actually functioning, rather than also incorporating time that the software is being restored and possibly even modified with corrective or emergency maintenance.

In other words, stakeholders assessing the performance of software may be asking the basic question, "How long was your software running *this time* before it failed?" In posing the question in this manner, stakeholders are requesting the exclusive MTBF—measured from the last recovery point to current failure—because they are not concerned with how long it took to modify and recover software following the *last* failure. Whichever method is selected—internal or external MTBF—it should be defined leaving no room for ambiguity or misinterpretation.

WHAT'S NEXT?

Reliability is the most critical dynamic performance attribute and, as described, incorporates availability, which measures performance against failure. For this reason, high availability requires swift software recovery, facilitated by recoverability principles. In the next chapter, recoverability—a subcomponent of reliability in the ISO software product quality model—is introduced and demonstrated.

NOTES

1. ISO/IEC 25010:2011. *Systems and software engineering — Systems and software Quality Requirements and Evaluation (SQuaRE)—System and software quality models*. Geneva, Switzerland: International Organization for Standardization and Institute of Electrical and Electronics Engineers.
2. IEEE Std 982.1-2005. *IEEE standard dictionary of measures of the software aspects of dependability*. Geneva, Switzerland: Institute of Electrical and Electronics Engineers.
3. IEEE Std 982.1-1988. *IEEE standard dictionary of measures to produce reliable software*. Geneva, Switzerland: Institute of Electrical and Electronics Engineers.
4. IEEE Std 610.12-1990. *IEEE standard glossary of software engineering terminology*. Geneva, Switzerland: Institute of Electrical and Electronics Engineers.
5. ISO/IEC 25000:2005. *Software engineering—Software product Quality Requirements and Evaluation (SQuaRE)—Guide to SQuaRE*. Geneva, Switzerland: International Organization for Standardization and Institute of Electrical and Electronics Engineers.
6. IEEE Std 982.1-2005.
7. IEEE Std 1044-2009. *IEEE standard classification for software anomalies*. Geneva, Switzerland: Institute of Electrical and Electronics Engineers.
8. Id.
9. Id.
10. Id.
11. "Types of Errors in SAS." *SAS® 9.4 Language Reference: Concepts, Fifth Edition*. Retrieved from http://support.sas.com/documentation/cdl/en/lrcon/68089/HTML/default/viewer.htm #n1g8q3l1j2z1hjn1gj1hln0ci5gn.htm.
12. "%GLOBAL Statement." *SAS® 9.4 Macro Language: Reference, Fourth Edition*. Retrieved from http://support.sas.com/documentation/cdl/en/mcrolref/67912/HTML/default/viewer.htm #p1lhhti7fjxgb1n1fuiubqk11h4d.htm.
13. "Chipotle Local Grower Support Initiative." *Chipotle* website. Retrieved from www.chipotle .com/localgrowersupport.
14. IEEE Std 982.1-1988.
15. "Amazon EBS Product Details." *Amazon.com* website. Retrieved from https://aws.amazon.com/ ebs/details/5.
16. ISO/IEC/IEEE 24765:2010. *Systems and software engineering—Vocabulary*. Geneva, Switzerland: International Organization for Standardization, International Electrotechnical Commission, and Institute of Electrical and Electronics Engineers.
17. Id.
18. Id.
19. Id.

CHAPTER 5

Recoverability

I'd worn the same pair of button-fly 505 Levi's for a month while backpacking through Central America and, aside from some pupusa stains and the subtle accumulation of chicken bus grime, they'd been exceedingly durable and reliable.

Busing and hitchhiking in pickup beds through Columbian FARC country, I'd finally arrived at San Andrés de Pisimbalá, a remote, mountainous pueblo bordering the Tierradentro archaeological and UNESCO World Heritage Site. Remote and inaccessible except by horseback, the seldom-seen Tierradentro features pre-Columbian tombs carved into hilltops with 7th-century ceramics and polychromatic paintings in situ.

Having serendipitously met my guide's wife while hitchhiking the gravelly road to San Andrés, I saddled up and followed Pedro on horseback, dismounting at numerous tombs to descend down rock-hewn spiral steps into spaciously carved caverns and indescribable antiquity. As the day wore on and the equatorial sun beat down, I was thankful for the cool breeze but, while dismounting at one site, the breeze suddenly intensified and I realized that amid the jolting, jarring hours on horseback I had completely ripped out the crotch of my Levi's. *¡Ay caramba!*

The front-to-rear tear was catastrophic and career-ending for the Levi's; the guide and his wife were amused when, back at the hostel, I learned that she could not repair the rip with a patch. I'd be in Quito, Ecuador, in a couple days, so I was resolved to modesty until then as I left San Andrés and continued onward.

Having lived in Ecuador previously, I reached out to Andres, a friend in Quito who, despite laughing at my crotchless travels, was happy to assist with my latest backpacking debacle. We searched for hours, going from mall to mall and, even with my interpreter, it took four hours to find a single pair of jeans that I could squeeze my gringo-sized body into.

At long last, Tommy Hilfiger came to the rescue; despite a ridiculous price of US$124, the jeans have been equally reliable, surviving South America and beyond.

■ ■ ■

If you start with a quality product—such as Levi's or Tommy Hilfiger jeans—you'll be less concerned with failure because the product is inherently more reliable and durable. But even 505s or extremely reliable software will fail at some point, and when this occurs, recoverability principles dictate how quickly recovery can occur.

My jeans theoretically could have been patched, but they were five years old, had had a good life, and were becoming threadbare elsewhere. Software, however, doesn't weaken or wizen over time, so failure typically doesn't denote end of product life. For example, when a hapless SAS administrator accidentally turns off the server, causing all SAS jobs to fail, the SAS server can be turned on again and the SAS jobs will continue to execute.

But recovery can be delayed where infrastructure outside the control of SAS practitioners has failed, such as power or network failures. In the remote Colombian mountainside, no amount of effort or time could have repaired or replaced my torn jeans—I could only close my legs and be patient while transiting to Quito. SAS practitioners, too, will encounter external causes of software failure that cannot be remedied and must be waited out; recoverability principles cannot lessen these delays.

With infrastructure restored, however, or in the case of failures that were programmatic in nature, the software should be able to be restarted easily and efficiently, but sometimes that is not the case. As with driving around Quito for hours, running from one mall to the next, developers can incur inefficiency and tremendous expense of time if software has to be jumpstarted through convoluted procedures or emergency maintenance to restore functionality.

Inefficiency is especially likely when failure was not predicted and no recovery strategy exists, or recovery efforts compete against other objectives. I knew I would eventually find jeans in Quito—at over 2.5 million *Quiteños* strong, plenty of shopping venues exist—but I didn't know the malls, the brands, the sizes, or the language, and all of these factors contributed to inefficient shopping. When software fails unexpectedly on a Tuesday, developers may have to spring into action, suppressing other activities and inefficiently prioritizing maintenance over scheduled work.

Recoverability principles thus seek to increase the overall reliability and availability of software by minimizing the amount of time required to restore software and data products following a failure.

DEFINING RECOVERABILITY

Recoverability is "the degree to which the software product can re-establish a specified level of performance and recover the data directly affected in the case of a failure"[1] In the case of failed software, the failure could have been caused by the SAS software itself or by any element of its infrastructure, such as the SAS application, SAS server, network, hardware, or even power supply. Recoverability is a critical component of reliability, especially in high-availability systems that measure time in a failed state against operational time.

Software recovery cannot prevent external sources of failure such as power outages from felling SAS software, but when the cause of the failure has been eliminated and the SAS server is operational, recoverability principles minimize the amount of time required to restore SAS software and, in some cases, the resultant SAS data products. Software recoverability principles follow the TEACH mnemonic: software recovery should be *timely, efficient, autonomous, constant,* and *harmless.*

In other cases, software fails because of internal causes, such as software defects or other errors that require corrective or adaptive maintenance. Software must be modified before it can be restored, which can be both time consuming and risky when unplanned emergency maintenance must be performed. Maintainability principles demonstrated throughout Chapter 13, "Maintainability," include best practices to support efficient and effective modification. When emergency maintenance occurs after a failure, however, this maintenance contributes to the recovery period and detracts from overall software reliability. Therefore, where software modification must occur during recovery operations, maintainability principles as much as recoverability principles will guide software recovery. This chapter demonstrates TEACH principles across discrete phases of the recovery period that follow software from failure to recovery.

RECOVERABILITY TERMS

Recoverability "The degree to which the software product can re-establish a specified level of performance and recover the data directly affected in the case of a failure."[2]

Failure "Termination of the ability of a product to perform a required function or its inability to perform within previously specified limits."[3]

Recovery "The restoration of a system, program, database, or other system resource to a state in which it can perform required functions."[4]

Recovery Period Synonymous with *downtime*, "the period of time during which a system or component is not operational or has been taken out of service."[5]

Mean Time to Recovery (MTTR) Synonymous with *mean time to repair*, "the mean time the maintenance team requires to implement a change and restore the system to working order."[6]

Maximum Tolerable Downtime (MTD) The amount of time a business can tolerate failure without loss of business value or greater failure.

Recovery Time Objective (RTO) The maximum allowable time that software can be failed before recovery occurs. The RTO should always be less than the MTD.

RECOVERABILITY TOWARD RELIABILITY

Many products, such as jeans, must be disposed of when they fail because failure represents end of product life or, in less extreme cases, the need to at least repair the product. In this sense, the mean time to *recovery* (MTTR) for software may in some environments be referenced instead as mean time to *repair,* reflecting terminology typically reserved for physical products. In physical products, mean time to failure (MTTF) describes the product life before it fails, typically with the expectation that the product will be discarded at failure. In describing software, however, mean time between failures (MTBF) rather than MTTF is used because software failure is presumed not to denote end of life but rather the start of the recovery period, after which software can continue operation.

In many cases, a performance failure may cause software to fail in the sense that it fails to meet performance requirements, but software neither terminates nor produces invalid data or data products. For example, SAS software that occasionally completes too slowly can cause performance failures when execution time exceeds technical requirements. Unless software is terminated after a performance failure to allow SAS practitioners to perform corrective maintenance, no recovery period or recovery exists, so recoverability principles are irrelevant. For this reason, failure logs—demonstrated in the "Failure Log" section in Chapter 4, "Reliability"—typically distinguish functional from performance failures so they can be tracked independently.

Therefore, recoverability represents an aspect of quality that, while critical to software development environments, is uncommon in most physical product types. Figure 5.1 compares MTTF patterns for a physical product like a pair of jeans to MTBF patterns for software. Note that no recovery period exists for the software experiencing only performance failures from slowed execution.

Described as a subcomponent of reliability in the International Organization for Standardization (ISO) software product quality model, recoverability contributes to reliability because it minimizes the recovery period during which

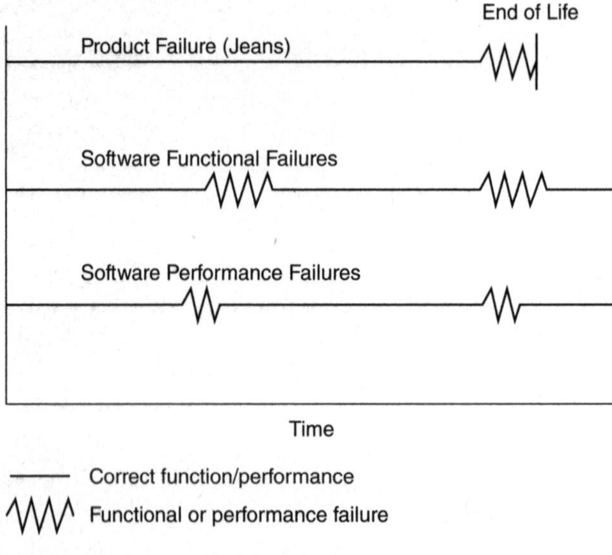

Figure 5.1 Comparing MTTF and MTBF

software is in a failed state, providing no or limited business value. Achieving reliable software is the most crucial step in maximizing software availability; code that doesn't fail in the first place won't need to recover. But because software failure is an inevitability, software that requires high reliability and availability must also inherently espouse recoverability principles to facilitate these goals.

Environmental, network, system, data, and other external issues can cause lapses in connectivity, memory errors, and other failures that result in unavoidable program termination. Planned outages, including upgrades and modifications to SAS software, its infrastructure, network, and hardware also require halting operational SAS programs. While *recoverability* reflects the ability to restore functionality or availability of a system or software following a failure or planned outage, *recovery* represents the state of restored functionality or availability, and the *recovery period* the time and effort required to restore the system or software.

The "Reliability Growth Model" described in Chapter 4, "Reliability," demonstrates the increasing reliability of software that is appropriately maintained over time. Figure 5.2 demonstrates the decreasing failure rates and increasing MTBF associated with the reliability growth curve, representing that software reliability and maintenance have been prioritized. In this example,

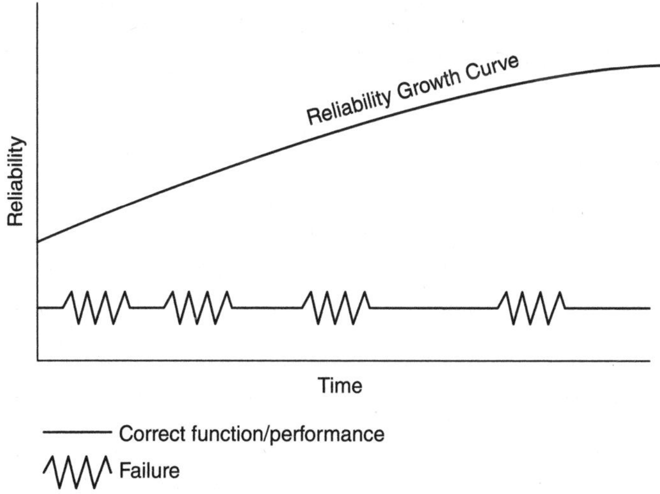

Figure 5.2 MTBF Influencing Reliability Growth Curve

however, rather than merely indicating when each failure occurred, the duration of the failure also is shown.

While the reliability growth curve demonstrates decreasing failure rates, it says nothing about the recovery period or duration of each failure. Thus, while higher availability and decreased failure rates can be achieved simply by following the reliability growth curve, the recoverability growth curve shows that availability can be further increased by decreasing the recovery period through the implementation of recoverability principles and phases.

Recoverability Growth Model

The recoverability growth model is a corollary to the reliability growth model, effectively demonstrating that where recoverability principles are prioritized in software, the software recovery should decrease asymptotically over time toward theorized automatic, immediate recovery. The recoverability growth curve is demonstrated in Figure 5.3, in which lower recoverability denotes longer recovery periods and higher recoverability denotes shorter recovery periods.

The recoverability growth curve often occurs in conjunction with the reliability growth curve. Thus, as SAS practitioners hone software and make it more resilient and robust, they often additionally seek to improve recovery because they learn ways to eliminate or make recovery phases more efficient.

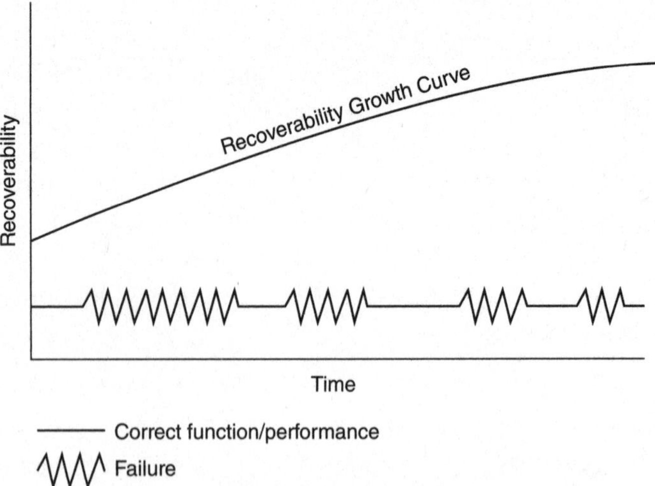

Figure 5.3 Recoverability Growth Curve with Decreasing Recovery Period

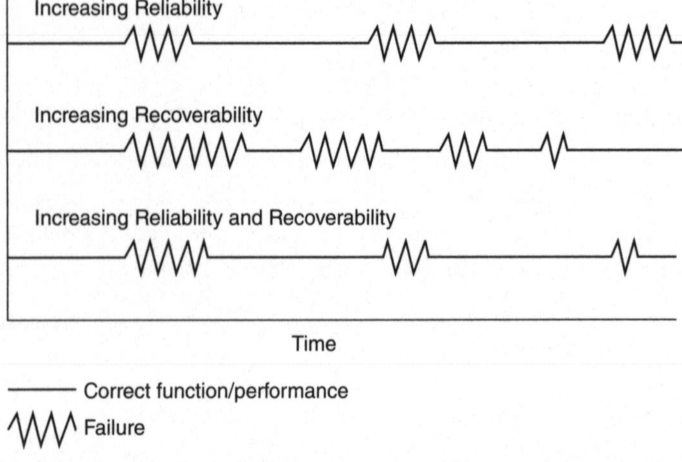

Figure 5.4 Interaction of Reliability and Recoverability

The recoverability curve is irrespective of the reliability curve, as Figure 5.4 demonstrates both the combined and individual effects of reliability and recoverability in relation to each other.

Availability is demonstrated by the percent of time that software is not in a failed state. As demonstrated in Figure 5.4, while the reliability growth curve and recoverability growth curve can occur separately over the software lifespan, the highest software availability occurs when both reliability and recoverability are maximized.

RECOVERABILITY MATRIX

In 2015, I introduced the Recoverability Matrix, which describes guiding principles that facilitate software recoverability as well as discrete phases that software passes through en route to recovery.[7] By incorporating these principles into code, SAS practitioners can facilitate code that minimizes and often eliminates phases entirely, thus tremendously improving the recovery experience.

Recoverability principles follow the TEACH mnemonic and state that software recovery should be:

- *Timely*—The recovery period should be minimized to maximize availability.
- *Efficient*—Recovery should be efficient, including an efficient use of software processing (i.e., system resources) as well as developer resources if maintenance is required during recovery.
- *Autonomous*—From failure detection to functional resumption, recovery should minimize or eliminate human interaction and recover in an intelligent, automated manner.
- *Constant*—Quick-fixes, jumpstarts, and one-time code bandages that facilitate recovery should be avoided in lieu of strategic, repeatable, enduring solutions.
- *Harmless*—The recovery process should not damage the data products or other aspects of the environment.

Each of these principles should be demonstrated throughout the following discrete phases that can occur during the recovery period after failure has been detected. The SPICIER mnemonic describes recovery phases that may exist but, in many cases, should be minimized or eliminated through recoverability principles.

- *Stop Program*—Because failure does not necessarily denote that software has stopped running, the software may need to be stopped.
- *Investigation*—Gather information about the source, cause, duration, impact, potential remedies, and other aspects of the failure.
- *Cleanup*—Remove faulty data, data products, or other results that may have been erroneously created due to the failure.
- *Internal issues*—Corrective, adaptive, or emergency maintenance may be required to restore software functionality.

- *External issues*—Issues beyond the immediate control of SAS practitioners may have caused the failure, but these may need to be researched or have quality controls applied to improve or automate future failure detection.
- *Rerun code*—The final phase in the recovery period and, after especially heinous failures that have crippled critical infrastructure, the point at which SAS developers tap that keg and finally celebrate.

TEACH RECOVERABILITY PRINCIPLES

An ideal recovery is *timely, efficient, autonomous, constant,* and *harmless*—tightly intertwined principles depicted through the TEACH mnemonic. Timeliness reflects the speed with which the recovery occurs, including discovery of the failure, how long it takes to restart software, and how long it takes to surpass the point of original failure within the code. Efficiency demonstrates code that does not require excessive processing power or developer effort, the latter of which can be eliminated through autonomous code. Autonomy reflects software that identifies failure, terminates automatically if necessary, and can be restarted with the push of a button without further developer intervention. Software constancy demonstrates that a recovery strategy is enduring, not an expedient "quick fix" that will need to be later improved or undone. Harmless software should both fail and recover without damaging its environment, such as causing other processes to fail or producing invalid data products.

Timely

Just as software speed is often used as a proxy to represent overall software performance, timeliness of recovery often acts as a proxy for other recoverability principles. When software recovery occurs quickly or instantaneously, the assumption is that other principles must have been followed. Thus, other recoverability principles often are not prioritized unless recoverability or reliability performance metrics are not being achieved.

At a higher level, recovery is viewed as a single event that restores functionality and terminates the recovery period. The RTO, if present in requirements, specifies the time period within which the average recovery must occur. Thus, the actual MTTR achieved over time should always be less than the RTO although individual recovery periods may exceed the RTO. In more restrictive environments, the RTO is interpreted as a *maximum* recovery period that no

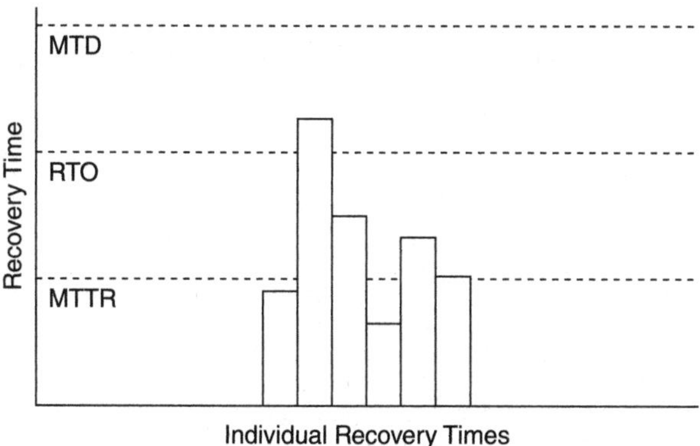

Figure 5.5 MTTR, RTO, and MTD

single recovery period should ever exceed. In this latter sense, implementation of the RTO more closely approximates the MTD. Figure 5.5 demonstrates six software failures whose recovery periods have been measured and recorded against recoverability requirements. While the MTTR fell below the RTO, one failure did take longer to recover than the RTO, which represents a performance failure. All recoveries did occur far below the MTD that stakeholders had established.

In the case of Figure 5.5, in which the RTO was not achieved because of one performance failure, attention should shift from higher- to lower-level recovery, including discrete phases that can occur during the recovery period. For example, SAS practitioners might discover that during the single failure, the cleanup phase of recovery took longer because invalid data products had been created and had to be corrected. Thus, in seeking to deliver more consistent recovery in the future, timeliness of the SAS software could be improved by investigating and improving individual phases of recovery.

Efficient

Efficiency can be represented by the effort or resources that a person or process must expend to facilitate software recovery. For example, an inexperienced SAS practitioner who is unfamiliar with specific software would be less efficient in investigating and remedying a failure than a more experienced developer who had authored the software. Code that is difficult to read or poorly documented can also unnecessarily delay recovery because it is not intuitive and thus more difficult to maintain and modify. A failure that requires substantial cleanup of the SAS system because the code produced spurious data, metadata, or data products can also be an inefficient use of a developer's time, because that cleanup could likely have been eliminated or could have occurred automatically, were more rigorous software development practices enforced. While software failure is never a pleasant experience, through proactive design and development, developers can ensure that their time is well utilized during the recovery period.

Development time, however, is not the only resource that must be taken into account. Software itself should recover efficiently, demonstrating a competent use of system resources and execution time. If SAS code fails to recognize that an internal failure has occurred and the failure goes unnoticed for hours or days, this is an inefficient use of the SAS processor, which was effectively churning without purpose for days while producing invalid results that must be reproduced after recovery. When a program must be rerun at the close of the recovery period, efficient code should also intelligently bypass any unnecessary steps. *Checkpoints* can demonstrate code modules that have completed successfully prior to failure, thus enabling process flow to skip these and proceed directly to the point of failure. This technique is demonstrated in the "Recovering with Checkpoints" section later in the chapter.

Autonomous

Autonomy typically includes and supersedes automation, specifying that a system is not only free from human interaction but also able to make decisions intelligently. Fuzzy logic SAS routines can guide processes to detect failure automatically and, based on metadata collected about the failure, to recover automatically in many cases. Process metrics collected during software failure can demonstrate the type and location of exceptions or errors, especially if they can be compared with baseline metrics that demonstrate previous process success. By predicting and detecting specific error types in exception handling routines, software can often terminate gracefully, thus setting the stage for a cleaner, more efficient recovery.

Consider a SAS process that fails because a required input data set is not available when the software is run. Autonomous recovery begs the question, "What steps would a skilled SAS practitioner follow in this situation?" and recreates this logic in software. For example, when alerted to the exception of the missing data set, the software might automatically send an email to stakeholders to inform them while entering a busy-waiting loop for a number of hours to give developers the chance to rectify the missing data set. After a parameterized amount of time, the software could gracefully terminate if the data set still could not be located. However, if stakeholders were able to restore the availability of the input data set, recovery could occur quickly and efficiently without a single runtime error.

This scenario underscores the importance of autonomy and automation throughout all recovery phases. For example, the obvious bottleneck in the proposed solution is the intervention of SAS practitioners, who must first receive an email and then perform some action. A single bottleneck that relies on human interaction can cause significant delays in recovery systems that are otherwise streamlined and automated, but in many cases it's unrealistic to expect an automated process to transfer the necessary data set to its expected location, so total automation may not be achievable. Notwithstanding, even adding processes that automatically detect and communicate software failures will provide a tremendous advantage for software recovery.

Constant

Constancy or consistency reflects stability in code over time, including the manner in which software is executed and the steps taken to revive software following failure. Stable software should be dynamic enough to determine how and where to restart after a failure. Thus, rather than highlighting sections of code to be run manually or creating quick fixes to be used to jumpstart code, developers should be able to run the same code in its entirety every time—whether following a successful or failed execution. Temporary solutions that can revive and run software *today* may be timely but, if they introduce new defects or technical debt that must be dealt with *tomorrow*, this added cost is typically not worthwhile.

For example, suppose that a serialized extract-transform-load (ETL) process is executing that ingests and transforms 20 SAS data sets in series. If the process fails on the 16th data set after successfully processing 15 data sets, some SAS practitioners might be tempted to rerun the software with manual modifications that specify that the software should start at the 16th rather than the 1st data set. This type of rogue emergency maintenance is abhorred in production software and should be avoided at all cost! Another tempting but terrible idea would be to rerun the software in its entirety, thus needlessly reprocessing the first 15 data sets and obviously violating the principles that recovery should be timely and efficient.

A consistent software solution, on the other hand, might instead consult a control table to interrogate performance metrics, which would demonstrate that the first 15 data sets had been processed, thus dynamically enabling the software to begin processing on the 16th data set rather than the 1st. As a result, software can remain constant by implementing dynamic processes that effectively utilize checkpoints, thus eliminating the poor practices of rogue software modification or redundant processing. This revised solution, which dynamically assesses past successes and failures of the software, can be utilized to execute software every time it is run, regardless of whether the previous execution succeeded or failed. The benefits of consistent, stable software are demonstrated throughout Chapter 17, "Stability."

Harmless

When harmless software fails, it should not adversely affect its environment, such as by producing faulty, incomplete, corrupt, or otherwise invalid data, metadata, data products, or output. Harmless code should automatically detect a failure when it occurs and handle it in a predictable, responsible manner, because only through predictability comes the assurance that dependent processes and data products are not negatively affected. In most cases, it is far more desirable to have data products that are unavailable or delayed (due to detected failure) than those that are dead wrong (due to undetected failure).

Not only failure but also recovery should be harmless, so the environment should not be altered unfavorably in an attempt to restore functionality. For example, if a developer modifies SAS library names in a production environment as an expedient solution to rerun code that has failed (because it erroneously referenced library names in the development environment), this could cause unrelated processes to fail. Another scenario often occurs in which software fails because of an exception or error within a macro shared by the development team. The temptation may exist to alter the macro so that it works in the current software that is failing, but if this modified macro is reused in other production software, the change cannot be made until the implications of the proposed modification are understood. A software reuse library, demonstrated in the "Reuse Library" section in Chapter 18, "Reusability," can facilitate an awareness of code use and reuse and ultimately backward compatibility when maintenance is required.

SPICIER RECOVERABILITY STEPS

While TEACH depicts recoverability principles that should be applied during software design and development, the SPICIER mnemonic identifies specific phases of the recovery period, each of which can be optimized through recoverability principles. In many cases, many if not all recovery steps can be automated, and some steps, such as cleanup, should often be eliminated in production software.

Stop Program

Failure comes in all shapes and sizes but does not necessarily denote program termination or even that runtime errors were produced. Failure can occur not only when software produces runtime errors or terminates abruptly, but also when it enters a deadlock or infinite loop, or continues to execute albeit producing invalid data, data products, or other output. A common goal of exception handling routines is to channelize software outcome toward two distinct paths, the *happy trail* and the *fail-safe path*.

The happy trail (or happy path) is the journey that software takes from execution to successful completion and delivery of full business value. When exceptions or errors are encountered, however, robust software aims to get back on the happy trail toward restoration of full or partial functionality. In other words, something bad happens during execution, but the software limps along or is fully restored, ultimately delivering either full or partial business value. Software that lacks robustness will still have an underlying happy trail, but once exceptions occur, no methods exist to rejoin that trail, so the software can only fail.

The fail-safe path is the alternative path in which software fails gracefully due to exceptions or errors that were encountered. Functionality halts and no or minimal business value is delivered, but the environment is undamaged and remains secure. An example of unsecure termination would be abruptly and manually aborting SAS software because invalid output had been detected by analysts. Secure termination, on the other hand, might automatically detect the abnormalities in output, update a control table to reflect the exception or error, and instruct SAS to terminate. Thus, automatic software termination must always begin with automatic failure detection.

Although production software should fail gracefully when errors are encountered, if an unexpected error is encountered for the first time, software may fail in an unsafe or unpredictable manner. In these cases, if software is still executing but errors have been detected, it must be terminated manually. For example, if an ETL program that performs daily data ingestion requires tables to have 40 fields, but on Monday it encounters a table that has only 37 fields, the process flow should be robust enough to detect this exception automatically. Business rules would direct the exception handling, but might include immediate program termination with notification to stakeholders, as well as preventing dependent processes from executing. If the system is not robust and does not detect the 37-field discrepancy, subsequent code could continue to execute and create invalid results, even if no runtime errors were produced.

Timely, efficient, and autonomous termination can often be achieved through exception handling routines that detect exceptions and runtime errors and provide graceful program termination. For example, the following SAS output demonstrates an error that is produced when the file Ghost does not exist:

```
data temp;
   set ghost;
ERROR: File WORK.GHOST.DATA does not exist.
run;

NOTE: The SAS System stopped processing this step because of errors.
WARNING: The data set WORK.TEMP may be incomplete. When this step was
stopped there were 0 observations and 0 variables.
WARNING: data set WORK.TEMP was not replaced because this step was
stopped.
%put &SYSCC;
1012
```

Testing the automatic macro variable &SYSCC after each procedure, DATA step, or other module offers one way to determine programmatically whether a runtime error occurred and can be utilized to alter program flow dynamically thereafter. The use of SAS automatic macro variables for use in exception detection and handling is described extensively in Chapter 3, "Communication," and in the "Exception Handling" section in Chapter 6, "Robustness."

Software failures, however, are not always signaled by runtime errors. The 37-field data set described earlier was structurally unsound and would produce invalid results yet possibly yield no obvious runtime errors. As another example, if the following code is run after the previous code, it runs without producing a runtime error but is still considered to have failed because an empty data set is produced:

```
data temp2;
   set temp;
run;

NOTE: There were 0 observations read from the data set WORK.TEMP.
NOTE: The data set WORK.TEMP2 has 0 observations and 0 variables.
```

It turns out that despite encountering an error, the first DATA step creates the Temp data set even though it has zero observations and zero variables. Thus, the second DATA step executes without warning or runtime error, despite having invalid input. In testing for conditions that should direct a program to abort, software must examine more than warning and error codes and must validate that business rules and prerequisite conditions are met. The following code first tests to ensure that the data set contains at least one observation and, if that requirement is not met, the %RETURN statement causes the macro to abort:

```
%macro validate(dsn=);
%global nobs;
%let nobs=;
data _null_;
   call symput('nobs',nobs);
   stop;
   set &dsn nobs=nobs;
run;
%mend;

* inside some parent macro;
%validate(dsn=work.temp);
%if &nobs=0 %then %return;
* else continue processing;
```

Despite diligence in software exception management, on occasion software will still need to be terminated manually. Terminating software manually reinforces the reality that software can fail inexplicably at any line of code. The hapless developer who trips over a cable and disconnects his machine from the server could do so while *any* line of code is executing. Recoverability requires developers to espouse this mind-set and ask themselves, after every line of code, "What will happen if my code fails *here*?" This mind-set reflects that code should always fail safe, in a harmless, predictable manner. This aspect is discussed more fully in the "Cleanup" section later in this chapter, because code that fails safe and is harmless will not require cleanup, thus avoiding this sometimes agonizing aspect of the recovery period.

Investigation

Where program termination has resulted from a scheduled outage due to maintenance or other program modification, this step won't be necessary because failure was planned. But in cases in which unplanned failure has occurred in production software, stakeholders will be clamoring not only for immediate restoration but also for answers. From a risk management perspective, higher-level stakeholders are often not concerned with great technical detail, but rather with more substantive facts surrounding the failure, such as "What caused this failure?" and "Will it happen again?" A best practice is to record these and other qualitative information about the failure in the failure log, as discussed in the "Qualitative Measures of Failure" section in Chapter 4, "Reliability."

Log files are a recommended best practice for production software because they can help validate program success through a baseline (i.e., clean) log and facilitate the identification of runtime or logic errors. When the failure source is captured in the log file and through continuous quality improvement (CQI), developers are able to refactor code to improve robustness, eliminating one failure at a time from occurring in the future, as depicted in the "Reliability Growth Model" section in Chapter 4, "Reliability." In other cases, the log file may exist but may not contain sufficient information to debug code. This also occurs when errors in business rules or logic may not be immediately perceived, resulting in the deletion of relevant log files before they can be analyzed. When this occurs, the code may require extensive debugging and test runs to reproduce and recapture the failure with sufficient information to understand the underlying defect or error.

Maintenance of a risk register is another best practice for production software because it describes known defects, faulty logic, and other software vulnerabilities that can potentially cause program failure. After encountering a failure, SAS practitioners can turn to the risk register to investigate whether associated threats and vulnerabilities had been identified for the failure and, in many cases, whether maintenance or solutions had been proposed. This is not to suggest that action will be taken once the source of failure is identified, but elucidation of the vulnerability is critical to assessing risk in reliable software. Use of the risk register is discussed in "The Risk Register" section in Chapter 1, "Introduction."

Cleanup

After investigation of a failure, the next step is either to rerun the software or perform maintenance if necessary. Especially in cases in which no defect exists and software failed due to an external threat such as a power outage, software should be able to be restarted immediately without further intervention. This pushbutton simplicity can be thwarted, however, when failures produce wreckage in the form of invalid data products—including data sets, metadata, control tables, reports, or other output. The principle that recovery be harmless dictates that carnage not be produced by production software; thus the site of failed software should not resemble an accident scene, with bits of bumpers, broken glass, and the fetid stench of motor oil and transmission fluid seeping into the asphalt.

Cleanup is typically required when some module or process started and did not complete, but left the landscape nevertheless altered. Some examples include:

- A data set is repeatedly modified through successive processes. After the failed process, the original data set is no longer available in its original format because some (but not all) transformations have occurred.
- A data set was locked for use with a shared or exclusive lock, after which the process failed before the data set could be unlocked. The data set may be unavailable for future use until the SAS session is terminated.
- The structure or content of a directory is modified, after which a failure occurs. For example, if raw data files were moved from an Original folder to a Processed folder, they may need to be moved back if the ingestion process was later demonstrated to have failed.
- A control table was updated to reflect that a process started but, because the process failed, the control table may never have been updated with this information. In many cases, SAS practitioners must manually edit the control table to remove these data, which can be risky when control tables contain months of precious, historical performance metrics and other metadata.

The first example often occurs in serialized ETL processes, in which each DATA step or procedure is dependent on the previous one. A DATA step may undergo successive transformations that, to save disk space or avoid complexity, are performed on the same data set without modifying its name. In the following example, the Final data set seems to indicate some finality, but instead is modified several times. Thus, if the code fails, it may be difficult to determine which transformations have occurred successfully (and should not be repeated during recovery). Moreover, where transformations permanently delete, modify, or overwrite data, it may be impossible to recreate the initial data set if necessary:

```
data final;
   length char1 $15;
   char1='taco';
run;

data final;
   set final;
   length char2 $15;
   char2='chimichanga';
run;

data final;
   set final;
   length char3 $15;
   if char1='taco' then char3='flauta'; * theoretical failure
      occurs here;
run;
```

Cleanup can also be required when explicit locks are not cleared because software fails before this can be done. Explicit locks are beneficial because they can ensure that a data set will be available before software attempts to utilize it. When an exception is encountered in which the data set is either in use or missing, the process can enter a busy-waiting cycle to wait for access or it can fail gracefully, thus avoiding runtime errors. The LOCK statement explicitly invokes an exclusive lock and was historically used in busy-waiting cycles, but its use has been deprecated (outside of the SAS/SHARE environment), as discussed in the "&SYSLCKRC" section in Chapter 3, "Communications." Rather, FOPEN can be used to secure an exclusive lock on a file or data set, and OPEN to secure a shared lock. These functions, however, require the FCLOSE and CLOSE functions, respectively, to ensure file streams are closed, as demonstrated in the "Closing Data Streams" section in Chapter 11, "Security."

Another example of cleanup describes changes in folder structure or content that often occur as part of an ETL process. Consider an ETL process that I once inherited from another SAS practitioner. Raw text files were continuously deposited into a folder Inbox and, once a day, the ETL process would first create a folder named with the current date, then copy all accumulated text files from Inbox to the new folder, then ingest and process the files.

The software was fairly reliable but, when it did fail, it failed miserably and required extensive cleanup.

When the text files were copied from Inbox to the respective daily folder, this process created an index that was later used to ingest each file. So, if the process failed sometime after files had been copied during ingestion, when the software was restarted, no files would be in the Inbox, so no new index would be created. Thus, to restart the software after a failure, the files had to be moved back to the Inbox, the index deleted, and a separate control table updated to reflect that the process was being rerun. While moving data sets or other files on a server can be a regular part of data maintenance and should be automated to the extent possible, SAS practitioners must also consider the implications of whether those actions and automation could cause later cleanup if software were to fail.

In all cases, SAS practitioners should strive to design and develop processes that require no cleanup, which not only can be costly but also can introduce the likelihood of further errors. Having to copy or rename files, modify data sets, edit control tables, or unlock data sets manually is risky business and should be avoided on production software.

Internal Issues

Once software has failed and stopped, the cause and potential remedies have been investigated, and any necessary cleanup has occurred, the next step is often to rerun the software. For example, this occurs when the software failed because of a power or server outage and no software maintenance is necessary. In the most straightforward failures, internal issues do not exist, and this phase is skipped entirely.

But what happens when the power fails and software isn't able to start immediately? Perhaps extensive cleanup of the landscape is required, or numerous software processes that had already completed need to be rerun, causing inefficiency. These all represent internal issues that should be corrected if recoverability has been prioritized, despite the actual failure occurring due to external causes. Thus, internal issues primarily describe corrective, emergency, or adaptive maintenance that must be performed on software either to restore functionality or to improve recoverability when future failures occur.

Some internal issues may not require software maintenance but may require SAS practitioners to resolve other issues in their environment. For example, if ETL software fails because a data set is locked by another process, the second process may need to be stopped (or waited for) so that the ETL software can be rerun after the data set is made available. In a similar example, memory-intensive SAS software can cause an out-of-memory failure, sometimes in the software itself and sometimes in other SAS software running concurrently on the server. Thus, when memory failures occur, other processes may need to be stopped (or waited for) so that the primary software can be restored and run with sufficient memory.

But if vulnerabilities or defects were identified in the code, or if developers or other stakeholders have decided that the recoverability or performance needs to be improved through maintenance, then the software itself must be modified in this phase. Maintenance types such as adaptive, perfective, corrective, and emergency are described in Chapter 13, "Maintainability," and by embracing maintainability principles in software design, maintenance can be performed efficiently after a failure, minimizing the recovery period.

External Issues

External issues reflect anything that stands between SAS practitioners and the recovery of their software—anything over which they have no control. The cause of failure may have been external, such as a failed connection to an external database administered by a third party. In this case, SAS practitioners would begin making phone calls to attempt to ascertain the cause of the failure and its expected duration. In many cases when an external cause is identified during investigation, a development team can only sit around and wait for the external circumstance to be rectified.

So what else can SAS practitioners do once software has crashed while they're twiddling their thumbs waiting for someone to restore a data stream? My team once lost a critical data connection to Afghanistan, and team members kept asking one of our developers, "Is the connection up yet? Do we have data?" And each time he was asked, the developer would disappear to his desk and rerun the software, producing hundreds of runtime errors that demonstrated not only the single exception—the broken data stream—but also cascading failures resultant from a total lack of exception handling. As it turns out, when external issues are preventing software from executing, this is the perfect time to build data connection listeners and other quality controls that test negative environmental states.

A negative state could represent a missing data set, a locked data set, a broken data stream, or other types of exceptions. Exception handling frameworks can be constructed to identify theoretical exceptions, but they can be difficult to test without recreating the exception in a test environment. For example, the following code tests to determine if a shared lock can be gained for a data set; only if the lock is achieved will the MEANS procedure be executed. This exception handling avoids the runtime error that occurs if another process or user has the data set exclusively locked when the MEANS procedure attempts to gain access:

```
data perm.original;
   length char1 $10;
run;

%macro freq(dsn= /* data set in LIB.DSN or DSN format */,
   var= /* single variable for frequency */);
%local dsid;
%local close;
%let dsid=%sysfunc(open(&dsn));
%if %eval(&dsid>0) %then %do;
   proc freq data=&dsn;
      tables &var;
   run;
   %let close=%sysfunc(close(&dsid));
   %end;
%mend;

%freq(dsn=perm.original, var=char1);
```

This exception handling will prevent failures from occurring from a locked or missing data set but, because exception handling is only as reliable as the rigor with which it has been tested, each of these states must be tested individually. A missing data set is easy to test—just delete Original and invoke the %FREQ macro as indicated, and the FREQ procedure will be bypassed. To test how exception handling detects and handles shared and exclusive file locks on Original, a separate session of SAS must be initiated in which Original is first locked with a shared lock and subsequently with an exclusive lock while the %FREQ macro is run from the first SAS session. This process is more complicated, but it is well worth the peace of mind to know that the framework handles negative states as expected.

However, exception handling designed to detect external exceptions or errors can be more difficult to test because SAS practitioners inherently don't have control over external data streams to test negative states. When a data stream is down—and down long enough to write and test some code—this presents the perfect opportunity to ensure that future, similar exceptions will be caught and handled programmatically and automatically. When our Afghanistan data stream was down for several hours, we built a robust process that could detect the exception—the failed data stream—and enter a busy-waiting cycle that would repeatedly retest every few minutes until the stream was recovered. Moreover, we directed these test results to a dashboard so all stakeholders could monitor data ingestion processes.

The refurbished solution was superior for several reasons and drastically improved recoverability of the software. By detecting the exception, the exception handling framework prevented cascading failures and invalid data sets that would have had to be deleted. The solution enabled the team to be more efficient during recovery because no one had to retest the software every time a customer asked for an update—this had been automated. Moreover, the recovery itself was made more efficient because, when the connection eventually was restored, the busy-waiting cycle identified this and ran the software only a few minutes after the connection was restored. And, finally, the solution promulgated the connection status—failed or active—to all stakeholders through the dashboard, thus helping us to more quickly identify and resolve similar negative states in the future.

This example demonstrates dichotomous states in which the external connection is either active or failed, and in which failure denotes total loss of business value and an active connection denotes full business value. But external issues often lack this dichotomy, especially where partial business value can still be delivered. For example, a similar database connection linked my team to a third-party Oracle database maintained overseas. Sometimes the entire database would be down due to external failure; in other cases, only some of the Oracle tables were unreachable. At this point, we had to make the business decision of whether to reject the entire database or pursue partial business value—we opted to take what we could get.

In this scenario, partial business value was deemed better than no business value, but we had to manipulate our software to detect and respond to the exception of individual missing tables. In so doing, we took a SAS process that had been consummately failing when only one table was missing and developed a solution that identified and bypassed missing tables and that conveyed available data to subsequent analytic reporting. For missing tables, the software essentially entered a busy-waiting cycle (similar to the prior data stream example) in which it tested the availability of those specific missing tables until they could be ingested. In delivering this solution, we ensured that business value was maximized despite the occurrence of external failures beyond our control. Furthermore, we received kudos from the data stewards because, in many cases, we were able to inform *them* that their database or tables were down before they were even aware!

Rerun Program

Once any necessary modifications to code, hardware, or other infrastructure have been made, the SAS software should be tested, validated, and executed. Because the cause of the failure was identified and remedied (or at least evaluated and included in the risk register), in an ideal scenario, the developer hits *Submit* or schedules the batch job and confidently walks away. Execution often is less straightforward, however, and unfortunately can involve highlighting bits and pieces of code to run in an unrepeatable fashion following a failure. As described previously in the "Constant" section, quick fixes should be avoided and the software that recovers best is the software that can be rerun in its entirety and autonomously skip processes that do not need to be rerun.

When SAS software fails in an end-user development environment, developers often make a quick fix, reattempt the offending module, and then run the remaining code from that point. End-user developers are well placed to make these modifications on the fly because they are, by definition, the primary users of the software. While this "code-and-fix" mentality is appropriate for end-user development, production software should never be maintained in a code-and-fix fashion. A more robust solution is to engineer SAS software that intelligently determines where it should begin executing, thus enabling both autonomous and efficient recovery. In addition to facilitating recovery after failure, this design further facilitates the recurrent software recoveries necessary during testing and development phases. In other words, the extra development effort expended to facilitate autonomous, efficient recovery will more than pay for itself over the course of the software lifespan.

To continue the scenario from the prior "Constant" section, consider that you have an ETL process that ingests and transforms 20 data sets in series. A failure occurs after the 15th data set has completed processing. SAS practitioners in end-user development environments might be tempted to modify the code so that when the software recovers, only the last five data sets are specified to be processed. This short-term solution should be avoided, especially where it would involve having to remodify the software afterward so it could again process all 20 data sets in the future.

A more autonomous yet less timely solution would be to run the entire process again, thus also redundantly processing the first 15 data sets. This duplication of effort would be neither timely nor efficient (for the SAS system), but it would at least be efficient for SAS practitioners because they wouldn't be stuck babysitting code during the recovery period. However, if ultimate business value is conveyed through analytic reporting that can be produced only once all 20 data sets have been processed, it would represent a tremendous inefficiency for the business analysts or other stakeholders who might be waiting on the other end of the SAS software for data products so that they can generate their reporting. Clearly a solution is required that is both efficient and autonomous.

A superior solution is to design modular code that intelligently identifies where it has failed and succeeded, and records process metrics in a control table that can be used to drive processing both under normal circumstances as well as under exceptional circumstances, including during and following catastrophic failure. By implementing checkpoints (demonstrated in the following section), the software can start where it needs to—where the failure occurred—rather than inefficiently reprocessing steps that had completed successfully. Through these best practices, timely, efficient, autonomous, constant, and harmless recovery can be achieved. Other benefits of control tables are demonstrated in the "Control Tables" section in Chapter 3, "Communication."

RECOVERING WITH CHECKPOINTS

A *checkpoint* (or *recovery point* or *restore point*) is "a point in a computer program at which program state, status, or results are checked or recorded."[8] One of the most successful methods to implement TEACH principles in software design is to utilize checkpoints that track successful and failed processes through control tables. The following code demonstrates the %BUILD_CONTROL macro that creates a control table Controller if it does not exist. It also improves autonomy because if the control table is accidentally deleted or corrupted, it will regenerate automatically once deleted. Note that the %SYSFUNC macro function only tests for the existence of the control table, so if another data set already exists with the same file name, the code as currently written will fail:

```
%macro build_control();
%if %sysfunc(exist(controller))=0 %then %do;
   data controller;
      length process $40 date_complete 8;
      format date_complete date10.;
      if ^missing(process);
   run;
   %end;
%mend;
```

The %UPDATE_CONTROL macro updates the control table with the date of completion for each module that completes without error:

```
%macro update_control(process=);
data control_update;
   length process $40 date_complete 8;
   format date_complete date10.;
   process="&process";
   date_complete=date();
run;
proc append base=controller data=control_update;
run;
%mend;
```

The primary business rules are located in the %NEED macro, which specifies that a process should not be executed if it has already been run on the same day. The %NEED macro is called immediately before each module in the ETL process to determine if that module needs to be executed or if can be skipped. Thus, if code failed after successfully completing one module, upon recovery, that first module would automatically be skipped. If the code is executing for the first time of the day under normal circumstances, all modules will be required, so all modules will be executed. This logic exemplifies constancy, because it allows the same code to be executed both under normal circumstances as well as following program failure:

```
%macro need(process=);
%global need;
%let need=YES;
%let now=%sysfunc(date());
data _null_;
   set controller;
   if strip(upcase(process))="%upcase(&process)" and date_complete
      =&now then call symput('need','NO');
run;
%mend;
```

Two modules, %INGEST and %PRINT, are included here as representations of much larger modules that would exist in an actual ETL process flow. Most importantly, the automatic macro variable &SYSERR tests module completion status and, only if no warnings or errors are encountered, the %UPDATE_CONTROL macro is called to record this successful completion in the control table. Note that in an actual infrastructure, additional information about process failures would likely be collected during this process and passed to the control table. In addition, failure of a prerequisite process (such as %INGEST) would necessitate that dependent processes (such as %PRINT) not initiate, but to improve readability, this additional business logic is omitted. Thus, while this code does not fail safe, it does eliminate redundant processing that often occurs during the recovery period.

```
%macro ingest();
%let process=&SYSMACRONAME;
data ingested;
   set incoming_data;
run;
%if &SYSERR=0 %then %update_control(process=&process);
%mend;

%macro print();
%let process=&SYSMACRONAME;
proc print data=ingested;
run;
%if &SYSERR=0 %then %update_control(process=&process);
%mend;
```

The %ETL macro represents the engine (AKA controller or driver) that drives processing, by first building the control table if necessary and by subsequently testing the need for each module before calling it. In an actual production environment, this code might be scheduled to run once a day to ingest and process new transactional data sets. To prevent the code from inefficiently performing redundant operations, successful execution of individual modules is tracked via the control table. If one of the modules does fail during execution, prerequisite processes are known to have completed, and data-driven processing will immediately direct execution to start at the specific checkpoint for the correct module.

```
%macro etl();
%build_control;
%need(process=ingest);
%if &need=YES %then %ingest;
%need(process=print);
%if &need=YES %then %print;
%mend;

%etl;
```

Checkpoints can improve software recoverability, but their implementation must be supported by modular code as well as a clear understanding of module prerequisites, dependencies, inputs, and outputs. *False dependencies*, discussed in the "False Dependencies" sections in Chapter 7, "Execution Efficiency," can occur during recovery, in which later processes are inefficiently waiting for earlier processes to redundantly run, despite earlier successful completion. Checkpoints effectively eliminate this type of false dependency by enabling software to recognize that prerequisite processes have completed successfully, even after a failure has later occurred. Without modular software design, however, discussed throughout Chapter 14, "Modularity," checkpoints will have limited benefit because process boundaries, prerequisites, and dependencies will be largely unknown or buried deep within monolithic code.

RECOVERABILITY IN THE SDLC

Recoverability is not often discussed until a high-availability threshold appears in reliability requirements. Because high availability environments minimize software downtime, only through a mix of both reliability and recoverability principles, as well as through nonprogrammatic endeavors, can high availability be achieved. As a frequent afterthought, recoverability too often is discussed only after software is in production, a failure has occurred, and through emergency maintenance developers have subjectively taken "too long" to restore functionality. While recoverability principles and methods can be infused into software after production, checkpoints and other recoverability techniques are far easier to design into software than to sprinkle on top as an afterthought.

Requiring Recoverability

Recoverability is less commonly prioritized in software requirements than reliability even though the recovery period directly detracts from software availability, the key reliability metric. Stakeholders too often want to increase reliability by preventing failures alone rather than by also reducing the impact that each failure has on the system. Incorporating recoverability metrics into software planning and design is beneficial because it enables SAS practitioners to further understand the quality that their software must demonstrate. Moreover, when failures do occur, all stakeholders have common benchmarks by which to assess whether recovery is occurring as expected or whether it is delayed, chaotic, or otherwise failing to perform to standards.

While the five principles of recoverability are closely intertwined, timeliness is the most quantifiable metric; it has been mentioned that timeliness often acts as a proxy for other principles. By specifying that software recovery additionally must be efficient, autonomous, constant, and harmless, software requirements set the stage for a timely recovery. In technical requirements, it's common to simply state the MTTR, RTO, and possibly MTD that are required of software performance. These terms are defined in the "Defining Recoverability" section earlier in this chapter and further described in the "Timely" section.

Redefining Failure

Both *failure* and *recovery* are technically defined in the "Recoverability Terms" text box earlier in the chapter, but definitions are worthless if they can't be interpreted within real-world scenarios. Moreover, it's important to understand that from different stakeholder perspectives, these seemingly straightforward definitions can be interpreted differently when not explicitly defined within an organization or team. *Failure* is, again, defined as "termination of the ability of a product to perform a required function or its inability to perform within previously specified limits."[9] Some of the more commonly held interpretations of failure and their applications to failed software follow:

Four Interpretations of Failure

- *"When the software actually fails."*—If software terminates with a runtime error, failure occurs at that point, whereas if software completes but does so too slowly, failing to meet performance requirements, the failure occurs at the software completion (or when it exceeds a requirements threshold). Because software failures are not necessarily immediately known to stakeholders, this definition may require researching logs or historical data products to determine when the failure occurred.
- *"When the software initially fails."*—The key to this common definition is *initially,* which describes that only the first failure in a series of successive same-type failures is counted. Thus, this definition of failure approximates the failure rate definition demonstrated in the "Failure Rate" section in Chapter 4, "Reliability," in which only the first failure is counted, even if the software cannot be repaired for days and thus exists in a failed state unable to deliver business value.
- *"When the software failure is discovered."*—If a tree falls in the forest and no one hears it, does it make a sound? In some environments, SAS practitioners are not penalized for failures that go unnoticed, since the situation represents an inability to identify the failure rather than an inability to act to correct it. Developers often may want to start the recovery clock when they or other stakeholders detect the failure, but should it actually

be started when the failure first occurred? For example, when software fails overnight and cannot possibly be discovered until morning, when should the failure actually be described as having occurred?

- *"When business value was lost."*—It's demonstrated throughout this text that ultimate business value in data analytic development environments is conveyed not through software products but through data products. Therefore, software might be expected to complete by 1 PM so that analysts can retrieve resultant data products by 2 PM. If the software failure occurs at noon, is the failure recorded at noon (the time of the failure) or 2 PM (when business value is lost)?

Continuing the scenario in the final definition of failure, imagine that the software failure had not only occurred but also been detected by SAS practitioners at noon. They worked feverishly for an hour to perform corrective maintenance to restore the software and reran it at 1 PM, and it completed seconds before analysts needed the data products at 2 PM. Developers and analysts were ecstatic that the deadline was met and no business value was compromised, so should this even be recorded as a failure? Further, if there were no failure, how could there be a recovery or a recovery period?

In this example, although business value was not diminished, developers did experience a recovery period because they hastily modified and tested SAS code for an hour, shirking other responsibilities. The event should be recorded in a failure log, because it incurred the expense of several developers' time for an hour. But it also should be noted that no business value was lost, thus diminishing the severity of the failure. But these decisions and all their complexity are best left to stakeholders to determine. This scenario does illustrate that especially where contractually imposed performance requirements exist, there should be no ambiguity in whether or when failures have occurred. The next section discusses the similar conundrum that can exist in attempting to clearly define recovery for all stakeholders.

Redefining Recovery

Recovery, like failure, has a clearly stated and accepted technical definition: "the restoration of a system, program, database, or other system resource to a state in which it can perform required functions."[10] This doesn't prevent stakeholders, however, from interpreting recovery differently based on their unique role, on their visibility to software failures, and possibly on whether they have some bias toward either minimizing or maximizing the recovery period. The following list demonstrates some of the more commonly held interpretations of recovery, and further illustrates the importance of unambiguously defining what is meant by recovery in requirements documentation.

Four Interpretations of Recovery

- *"When the software developer finishes corrective maintenance that was required, and hits Submit."*—This definition is common in end-user environments, in which a SAS practitioner tells a customer pointedly, "Yes, the software is fixed. It's running now." But was it sufficiently tested and validated? Has it achieved the necessary outcome? Has it even surpassed the point at which the software initially failed? This is an exceedingly weak and often inaccurate definition of recovery, yet one which unfortunately is tacitly conveyed far too often.
- *"When the software has been restarted and has surpassed the point of failure."*—For example, if ETL software failed when ingesting the 16th of 20 data sets, recovery would be defined as having occurred when that 16th table was successfully processed, thus ensuring that the same failure will not occur.
- *"When the software has been restarted and completed."*—A clean run and SAS log are the best indication of a successful process, especially when additional quality control measures are able to validate the data products or other solutions that are produced. Thus it's common to wait until software has completed to declare it recovered. However, in the case of some lengthy, serialized processes, and especially where the failure occurred in early modules, it might be more appropriate to declare recovery when the point of failure has been surpassed, especially where corrective maintenance has been implemented to prevent that failure in the future.
- *"When lost business value is restored."*—Just as failure is sometimes defined as the loss of business value, recovery can be defined as its restoration. For example, non-technical customers, analysts, or other stakeholders are less likely to be concerned with the functional status of data analytic software and more likely to be concerned with the accuracy and timeliness of resultant data products. If a failed SAS process overwrites a permanent analytic data set or report with invalid data, recovery might not be considered to have been achieved until a data product is again available.

These complexities of failure and recovery definitions can be subtle, but when met unexpectedly in an operational environment, can lead to tumultuous discussions as various stakeholders express individual interpretations of these complex constructs. A far more desirable outcome is to have had these discussions and collectively defined failure and recovery among stakeholders before software is executing and has had a chance to fail.

Measuring Recoverability

Consider that a SAS process runs every hour on the hour 24×7 for March and April but has experienced five failures, each of which is recorded in the failure log. Table 5.1 presents this failure log in an extremely abbreviated format, reprising the failure log demonstrated in the "Measuring Reliability" section in Chapter 4, "Reliability." Please visit that section to understand the difference between reliability and recoverability metrics.

The "Redefining Failure" and "Redefining Recovery" sections earlier in the chapter delineate a host of issues involving the interpretation of these two terms; however, for the purposes of this demonstration, failure and recovery are represented in the failure log but not further defined. Thus, the failure occurring on March 16 incurred a recovery period of 18 hours, with subsequent recovery periods including 20, 1, 1, and 4 hours. Note that unless otherwise specified, the cause of the failure—internal or external—is irrelevant, as is severity of the failure, so every failure and its corresponding recovery contributes to recoverability metrics. In this general sense, the MTTR is (18 + 20 + 1 + 1 +4) / 5 or 8.8 hours.

Well, this doesn't seem too fair. The first 18-hour outage was caused by a power failure over which the data analytic team likely had no control. For this reason, team- and organizational-level documentation such as SLAs

Table 5.1 Sample Failure Log for SAS Software Run Daily for Two Months

Error Number	Failure Date	Recovery Date	External/ Internal	Severity	Description	Resolution
1	3-16 3 PM	3-17 9 AM	Ext	1	power outage	waited
2	3-18 2 PM	3-19 10 AM	Int	1	failure related to power outage	made more restorable
3	3-25 10 AM	3-25 11 AM	Ext	1	network outage	waited
4	4-5 3 PM	4-5 4 PM	Int	1	out of memory SAS error	restarted software
5	4-22 9 AM	4-22 1 PM	Int	2	minor data product HTML errors	modified HTML output

often explicitly state the software or infrastructure for which a team will have responsibility—including incurring risk of failure—as well as external infrastructure for which they will not be held responsible. Thus, in a slightly modified example, if the team explicitly is not held responsible for power or network outages, then the first and third items in the failure log should not be included in recoverability metrics. In this example, the MTTR is (20 + 1 + 4) / 3 or 8.3 hours, only negligibly better than the previous example.

The first and second failures in the log also illustrate an important point: the recovery period reflects the duration of recovery but not necessarily the effort involved. When an externally caused failure occurs, the majority of the recovery period can be the team performing unrelated tasks while they wait for power to return or a system to come back online. Even when SAS practitioners work diligently to repair a failure like the one experienced on March 18, it's unlikely that they literally worked around the clock until 10 AM to recover the software. More likely, they stayed late but still made it home in time for dinner or at least to sleep. Yet, because they had not yet successfully recovered the software, the time which they slept counted for recovery metric purposes. Thus, the critical point to understand is that when stakeholders are discussing whether and how to implement recoverability metrics such as RTO or MTD, realistic estimates and thresholds must be generated that take into account variability such as software that remains in a failed state overnight.

WHAT'S NEXT?

Reliability aims to achieve software that produces required functionality and performs consistently for some intended duration. And if variability is said to be the antithesis of reliability, how can that variability be identified, predicted, detected, and handled to prevent or minimize failure while promoting reliability? The objective of robust software is to predict variability in the environment, in data, and in other aspects of software to expand the ability of software to function despite that variability. And, if software cannot function, a hallmark of robust software is that it fails gracefully, thus failing safe without damaging itself, data products, other output, or other aspects of the environment.

NOTES

1. ISO/IEC 25010:2011. *Systems and software engineering—Systems and software Quality Requirements and Evaluation (SQuaRE)—System and software quality models*. Geneva, Switzerland: International Organization for Standardization and Institute of Electrical and Electronics Engineers.
2. Id.

3. ISO/IEC 25000:2005. *Software engineering—Software product Quality Requirements and Evaluation (SQuaRE)—Guide to SQuaRE*. Geneva, Switzerland: International Organization for Standardization and Institute of Electrical and Electronics Engineers.
4. ISO/IEC/IEEE 24765:2010. *Systems and software engineering—Vocabulary*. Geneva, Switzerland: International Organization for Standardization, International Electrotechnical Commission, and Institute of Electrical and Electronics Engineers.
5. Id.
6. Id.
7. Troy Martin Hughes, "When Reliable Programs Fail: Designing for Timely, Efficient, Push-Button Recovery." Presented at PharmaSUG, 2015.
8. ISO/IEC/IEEE 24765:2010.
9. ISO/IEC 25000:2005.
10. ISO/IEC/IEEE 24765:2010.

CHAPTER 6
Robustness

More coca, please.

Off-roading at 16,000 feet across the Bolivian *Altiplano* (high plane), your head is literally in the clouds. Mine was also figuratively in the clouds as I munched on a Diamox and coca candy cocktail, trying to stave off altitude sickness while taking in the extraterrestrial landscape.

While my head wasn't prepared for the three-day trek through the heavens, my guides fortunately had thought of everything. Those old black-and-white photos of "Okies" escaping the Dust Bowl, setting out with high hopes and humble means on Route 66 for a new life in California, their Model Ts overburdened, all their earthly wares tied haphazardly to anything that would hold a rope—our two Land Cruisers looked no different as we departed from Uyuni, Bolivia, overflowing with bags and backpacks, and crowned with tires and tarps. The only discernible difference was that we were driving *into* our Dust Bowl.

Carving trails into the famed Salar de Uyuni (at more than 4,000 square miles, the largest salt flats in the world), we raced across in blinding sunlight that reflected off the brilliant white landscape. You touched the brim on your polarized sunglasses every couple minutes just to prove to yourself they were still in place despite the necessity to squint. I had trekked to Badwater Basin, Death Valley, to visit the California salt flats and lowest point in North America—yet Bolivian brine was like nothing I had ever seen. You can succumb to sunburn in minutes here, so when we disembarked for an hour to go play in the salt, our guides fortunately provided sunscreen for the tourists who had neglected to bring any.

The caravan continued across the salt flats, crushing the hexagonal plates of salt that burgeoned forth beneath our tires. The next destination was Isla Incahuasi, a rugged, rocky "island" outpost that perforates the otherwise bleak incandescence. For the guys—all over 6 feet and crammed into the Beverly Hillbillies roadster—it offered a chance to unfold our bodies and, for our lone female companion, a *baño* (bathroom) provided a welcome relief.

Seconds after she had darted into the primitive cobblestone structure, she raced back to our SUVs, her look of utter dejection turning to elation as our guide's outstretched arm held toilet paper. They had fortunately thought of that, too.

By the appearance of the overladen Land Cruisers, you'd have thought we were pitching tents on the Salar, but no, we were booked into a hotel constructed entirely from salt blocks perched somewhere in the mountains. However, getting there would prove to be the true adventure.

As we continued on, the salt dissipated, giving way to mud, then sand, then gravel, then outright rocks, as driving conditions continued to deteriorate. The first flat tire—exposing the more infamous meaning of the salt *flats*—came in the afternoon. Our guides were able to repair it quickly with an air compressor they had fortunately brought.

As we departed the basin and the terrain turned more mountainous, several other flats would follow which were met with more compressor sessions, aerosol tire sealant, and finally full replacement, all of which were adroitly handled by the guides. We made it to our briny, barebones hotel by dusk to dine sumptuously on a salt table and sleep on salt beds.

Day two was a trek through the *Reserva Nacional de Fauna Andina Eduardo Abaroa*, a federal nature preserve of untold beauty and often austerity. We chased flamingos, jumped through sulfuric clouds of off-gassing geysers, relaxed with cold beers in geothermal hot springs, and of course made countless stops for recurring flat tires endemic to the rough terrain.

At one point, having limped as far as we could on a bandaged spare, the guides pulled into a remote service station, where the tire was professionally patched—of course by a crew with whom the guides were fortunately besties.

But throughout the adventure, at every turn the guides had our backs—and fortunately an endless supply of coca leaves to chew to stave off the ever encroaching altitude sickness.

■ ■ ■

I've intentionally overused "fortunately" throughout the scenario to illustrate that fortune in fact had nothing to do with our successful trek. The guides were pros. They'd taken this exact route hundreds of times. They knew the terrain. They knew their vehicles. They knew their equipment. The only variability seemed to be the occasional curveball that a tourist would throw, but after so many expeditions, even that variability was subtle and predictable.

Robust software should be equally equipped to detect and handle predicted threats and vulnerabilities. You don't go off-roading in the wilderness for days without a spare tire (or two) because you know the terrain is unforgiving and flat tires can occur. You similarly don't assume that a data set will be available when you try to run a MEANS procedure, because you can predict that it could be missing, locked, or possibly corrupted.

Over time, additional threats or vulnerabilities may be identified that weren't expected at the outset. The guides initially hadn't carried sunscreen but, after a couple early trips in which trekkers got horribly sunburned on the flats, the safety in a bottle became a small measure they could take to avoid a huge risk. Software exception handling routines similarly can be modified over time to address new or unpredicted risks to facilitate smooth performance.

The trek also demonstrated the flexibility that should exist in a risk management framework. When our tire initially went flat, they pumped it immediately. When leaks worsened, they applied tire sealant. When we'd actually found a patch of level ground, they took the opportunity to more closely examine and often exchange failing tires. Finally, the desert mechanics patched a tire when we'd nearly exhausted our nine lives.

SAS practitioners should also understand the full array of tools and techniques that can eliminate or mitigate risk to software. Exception handling aims not only to identify threats but also to enable software to continue despite encountering threats, thus maximizing business value.

The guides never had to use it, but they had one final Hail Mary in their pockets in the event of extreme injury, sickness, or other calamity—a satellite phone charging in the glovebox. Even the most robust software will encounter exceptions, errors, or environmental conditions that cannot be overcome and which signal imminent failure. The Hail Mary in software development should be a fail-safe path that facilitates graceful termination.

We never had to use the sat phone, but its presence was reassuring in case we ever ran out of tires … or coca.

DEFINING ROBUSTNESS

Robustness is "the degree to which a system or component can function correctly in the presence of invalid inputs or stressful environmental conditions."[1] This is distinguished from *fault tolerance*, defined as "the degree to which the software product can maintain a specified level of performance in cases of software faults or of infringement of its specified interface."[2] The terms are often commingled even in technical discussions, but robustness speaks to overcoming variability external to software, whereas fault-tolerance aims to overcome unwanted (and sometimes unpredictable) variability within software.

The main objective of robust and fault-tolerant software is to detect and overcome variability so that business value can be delivered. A secondary objective, however, is to enable software to *fail safe* when software execution cannot continue. *Fail safe* is defined as "pertaining to a system or component that automatically places itself in a safe operating mode in the event of a failure."[3] Failing safe is colloquially known as *failing gracefully* or *graceful termination* while the *fail-safe path* terminates program flow and safely and securely shuts down software when all hope has been lost. In other words, software will do its best to overcome challenges but, when faced with insurmountable obstacles, will exit in a manner that doesn't damage itself, its data products, other output, or other aspects of the environment.

This chapter introduces defensive programming techniques that require the creative wherewithal to imagine potential vulnerabilities in and threats to software, as well as the technical prowess to design solutions that mitigate or eliminate those vulnerabilities or threats before they occur or as they are occurring. The chapter also introduces exception handling, which enables software to detect variability such as exceptions, runtime errors, or other faults and to respond dynamically based on prescribed business rules. In some cases, full business value can still be delivered but, in cases in which only partial business value can be delivered, it is nevertheless done in a planned manner. Through this chapter, SAS practitioners will gain an understanding of how to build robust software that benefits from exception handling and facilitates more reliable software performance.

ROBUSTNESS TOWARD RELIABILITY

The journey toward robustness is ultimately one toward reliability. The intent of equipping your Land Cruiser with off-road shocks and all-terrain tires is not to impress the neighbors, but ultimately to get somewhere reliably despite the adversity or obstacles you may face along the way. Making software more robust does not remove the variability that reliability loathes, but instead improves the adaptability of software by allowing it to deliver consistent performance despite inconsistent inputs or environmental elements.

Consider the example of SAS extract-transform-load (ETL) software designed to ingest third-party data from an external source. Because SAS practitioners cannot control the quality of third-party data input, they must validate data quality when data are ingested. Thus, robust ETL software is able to continue executing even if some values, observations, or data sets must be deleted or modified because they are missing, duplicate, corrupt, or otherwise invalid. For example, if 19 data sets were received correctly, but one was corrupt, robust ETL software should be able to process the valid data when ingestion can still deliver partial business value, thus maximizing reliability to the extent possible.

In another example, software intended to be run in different environments (such as in both SAS Display Manager and the SAS University Edition) will need to be robust to variability in those environments. File naming conventions, file structure, system options, and other functionality do subtly differ between the environments, and the inability to recognize these differences could result in failure. In this sense, the aim of achieving software portability is to ensure that the software is robust enough to function and perform equivalently across SAS environments in which the SAS interface, installation type, or other aspects vary. Thus, software portability, discussed in Chapter 10, "Portability," aims to make software robust and reliable despite specific sources of environmental variability.

While the primary function of robustness is to enable software to continue functioning despite the bumps in the road (and thus deliver full or partial business value), the secondary function is the facilitation of fail-safe termination. In data analytic development, the most important aspect of software reliability is the generation of an accurate solution or data product, without which the software is worthless. Robust software can prevent inaccurate or invalid results by ensuring that software terminates when it encounters an environment or condition that would cause functional failure, just as experienced off-roaders can read the terrain and realize when it's time to turn back toward safer ground. Thus, as a last resort, even the most elaborate exception handling routines should provide a fail-safe path to graceful software termination.

DEFENSIVE PROGRAMMING

Defensive programming is another name for developing robust software. By identifying known threats and vulnerabilities, developers *defend* against failure by making software more robust and reliable. Defensive programming is greatly facilitated by an awareness of specific threats and vulnerabilities, whether external (such as server outage) or internal (such as unexpectedly locked SAS data sets) that can cause runtime errors. These threats are unexpected in the sense that no one knew the data set would be locked at 10 AM or the server would fail at 3 PM, but they are nevertheless predictable because these exceptions are readily identifiable. Defensive programming assumes the mind-set that every data set is locked or missing until proven otherwise and every SAS function, statement, and procedure has failed until otherwise validated.

Risk identification is the first step in defensive programming. Risks typically include software that terminates early with a failure, or software that completes but with functional or performance failures. Each of these situations reduces or eliminates business value. Thus, the second step is to identify specific vulnerabilities within software (or requirements) that could be exploited to produce a failure. These vulnerabilities comprise the risk register, described in the "Risk Register" section in Chapter 1, "Introduction," and should be used to identify which vulnerabilities should be eliminated or mitigated through exception handling. In software that does not demand high reliability or robustness, the decision is often made to forgo exception handling and accept all risks; however, robust software should incorporate an exception handling framework that identifies and handles exceptions and runtime errors that occur.

Specific Threats

The following SAS code demonstrates the "data-set-run" mentality—the unfortunate reality of how most SAS practitioners are taught to code through instruction, in literature, and by their peers. This straightforward representation may suffice in end-user development environments or when code doesn't need to demonstrate robustness, but it lacks sufficient quality for production environments:

```
data final;
   set original;
run;
```

The code creates the Final data set from the Original data set and is the most basic implementation of the DATA step. However, in complex, variable environments, defensive programmers must ask themselves two questions about every line of code:

- What are the specific threats that could cause this line of code to fail?
- What occurs if the software fails for any reason at this line of code?

The first question addresses robustness while the second question addresses fault-tolerance and risk. Although the degree to which robustness and fault-tolerance are integrated into software should depend on the required level of software reliability, defensive programming techniques teach developers to eschew error-prone assumptions—except to assume that if it can fail, it will.

Some threats to the previous DATA step include:

- The Original data set may not exist.
- The Original data set may exist but be exclusively locked so it cannot be read from.

- The Original data set may exist but in fact be the wrong data set.
- The Original data set may exist but be corrupt.
- The Original data set may exist but be missing data, variables, or observations or otherwise lack completeness or accurate structure.
- The Original data set may be too large to duplicate in the SAS application.
- The Original data set may be too large to create in the WORK library.
- The Final data set may exist but be exclusively locked so it cannot be overwritten.
- The SAS application may run out of memory processing the DATA step.
- The DATA step may incur other unspecified warnings or errors.

The first threat (a missing file) is a specific software threat that can exploit a specific software vulnerability—the failure to validate data set existence. If SAS software requires reliable code, but fails to validate data set existence and availability, these vulnerabilities represent errors (human mistakes) as well as code defects that must be remedied to achieve the required level of robustness. When this vulnerability is exploited, the following failure is produced:

```
data final;
   set original;
ERROR: File WORK.ORIGINAL.DATA does not exist.
run;

NOTE: The SAS System stopped processing this step because of errors.
WARNING: The data set WORK.FINAL may be incomplete.  When this
step was stopped there were 0 observations and 0 variables.
```

While the previous threats are all possible, the risk they pose will vary substantially from one environment to the next and even from one DATA step to the next. For example, the DATA step reads the Original data set from the WORK library, which normally is inaccessible to other users and processes. Thus, while it is technically possible on some networks to manually assign a library reference to another SAS practitioner's temporary WORK library, in general, this constitutes an extremely low risk; thus data sets in the WORK library do not need to be protected against the risk of inaccessibility.

However, the following modified DATA step is inherently riskier than the previous because it references a shared library PERM that could be accessed by other developers; therefore, testing both the availability and the file lock status of PERM.Original before the DATA step would be more valuable than in the prior process.

```
data final;
   set perm.original;
run;
```

Separate threats do not necessarily require separate remedies. For example, to test for data set existence, the EXIST function (often operationalized with the %SYSFUNC macro function) is commonly used. The following code now first determines whether PERM.Original exists before continuing:

```
libname perm 'c:\perm';

%macro doit();
%if %sysfunc(exist(perm.original)) %then %do;
   data final;
      set perm.original;
   run;
   %end;
%mend;

%doit;
```

An unrelated threat occurs when the PERM.Original data set is exclusively locked and the SET statement fails. This vulnerability can be eliminated by first testing the lock status of the data set with the OPEN function, but as it turns out, the OPEN function will also fail if the data set is missing. Thus, the following code ensures that the data set is both existent and accessible before the DATA step execution, effectively killing two threats with one stone:

```
%macro doit();
%local dsid;
%let dsid=%sysfunc(open(perm.original));
%if %eval(&dsid>0) %then %do;
   data final;
      set perm.original;
   run;
   %let close=%sysfunc(close(&dsid));
   %end;
%mend;

%doit;
```

This code now defends against the first two enumerated threats, lack of existence and accessibility. However, in defending against failure, SAS practitioners must always remember to ask the second, general question: *What happens if my code fails here?* In other words, if an unknown exception or runtime error occurs (possibly due to a latent defect) on line 5, what will happen? Will the program continue in a safe, predictable manner? Or, if the power fails while line 5 is being processed, what will occur, and will data products or the environment be damaged in the process, either during failure or resultant recovery?

As demonstrated throughout later sections, exception handling can make software much more reliable and robust. However, it can also introduce additional, unintended errors if done improperly or incompletely, providing SAS practitioners and other stakeholders with a false sense of security. Some vulnerabilities can be introduced through exception handling routines and cannot be overcome, but SAS practitioners should at least be made aware of them. For example, the following code represents a slightly more complex implementation of the %DOIT macro, which contains an error (and latent defect) immediately after the DATA step.

```
%macro doit();
%local dsid;
%let i=1;
%let dsid=%sysfunc(open(perm.original));
%if %eval(&dsid>0) %then %do;
   data final;
      set perm.original;
   run;
   %if &i=1 %then growl; *defect;
   %let close=%sysfunc(close(&dsid));
   %end;
%mend;

%doit;
```

In this example, the macro variable &I simulates a loop or some other variable construct that was not thoroughly tested. Thus, when &I is not equal to 1, no runtime error occurs and the defect goes undetected, but when &I is 1, the SAS application doesn't know how to "growl" and fails. Central to the exception handling that has been implemented, when SAS fails to growl it also prevents the CLOSE function from executing, which in turn retains the data stream that was opened. Thus, after the initial failure, even after the growl line is removed, the code will perform incorrectly because it will open a second data stream to the Original data set with the OPEN function. This failure, while not producing a runtime error, will effectively prevent the software from closing the stream, which it will unnecessarily maintain until the SAS session is terminated.

Thus, in implementing use of %SYSFUNC(OPEN) to validate data set existence and accessibility, SAS practitioners must understand the new risk posed by a failure to close the associated data stream. Fault-tolerant software that fails safe would ensure that this stream is closed as part of terminating a SAS macro or program, as demonstrated in the "Closing Data Streams" section in Chapter 11, "Security." It's impossible to eliminate all risk from software, and accepting risk is a necessary component of all software development. So while developers might do nothing to mitigate or eliminate the vulnerability that exists when CLOSE fails to execute, they should still be aware of it. In general, however, software is significantly less risky and more reliable with exception handling in place.

General Threats

General threats represent unpredictable or unknown threats to software functionality. For example, occasionally SAS will unexpectedly throw a runtime error whose description references a Java error—something completely outside the Base SAS world—which must be tracked down. For example, what do you do when your program is executing confidently, and suddenly you're embroiled in this morass?

```
ERROR:  An exception has been encountered.
Please contact technical support and provide them with the following
traceback information:

The SAS task name is [Submit]
ERROR:  Read Access Violation Submit
Exception occurred at (04A3E47A)
Task Traceback
Address    Frame      (DBGHELP API Version 4.0 rev 5)
0000000004A3E47A  000000000737ED00  sasxkern:tkvercn1+0xBD43A
00000000049D1074  000000000737ED80  sasxkern:tkvercn1+0x50034
00000000049E382C  000000000737EF30  sasxkern:tkvercn1+0x627EC
00000000049E2AC0  000000000737F020  sasxkern:tkvercn1+0x61A80
00000000049E56B8  000000000737F290  sasxkern:tkvercn1+0x64678
00000000049E098C  000000000737F3C0  sasxkern:tkvercn1+0x5F94C
00000000049E053C  000000000737F450  sasxkern:tkvercn1+0x5F4FC
```

```
000000004DA2761  000000000737F458  sasxshel:tkvercn1+0x1721
000000004DA19D5  000000000737F610  sasxshel:tkvercn1+0x995
000000004DCAF20  000000000737F750  sasxshel:tkvercn1+0x29EE0
000000004DF4341  000000000737FAB0  sasxshel:tkvercn1+0x53301
000000004DF7BAC  000000000737FBA0  sasxshel:tkvercn1+0x56B6C
000000004DFA4A0  000000000737FBF0  sasxshel:tkvercn1+0x59460
00000000034F89DB 000000000737FBF8  sashost:Main+0x10EBB
00000000034FE62D 000000000737FF50  sashost:Main+0x16B0D
00000000770259BD 000000000737FF58  kernel32:BaseThreadInitThunk+0xD
000000007715A2E1 000000000737FF88  ntdll:RtlUserThreadStart+0x21
```

The reality is that the SAS application itself is software and susceptible to its own defects and errors, and, while rare, they inexplicably rear their ugly heads from time to time. Fault-tolerant code would not be expected to identify the specific threat of the previous error, but it should appropriately handle the general threat that SAS code at any point could experience a similar failure. For this reason, testing general code failures with automatic macro variables such as &SYSERR and &SYSCC is a critical component of fault-tolerant design.

A second interpretation of general threats includes any threat not explicitly identified and communicated within an exception handling framework. Failure to detect and handle specific threats can occur for several reasons, including:

- *Ignorance*—SAS practitioners unfamiliar with exception handling, or possibly with Base SAS in general, may not realize the importance of exception handling or understand the nature of a specific exception or error.
- *Negligence*—This can occur when robustness isn't coded into software (when requirements direct it should be) or when robustness is not prioritized in software requirements. Choosing not to implement robustness should always be an intentional decision based on calculated risk and knowledge of vulnerabilities.
- *Acceptance*—By accepting the risk of a specific threat, software robustness and reliability may be diminished. But when threats pose only negligible risk, they often are accepted. For example, the "data-set-run" mind-set accepts *en masse* all of the previously enumerated threats.

To illustrate the difference between specific and general threats, the simple DATA step is reprised from the "Specific Threats" section:

```
data final;
   set original;
run;
```

This time, instead of Original being missing or some other predictable runtime error occurring, Base SAS informs you "ERROR: An exception has been encountered." and throws some unfriendly hex your way—that's hexadecimal and not an actual curse, to be clear. But it may feel like witchcraft, because there's neither a way to predict nor recover from the failure. Fault-tolerant SAS software will instead test for any general failure (which includes the hex) by assessing &SYSERR immediately after a procedure or DATA step or &SYSCC before exiting a macro or program. The use of automatic variables as return codes is demonstrated throughout Chapter 3, "Communication."

The following revised code uses the &SYSERR automatic macro variable to detect any warnings or runtime errors that may have been produced by the DATA step:

```
%macro doit();
%let syscc=0;
%local dsid;
%let dsid=%sysfunc(open(perm.original));
%if %eval(&dsid>0) %then %do;
   data final;
      set perm.original;
   run;
   %let close=%sysfunc(close(&dsid));
   %end;
%if &syscc>0 %then %return;
* otherwise do other stuff;
%mend;

%doit;
```

The code now terminates the %DOIT macro using the %RETURN statement if warnings or runtime errors are encountered. This framework detects not only the threat of a missing or inaccessible data set but also other threats that cause warnings or runtime errors, including the previous hex. The %DOIT macro is not made more *robust* by this additional post hoc exception handling that checks &SYSCC, because robustness implies that the code is able to continue toward the happy trail and produce some ultimate business, but it is made more *fault-tolerant* because code that otherwise would follow the DATA step will be prevented from executing when errors are detected. However, the software *as a whole* could have been made more robust because the failure of the %DOIT module can now be detected and used to alter program flow dynamically to some other process to deliver full or partial business value.

Notwithstanding, a zero-value &SYSERR or &SYSCC does not necessarily denote process success, just as a clean log does not necessarily represent that software has not failed. Some failures occur but are displayed only in SAS notes within the log. In other cases, return codes such as &SYSFILRC or &SYSLIBRC must be individually assessed to determine if statements or functions failed. These automatic macro variables are discussed in Chapter 3, "Communication." Failures can also occur due to errors in business logic and produce neither warnings nor runtime errors. While exception handling represents the most important tool to facilitate robust software, it shouldn't replace common sense and code inspection.

EXCEPTION HANDLING

Exception handling is the primary *defensive programming* method to facilitate *robust* and *fault-tolerant* software. These three terms are sometimes used interchangeably, although exception handling always represents the technical implementation or tool to further robustness. *Exception handling* describes the identification and dynamic resolution of adverse, unexpected, or untimely events or environmental states during software execution.[4] Resolution doesn't imply that the exception is eliminated or overcome—only that it is handled. In software literature, exception handling is also referenced as *error handling, error processing, error trapping, event handling, event trapping, defensive programming,* or *fault-tolerant programming*.

The goal of exception handling is always to reroute program flow back to the happy trail—that is, the originally intended process path that delivers full business value. This is illustrated in the "Happy Trail" section later in the chapter. Thus, exception handling is always implemented to alter program flow dynamically. The "Exception Handling, Not Exception Reporting!" section distinguishes *exception reporting,* which only produces static reports that demonstrate exceptions. Thus, while exception handling is a quality assurance tool intended not only to detect but also to overcome exceptions and runtime errors, exception reporting is a quality control mechanism that alerts the developer or user to aberration or failure.

While exception handling is inherent in languages such as Java, Python, and even the SAS Component Language (SCL), no inherent exception handling functionality exists in Base SAS.[5] You have to fake it in SAS, which is unfortunate considering the tremendous role that exception handling plays in facilitating software robustness and reliability. Despite these numerous weaknesses, faking it is better than unhandled exceptions that can cause abrupt or unpredictable failures in production software.

EXCEPTION HANDLING

Exception Handling The identification and dynamic resolution of adverse, unexpected, or untimely events or environmental states during software execution.

A Priori Exception Handling Ensuring that known and detectable prerequisites are met for a process before its execution.

Post Hoc Exception Handling Executing a process and immediately assessing its success or failure.

Exception An adverse, unexpected, or untimely event that may or may not produce a warning, runtime error, or other system notification.

Event User inject, process occurrence, environmental state, or other dynamic attribute that is detectable.

Happy Trail The program flow path that leads to the delivery of full business value or, following one or more exceptions, to the maximum attainable business value.

Exception Handling Framework Comprehensive detection, handling, and inheritance of exceptions to ensure exceptions are dynamically conveyed to dependencies.

Exception Handling Elsewhere

To understand classic exception handling theory and function, you need to step outside the SAS world and explore languages that have inherent exception handling functionality. The following Python code uses the PRINT function to print the variable X:

```
print(X)
```

But what if the variable X doesn't exist? A runtime error results, the program terminates, and the following output is displayed:

```
print(X)
Traceback (most recent call last):
  File "<stdin>", line 1, in <module>
NameError: name 'x' is not defined
```

This outcome is unacceptable in robust software, because the software should either display a message to the user, prompt the user to initialize a value for X, continue to another process, or a host of other activities. But because the exception is unhandled, the code abruptly terminates. Without exception handling, that blue screen of death that 1990s software so frequently exhibited would today still be common. Exception handling allows software to encounter variability—including errors and faults—and to adapt through prescribed channels.

The following modified Python code demonstrates exception handing that catches the exception and transfers control to an exception handler—the EXCEPT statement—for further processing. The underlying functionality—the PRINT function—remains unchanged but only has been wrapped in an exception handling block:

```
try:
   print(X)
except NameError:
   sys.exit()
```

Now when the code executes, if the variable X is missing, the SYS.EXIT function terminates the program. This is not the only outcome of exception handling, but one of many that are discussed in the section titled "Exception Handling Pathways." By intercepting the exception, no runtime error occurs and the software can continue undaunted or, in this example, terminate gracefully. Note that because NameError (a specific type of error) is specified, the exception handling block only detects a certain type of exception, thus eliminating the specific threat of a missing variable. To eliminate all threats (as discussed previously in the "General Threats" section), NameError can be removed. The revised code, for example, also now detects and handles the syntax error that would occur if the PRINT function were misspelled:

```
try:
   print(X)
except:
   sys.exit()
```

Any number of Python statements can be included within an exception handling block, demonstrating the tremendous efficiency of implementing exception handling. In the following example, five successive PRINT functions attempt to print the variables A through E. If A and B exist but C does not, A and B will be printed, after which the exception—the missing variable C—will be detected and program flow will transfer to the EXCEPT block.

If all five variables exist, no exception occurs and the EXCEPT block is never executed:

```
try:
   print(A)
   print(B)
   print(C)
   print(D)
   print(E)
except NameError:
   print("Something funky here...")
   sys.exit()
```

In languages that offer inherent exception handling functionality, the full capabilities of exception handling extend far beyond this abridged introduction. In the following section, Base SAS exception handling is differentiated and demonstrated.

Faking It in SAS

Because Base SAS provides no inherent exception handling capabilities, SAS practitioners can only fake it by testing detectable exceptional states, warnings, and runtime errors. Thus, major limitations exist that distinguish exception handling contrived through Base SAS from the real deal:

- *Exception handling cannot be used within DATA steps or SAS procedures*—Thus, the two primary constructs of SAS software offer no ability to test or handle exceptions. For example, when the DATA step encounters most exceptions or errors, it immediately terminates with a runtime error, offering no ability to handle it dynamically. In SAS, while some exceptions can be avoided through a priori detection, many exceptions immediately produce runtime errors when they occur. There are few opportunities to catch exceptions because Base SAS typically produces only runtime errors, not exceptions. In other languages, a handled exception does not become a runtime error.
- *Faults cannot be detected through exception handling*—For example, real exception handling catches not only faulty or exceptional data or environmental states but also syntax errors and other defects that may lie dormant in code. When SAS encounters a syntax error, this triggers a runtime error, and no exception handling can occur. If Python code perceives a falling basket of eggs, it identifies the exception and swoops in to rescue the eggs before they can hit the floor; post hoc

detection handling in SAS instead provides a bucket that the basket can be dropped into to minimize the mess, but does nothing to stop the impact.

- *Exception handling can be invoked only during specific lines of code that constitute exception handling routines*—For example, to test exceptions over multiple lines of code, multiple exception handling routines are required. In other languages, a single exception handling block (like the TRY-EXCEPT block demonstrated previously) handles any exception immediately as it occurs.

Despite all the ways that Base SAS exception handling is limited, it is still beneficial when more robust and reliable software is required. This section recreates the functionality of the Python exception handling demonstrated throughout the "Exception Handling Elsewhere" section.

The following SAS code uses the %PUT macro function to print the macro variable &X:

```
%put &x;
```

But what if the macro variable &X doesn't exist? A warning results and the log displays the following output:

```
%put &x;
WARNING: Apparent symbolic reference X not resolved.
&x
```

This is unacceptable in production software because logs are not manually monitored, so some mechanism should detect the exception and do something about it. Two global macro variables—&SYSERR and &SYSCC—demonstrate warnings and runtime errors that occur and are essential to approximating exception handling functionality. These variables and their implementation within exception handling are described in Chapter 3, "Communication."

The following code now detects warnings and runtime errors and dynamically alters program flow to terminate the macro when they occur. While not all exception handling in SAS requires use of the macro language, in more complex code and logic this is nearly always a requirement:

```
%macro meh();
%let syscc=0;
%put &X;
%if &syscc>0 %then %return;
%mend;

%meh;
```

This code tests for warnings and runtime errors, rather than testing solely for a runtime error that occurs when a variable is missing, which Python can accomplish. Already, at this most basic level, SAS has deviated from exception handling functionality native in other languages. In SAS, while some errors such as the general memory error are separately distinguished, many runtime errors are lumped together and produce only a vague 1012 return code. All warnings, similarly, receive the return code 4 regardless of their nature, making them impossible to distinguish. But, at least Base SAS is able to detect and handle general threats, even if they often cannot be distinguished programmatically.

A second deviation already apparent in Base SAS is its inability to catch a general exception before it becomes a warning or runtime error. For example, in Python, once the exception of a missing variable is detected, program flow immediately shifts to the EXCEPT block. In Base SAS, however, the presence of a non-zero value in &SYSERR or &SYSCC reflects that a warning or runtime error has already occurred. For this reason, even with effective exception handling in place, the SAS log will often be littered with warnings and runtime errors.

In other languages, an infinite number of statements can be included within one exception handling block, but not so in Base SAS. Suppose you want to print five macro variables, &A through &E, but if any of these are missing, you want to immediately shift program flow elsewhere to handle this exception. The following SAS code attempts (and fails) to replicate this business logic:

```
%macro meh();
%let syscc=0;
%put &A;
%put &B;
%put &C;
%put &D;
%put &E;
%if &syscc>0 %then %do;
   %put Something funky here...;
   %return;
   %end;
%mend;

%meh;
```

When the missing macro variable &C is encountered, the code continues to attempt to print &D and &E rather than immediately transferring control to the exception block. Rather than assessing the value of &SYSCC automatically after each statement, &SYSCC is only assessed once and thus fails to halt execution in time. If &D were dependent on the successful execution of the previous line of code, then the &D line would also fail. These types of failures are discussed later in the "Cascading Failures" section.

A second attempt to deliver the necessary functionality and exception handling creates the EXCEPT label to which successive %GOTO statements can transfer program flow when an exception is detected. However, this is not only redundant, but also terribly faulty. Because SAS labels cannot be nested within %DO, %WHILE, or %UNTIL conditional logic, they essentially are always executed unless explicitly leapfrogged over through additional conditional logic. The following code demonstrates equivalent functionality and exception handling but is an utter disaster and demonstrates why procedural %GOTO logic is abhorred in software development.

```
%macro meh();
%let syscc=0;
%put &A;
%if &syscc>0 %then %goto except;
%put &B;
%if &syscc>0 %then %goto except;
%put &C;
%if &syscc>0 %then %goto except;
%put &D;
%if &syscc>0 %then %goto except;
%put &E;
%if &syscc>0 %then %goto except;
* other logic would go here;
%if &syscc=0 %then %return; * successful run;
%EXCEPT: %put Something funky here...;
%mend;

%meh;
```

Because the code has inherent dependencies—for example, &B should be printed only if &A printed successfully—another way to conceptualize the program flow is through nested conditional logic. The following code, while arguably as convoluted as the previous %GOTO logic, could be preferred because it doesn't require leapfrogging over SAS labels. Notwithstanding, it still requires redundant testing of &SYSCC.

```
%macro meh();
%let syscc=0;
%put &A;
%if &syscc=0 %then %do;
   %put &B;
   %if &syscc=0 %then %do;
      %put &C;
      %if &syscc=0 %then %do;
         %put &D;
         %if &syscc=0 %then %do;
            %put &E;
            %end;
         %end;
      %end;
   %end;
%if &syscc>0 %then %do;
   %put Something funky here...;
   %return;
   %end;
* other logic would go here;
%mend;

%meh;
```

Exception handling in Base SAS sadly doesn't get any better than this. And bear in mind that the above functionality was successfully demonstrated in Python in only nine lines of code without conditional nested logic! SAS exception handling will always make code more lengthy and convoluted while often failing to provide functionality achievable in other languages. For example, because exception handling plays no role inside DATA steps or SAS procedures, the scope and capability of SAS exception handling is tremendously reduced. Notwithstanding all these technical limitations, exception handling is so critical to achieving reliability, robustness, and fault-tolerance in software design that the remaining sections of this chapter demonstrate best practices for implementing exception handling routines within Base SAS.

Asking for Forgiveness versus Permission

While traditional exception handling requires an exception to have occurred, been detected, and handled, some equivalent robustness can be delivered in SAS when specific threats to software are known. The "Specific Threats" section, as discussed earlier, includes examples such as a missing data set or locked data set that are predictable, identifiable, and, by implementing exception handling, sometimes also preventable. Thus, where causes of failure can be detected before they occur, the corresponding failures often can be prevented through a priori exception handling. This contrasts with post hoc exception handling, which typifies traditional exception handling that detects and handles an exception that has actually occurred, not just a condition or state that would have caused an exception.

These two methods—a priori and post hoc—are often characterized as asking for permission and asking for forgiveness, respectively. Thus, traditional exception handling (as seen in object-oriented languages) is likened to the adage, *It's better to ask for forgiveness than for permission,* because post hoc detection can identify exceptions the instant they occur. However, because of the limitations of Base SAS, an exception handling framework in SAS typically consists of both a priori and post hoc exception handling routines, with significant emphasis on a priori detection to identify specific threats, more so than in software languages that provide inherent exception handling capabilities.

A priori exception handling is sometimes referred to in literature as *error proofing,* although this should not be misconstrued to represent that software is exception-free or error-free, only that it is made more resistant to exceptions or errors. Post hoc exception handling is conversely referred to as *error testing* because tests are conducted after some process to determine if exceptions or errors occurred. This highlights the important distinction that while both a priori and post hoc exception handling can improve the reliability of a process, neither can *guarantee* process success, and only post hoc routines can demonstrate or validate process success.

The differences between a priori and post hoc exception handling are further described in the following two sections in which exception handling routines in Base SAS and Python are demonstrated and contrasted.

A PRIORI VERSUS POST HOC IN SAS

A Priori Exception Handling	Post Hoc Exception Handling
asking for permission	asking for forgiveness
also known as *error proofing*	also known as *error testing*
cannot guarantee success	cannot guarantee success
cannot demonstrate/validate success	can demonstrate/validate success
more commonly used in Base SAS exception handling	more commonly used in OOP software development languages
used to make software more robust against specific threats	can identify both specific and general threats but only if they are programmatically detectable
more focused on robustness	more focused on fault-tolerance

Forgiveness

You've been trapped on an intercontinental flight to parts unknown, South America, for hours and, despite the turbulence and the insistence by flight attendants to "Please take your seats," you hasten down the aisle anyway, slam the tiny *Occupado* latch, and apologize afterward. Often it's just easier in life and software development to make an attempt and, if it fails, ask for forgiveness afterward. In software development, however, forgiveness requires apt, post hoc exception handling routines that detect exceptions and dynamically redirect software based on business rules. Because post hoc exception handling was demonstrated in the "Exception Handling Elsewhere" and "Faking It in SAS" sections, it is only briefly reprised here.

The following Python code attempts to print a variable but, if the variable does not exist, the program exits safely without runtime error:

```
try:
   print(X)
except NameError:
   sys.exit()
```

Similar SAS code also attempts to print a macro variable, but if the variable does not exist, the program exits safely without runtime error. The global macro variable &SYSCC is set to 4, representing the warning that occurs when the referenced macro variable &X does not exist:

```
%macro meh();
%let syscc=0;
%put &X;
%if &syscc>0 %then %return;
%mend;

%meh;
```

But what occurs when a variable is referenced in a DATA step and is missing? As it turns out, a note is printed to the SAS log, but no return code is generated that can be programmatically evaluated.

In the following example, the DATA step relies on third-party data that should (but erroneously does not) contain the variable VERYIMPORTANT:

```
data test;
   set thirdpartydata;
   a=veryimportant; * does not exist so this represents an exception;
run;

NOTE: Variable veryimportant is uninitialized.
NOTE: The data set WORK.TEST has 1 observations and 2 variables.
NOTE: DATA statement used (Total process time):
      real time           0.01 seconds
      cpu time            0.01 seconds

%put SYSERR: &syserr   SYSCC: &syscc;
SYSERR: 0   SYSCC: 0
```

Because the exception cannot be detected programmatically, it also cannot be handled programmatically, so SAS post hoc exception handling is powerless and not robust to this type of common data variability. However, because the exception represents a specific, predictable threat, it can be handled through a priori exception handling, as depicted in the "Permission" section.

Another common failure of a priori exception handling occurs in SAS because many exceptions can only be captured as runtime errors. For example, the following log demonstrates a DATA step that fails because the Doesnotexist data set does not exist:

```
%macro moremeh();
%let syscc=0;
%global moremehRC;
```

```
%let moremehRC=GENERAL FAILURE;
data test;
   set doesnotexist; * missing data set;
   * transformations go here;
run;
%if &syscc>0 %then %do;
   %let moremehRC=you broke me!;
   %return;
   %end;
%else %let moremehRC=;
%mend;

%moremeh;
ERROR: File WORK.DOESNOTEXIST.DATA does not exist.

NOTE: The SAS System stopped processing this step because of errors.
WARNING: The data set WORK.TEST may be incomplete.  When this step
was stopped there were 0 observations and 0 variables.
WARNING: Data set WORK.TEST was not replaced because this step was
stopped.
NOTE: DATA statement used (Total process time):
      real time           0.01 seconds
      cpu time            0.01 seconds

%put RC: &moremehRC;
RC: you broke me!
```

However, because automatic macro variables like &SYSCC and &SYSERR do not update until the boundary step (like a RUN statement), it's impossible to implement post hoc exception handling that catches the exception—the missing data set—before the exception is recognized as a runtime error. Paradoxical to the very intent of exception handling, as soon as the SAS processor recognizes the exception it is simultaneously too late to do anything to prevent the subsequent runtime error so the exception and runtime error occur as a single event.

Thus, while asking for forgiveness is more closely associated with fault-tolerant software design and forms the underpinning of exception handling in most software languages, to be tolerant of faults, those faults must be programmatically detectable. Many specific threats to SAS software either cannot be identified programmatically or will be identified too late (i.e., after failure) to do anything useful. The only remaining solution is to detect the specific cause of the exception—the missing data set—before it is referenced, which lies at the heart of asking permission through a priori exception handling.

Permission

Asking for permission squarely addresses specific threats to software execution by removing vulnerabilities in code. By detecting causes of exceptions before they occur, the resultant exceptions and runtime errors can be eliminated and robustness and reliability are improved.

The "Forgiveness" section demonstrates that in some cases, exceptions that occur in the DATA step produce SAS notes but provide no programmatic feedback to facilitate exception handling. In other cases, as soon as an exception is detected—like a missing data set referenced by a SET statement—a runtime error is produced, which also defeats post hoc exception handling methods. Given these limitations in the Base SAS language, a priori exception handing is required to detect specific threats and reroute program flow.

The "Forgiveness" section depicts the %MEH macro in which post hoc exception handling detects the warning that occurs when a nonexistent SAS macro variable is referenced. Equivalent functionality can be accomplished through a priori testing, which provides the additional benefit of not producing warnings or runtime errors. For example, although the macro variable &X should exist, when &X does not exist, the exception is detected and handled before the variable is referenced:

```
%macro test;
%if %symexist(X) %then %put &X;
%else %return;
%mend;
```

As demonstrated in the "Forgiveness" section, sometimes an exception in SAS—like referencing a nonexistent variable in a DATA step—displays a note to the log but offers no return code (like &SYSCC) that can be programmatically assessed. This challenge can be overcome with a priori exception handling and is demonstrated in the following "Happy Trail" section, which uses the ATTRN function to determine whether a variable exists in a data set.

A final weakness of post hoc exception handling in SAS is its inability to be used effectively within the DATA step when exceptions immediately cause runtime errors that terminate the DATA step. Because a missing data set is a specific threat that can be identified programmatically, a priori exception handling routines can be emplaced before the DATA step (or SAS procedure) and can validate whether all prerequisites have been met. The %MOREMEH macro from the "Forgiveness" section is reprised and modified so that it validates data set existence before attempted use:

```
%macro moremeh();
%let syscc=0;
%global moremehRC;
%let moremehRC=GENERAL FAILURE;
%if %sysfunc(exist(doesnotexist))=0 %then %do;
   %let moremehRC=missing data set;
   %return;
   %end;
%else %do;
   data test;
      set doesnotexist; * missing data set;
      * transformations go here;
   run;
   %end;
%if &syscc>0 %then %do;
   %let moremehRC=you broke me!;
   %return;
   %end;
%else %let moremehRC=;
%mend;

%moremeh;
%put RC: &moremehRC;
```

In the revised example, the post hoc exception handling is not removed. While the newly implemented a priori exception handling can ensure that the DATA step does not execute if the Doesnotexist data set is missing, as noted earlier, a priori routines cannot guarantee process success; they can prevent failure only from specific threats. For example, the DATA step could still run out of memory while executing, which would cause a runtime error detected by the post hoc analysis of &SYSCC. Thus, in SAS, robustness and reliability can often be achieved by asking for both permission and forgiveness.

Exception Handling Framework

Exception handling gets you around or through a specific exception (or exception cluster), but what happens next? The %MOREMEH macro in the "Permission" section avoids the DATA step with a priori exception handling if the required data set is missing and subsequently terminates the macro. Post hoc routines additionally terminate the macro if any warnings or runtime errors are detected after the DATA step. When exception handling is demonstrated in SAS literature, however, this is often where the demonstration concludes—either a return code is generated that reflects the exception, or, more commonly and significantly less useful in production software, an exception report is produced that prints the exception to the log or other static location.

Contrary to examples demonstrated throughout literature, a dynamic exception handling framework is required in robust production software and utilizes exception management to operationalize business rules and logic through exception handling routines. Exception handling frameworks are typically not demonstrated in software literature for the same reason that exception handling is absent unless it is the focus of the discussion—they add complexity, diminish readability, and require the added context of business rules to be fully understood. For example, to demonstrate robustness rather than only fault-tolerance, the %MOREMEH macro would need to show through an exception handling framework how full or partial business value was achieved despite encountering the exception.

Because of this added complexity (and necessary context) to demonstrate an exception handling framework, while exception handling is demonstrated throughout this text, exception handling frameworks are demonstrated only in the following sections. For example, in literature the %RETURN statement often depicts termination of a process flow, and return codes are often generated to imply—if not demonstrate—subsequent dynamic program flow based on their values. Where this text references "exception handling frameworks," however, these following sections can serve as a guide to demonstrate the complexity and control necessary to achieve robust software.

Exception Inheritance

Exception inheritance requires that exceptions are propagated from prerequisite processes to dependent processes and from child processes to parent processes. Without exception inheritance, subsequent processes can fail when previous processes encounter exceptions. This will typically cause cascading failures in which failure begets failure until the software terminates. Inheritance links the success or failure of one process to the next, providing validation from software outset to end.

The "Modularity" section in Chapter 18, "Reusability," demonstrates a DATA step that calls the %GOBIG macro, which in turn calls the %FINDVARS macro. When all three modules execute successfully, this provides a reusable solution that dynamically capitalizes every character variable in the parameterized data set. However, because the code includes no exception handling, a more robust example is demonstrated in this section and handles two specific exceptions that can occur. The concepts of black-box and white-box testing, referenced later in this section, are introduced in the "Functional Testing" and "Unit Testing" sections, respectively, in Chapter 16, "Testability."

The %ENGINE macro capitalizes all character variables in the Sample data set. Embracing modular design, however, %ENGINE actually calls %GOBIG to write the capitalization code dynamically; thus %ENGINE must check the &GOBIGRC return code to ensure that %GOBIG was successful:

```
data sample;
   length char1 $20 char2 $20 num1 8 num2 8;
   char1="I love SAS";
   char2="SAS loves me";
run;

%macro engine();
%global engineRC;
%let engineRC=GENERAL FAILURE;
data uppersample;
   set sample;
   %gobig(dsn=sample);
   num1=99;
run;
%if %length(&gobigRC>0) %then %do;
   %let engineRC=something broke! GOBIG: &gobigrc;
   %return;
   %end;
%else %let engineRC=;
%mend;

%engine;
```

Because %GOBIG is encapsulated (essentially inside a black box), %ENGINE is unaware of how %GOBIG creates the capitalization code for each variable. The %ENGINE macro relies on receiving two things from %GOBIG—the capitalization code and the return code &GOBIGRC—and has no idea that %GOBIG has in fact subcontracted out much of its work to its own child process, the %FINDVARS macro. Only through a white-box inspection of %GOBIG is this apparent:

```
* dynamically changes all character variables in a data set to
   upper case;
%macro gobig(dsn= /* old data set in LIB.DSN or DSN format */);
%global gobigRC;
%let gobigRC=GENERAL FAILURE;
%local i;
%findvars(dsn=&dsn, type=CHAR);
%if %length(&findvarsRC)=0 or &findvarsRC=no variables %then %do;
   %let i=1;
   %do %while(%length(%scan(&varlist,&i,,S))>1);
      %scan(&varlist,&i,,S)=upcase(%scan(&varlist,&i,,S));;
      %let i=%eval(&i+1);
      %end;
   %let gobigRC=;
   %end;
%else %do;
   %let gobigRC=FINDVARSRC: &findvarsRC;
   %end;
%mend;
```

While %GOBIG writes the dynamic code to capitalize variables (with the %SCAN statement), it relies on the %FINDVARS macro to provide the list of variables. The %FINDVARS macro is unaware of %ENGINE, and %ENGINE conversely is unaware of %FINDVARS, but because %ENGINE is indirectly dependent on %FINDVARS functionality (since %GOBIG directly relies on %FINDVARS), a method must exist to pass the %FINDVARS return code (&FINDVARSRC) directly to %GOBIG as well as indirectly to %ENGINE. This is inheritance.

The following %FINDVARS macro creates the space-delimited list of relevant variables (i.e., character variables, in this example) and, if an exception occurs, the macro passes it within the return code &FINDVARSRC:

```
* creates a space-delimited macro variable VARLIST in data set DSN;
%macro findvars(dsn= /* data set in LIB.DSN or DSN format */,
   type= /* ALL, CHAR, or NUM to retrieve those types of variables */);
%global findvarsRC;
%let findvarsRC=GENERAL FAILURE;
%local dsid;
%local vars;
%local vartype;
```

```
%global varlist;
%let varlist=;
%local i;
%local vartot;
%let dsid=%sysfunc(open(&dsn,i));
%if &dsid<1 %then %do;
   %let findvarsRC=missing or locked data set;
   %return;
   %end;
%let vars=%sysfunc(attrn(&dsid, nvars));
%do i=1 %to &vars;
   %let vartype=%sysfunc(vartype(&dsid,&i));
   %if %upcase(&type)=ALL or (&vartype=N and %upcase(&type)=NUM) or
         (&vartype=C and %upcase(&type)=CHAR) %then %do;
      %let varlist=&varlist %sysfunc(varname(&dsid,&i));
      %let vartot=%eval(&vartot+1);
      %end;
   %end;
%if &vartot=0 %then findvarsRC=no variables;
%else %let findvarsRC=;
%let close=%sysfunc(close(&dsid));
%mend;
```

Because Base SAS does not inherently provide return code functionality, return codes must be faked by initializing global macro variables within macros. While these global macro variables can be accessed from anywhere inside a SAS session, inheritance should be used to pass return codes so that the chain of custody for software performance can be demonstrated. In other words, while %ENGINE could directly assess the value of &FINDVARSRC, don't get lazy and allow it to—any feedback from %FINDVARS must be passed through %GOBIG. Due to encapsulation and loose coupling, which are discussed and demonstrated throughout Chapter 14, "Modularity," the %ENGINE macro should have no knowledge that the %FINDVARS macro (or its respective return code &FINDVARSRC) even exists!

To put the complexity of exception handling and inheritance in perspective, the original functionality that this code is producing is demonstrated in the following five lines of code:

```
data uppersample;
   set sample;
   char1=upcase(char1);
   char2=upcase(char2);
run;
```

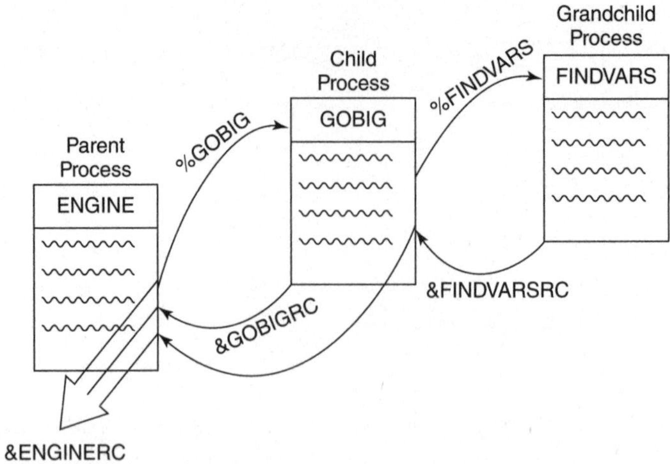

Figure 6.1 Exception Inheritance

Notwithstanding its additional complexity, the exception handling framework improves reliability and robustness by testing for known threats, and improves static performance by representing a more modular, reusable, extensible, and stable solution that can improve the efficiency of future software development and maintenance. However, because of the inherent complexities of modular software design and the necessity to propagate and inherit return codes, it's critical that modules not only return their own exceptions but also detect and return the exceptions of child processes upon which they rely. Figure 6.1 demonstrates inheritance of the two exceptions (missing data set and no relevant variables) from the %FINDVARS macro to the %GOBIG macro, and the subsequent inheritance to the %ENGINE macro.

The importance of inheritance becomes clearer when the intent of modular, reusable code is understood. For example, the %FINDVARS macro can be used for an infinite number of purposes that need to create a list of all variables (of either only character or only numeric variables) in a data set. For some purposes, such as the %GOBIG macro, if %FINDVARS produces an empty list—that is, no relevant variables exist in the data set—this could be valid and not represent an exception. This relationship is demonstrated in Figure 6.1.

However, because modules are intended to be reused in different situations and thus called by different parent processes, in other cases, if %FINDVARS finds no relevant variables, this could represent an exception in the parent process that would need to be subsequently handled. This more expanded relationship is demonstrated in Figure 6.2.

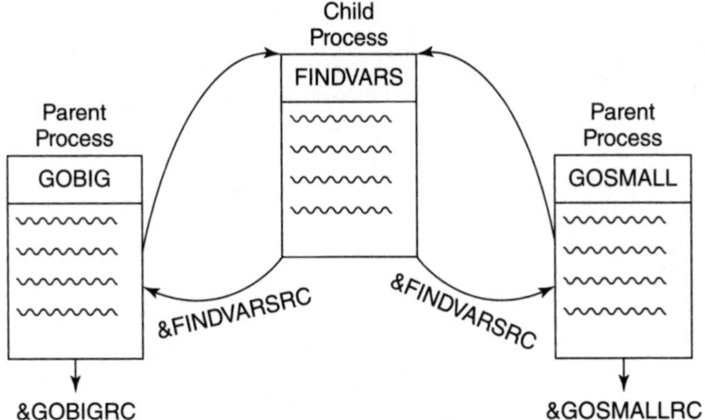

Figure 6.2 Exception Inheritance in Software Reuse

By implementing inheritance methods, processes can ensure that all prerequisite processes—including both child processes as well as previous serialized processes—have completed without error, or sufficiently to allow subsequent program flow to continue. Exception inheritance overcomes cascading failures that can occur in SAS when unhandled exceptions and runtime errors occur, as demonstrated in the following section. Exception inheritance is also demonstrated in the "Data Governors" section in Chapter 9, "Scalability."

Cascading Failures

Cascading failures are caused by unhandled exceptions, errors, or software defects and occur when failures in a child process are not communicated to the parent process, or when failures in a prerequisite process are not communicated to subsequent, dependent processes. The following code and output illustrate a DATA step that fails, which causes a cascading failure in the subsequent MEANS procedure:

```
data final;
   length char1 $10 num1 8;
   char1='sas is rad';
   num1=99;
   oopsy!;
         -----
         180
ERROR 180-322: Statement is not valid or it is used out of proper order.

run;

NOTE: The SAS System stopped processing this step because of errors.
```

```
WARNING: The data set WORK.FINAL may be incomplete.  When this step
was stopped there were 0 observations and 2 variables.
NOTE: DATA statement used (Total process time):
      real time           0.01 seconds
      cpu time            0.01 seconds

proc means data=final;
   var num1;
run;

NOTE: No observations in data set WORK.FINAL.
NOTE: PROCEDURE MEANS used (Total process time):
      real time           0.00 seconds
      cpu time            0.00 seconds
```

While the software defect (oopsy!) lies in the DATA step, it causes a cascading failure in the MEANS procedure. Recall that failures do not necessarily denote warnings, runtime errors, or software termination, but occur whenever an invalid solution or data product is generated or requirements are not met. Thus, although the MEANS procedure executes, it produces no output because the failed DATA step creates a spurious data set with two variables but no observations. Even more critical, because SAS notes cannot be programmatically detected, if the original defect is not detected and handled in the DATA step, it's more difficult to determine programmatically during the MEANS procedure that a failure has occurred. Validation could be performed by inspecting data produced with the OUT statement of the MEANS procedure, but it is much better to have caught the failure before the MEANS procedure is ever invoked.

The following revised code now uses post hoc exception handling to detect if warnings or runtime errors occurred in the DATA step and, if something goes awry, the MEANS procedure is skipped, preventing the cascading failure:

```
%macro oopsy();
%let syscc=0;
%global oopsyRC;
%let oopsyRC=GENERAL FAILURE;
data final;
   length char1 $10 num1 8;
   char1='sas is rad';
   num1=99;
   oopsy!;
run;
%if &syscc>0 %then %do;
```

```
   %let oopsyRC=data step had an OOPSY!;
   %return;
   %end;
proc means data=final;
   var num1;
run;
%if &syscc>0 %then %do;
   %let oopsyRC=proc means had an OOPSY!;
   %return;
   %end;
%else %let oopsyRC=;
%mend;

%oopsy;
%put RC: &oopsyRC;
```

This code makes no attempt to deliver partial or delayed business value, but rather is focused on detection of general failures to facilitate fault tolerance. A vulnerability in the above code still exists, however, because the Final data set is created despite the failure, as demonstrated in the following SAS log:

```
%oopsy;
NOTE: Line generated by the invoked macro "OOPSY".
data final;     length char1 $10 num1 8;     char1='sas is rad';
   num1=99;     oopsy!;

-----

180
1  ! run;

ERROR 180-322: Statement is not valid or it is used out of proper order.

NOTE: The SAS System stopped processing this step because of errors.
WARNING: The data set WORK.FINAL may be incomplete.  When this step
was stopped there were 0 observations and 2 variables.
NOTE: DATA statement used (Total process time):
      real time           0.01 seconds
      cpu time            0.01 seconds

%put RC: &oopsyRC;
RC: data step had an OOPSY!
```

Because failure of the DATA step generates a spurious, empty data set, the exception handling routine should embrace security (and recoverability principles) and delete Final so it is not accidentally utilized. Either the DELETE procedure or DATASETS procedure should be used to delete Final when warnings or runtime errors are detected during its creation.

Happy Trail

Exception handling has been introduced and demonstrated, but, in every example thus far, the goal has been to detect a failed state and terminate subsequent processes. While this describes the secondary goal of robustness—to gracefully terminate through a fail-safe path—it only scratches the surface of exception handling, and it fails to demonstrate the primary goal of robustness—maximization of business value in the face of software variability.

Thus, the goal of exception handling should always be to detect variability and reroute process control back to the happy trail—the path that delivers full business value or, after exceptions or errors have occurred, the maximum attainable business value. For example, in some cases, a missing data set signals that an entire SAS process or program should be terminated, but in other cases, functionality still can be salvaged. In the "External Issues" section in Chapter 5, "Recoverability," SAS software was modified so that it could flexibly adapt when individual tables were missing within a third-party database, thus eliminating runtime errors while improving software reliability.

Despite the challenges of implementing exception handling within Base SAS software, as a data analytic language SAS does have one *huge* advantage—data analytic software is inherently serialized. Where user-focused software often places a user at some central location from which endless program flow paths may follow (based on user input), data analytic development typically transforms data from one form to another over a series of sequenced processes. In many instances, the failure of earlier processes predicates failure of later processes, thus demonstrating clear dependencies that exist. In some cases, however, full or partial business value still can be achieved despite exceptions or failures in earlier processes, thus allowing program flow to return to the happy trail.

For example, the following SAS code simulates software that ingests and transforms a data set before performing some analysis, represented by the

MEANS procedure. The LIBNAME statement and DATA step precede the code to provide the necessary reference and input:

```
libname perm 'c:\perm';

data perm.original;
   length char1 $10 var1 8;
   var1=5;
run;

* PROCESS STARTS HERE;
data final;
   set perm.original;
   * transformations;
run;

proc means data=final;
   var var1;
run;
```

To understand the happy trail for specific software, it's first necessary to describe threats that can derail software from that path of full business value. Table 6.1 demonstrates an abbreviated risk register that lists processes, their specific threats, and respective exception handling pathways. For example, the business rule in the first line indicates that the risk of attempting to access the Final data set when it is locked should be accepted. In other words, because Final is maintained in the WORK library and the risk of other processes locking it is negligible, the decision has been made not to test for data set accessibility before creating the data set and to assume that it is unlocked.

The second business rule in Table 6.1 demonstrates that if the Original data set does not exist, program flow should use a backup data set that will have

Table 6.1 Abbreviated Risk Register

Process	Threats	Solution
DATA Final	Final is exclusively locked	ACCEPT
	Original does not exist	Switch to backup
	Original is exclusively locked	Switch to backup
PROC MEANS	Original has no observations	ACCEPT
	Original has no VAR1	Terminate program

to be maintained. Recall, however, that at this early planning phase in the SDLC, activities are *needs-focused* rather than *software-* or *solutions-focused,* so a specific method to achieve this backup won't necessarily have been selected or even discussed.

Where SAS practitioners either fail to identify threats to software or fail to implement solutions that eliminate vulnerabilities, they tacitly accept higher risks and may reduce the robustness and reliability of their software. Whereas a risk register typically records proposed solutions to specific known software vulnerabilities, Table 6.1 depicts business rules that have been accepted for implementation to make software more robust. Given these business rules, the flowchart in Figure 6.3 depicts program flow, including the happy trail.

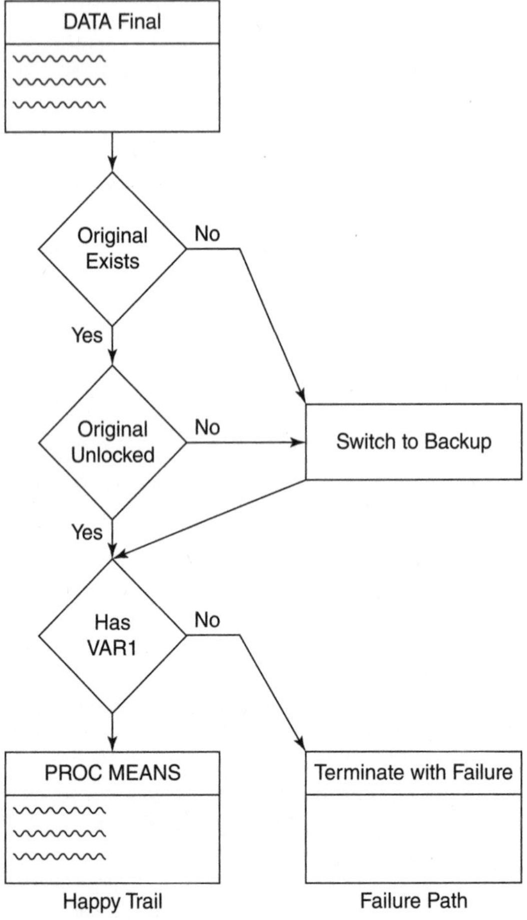

Figure 6.3 Program Flow and Happy Trail

Once business rules within the risk management strategy are established that specify how specific threats and vulnerabilities will be remedied, the actual technical solutions can be designed and developed. In this scenario, to make the process more robust, a mirrored SAS library PERM2 is created on a separate drive. When the PERM.Original data set is exclusively locked by a user or process, the program will automatically switch to the backup. While the backup may have some outdated data, in this scenario, day-old donuts are better than no donuts at all, and using day-old data can still maximize business value given the exception of a missing or locked primary data set.

The Final data set is also now tested before the MEANS procedure to ensure that the variable VAR1 exists in the data set. And, finally, post hoc exception handling is added to ensure that if either the DATA step or MEANS procedure produce warnings or runtime errors, the macro terminates via the %RETURN statement:

```
%macro makemeans();
%let syscc=0;
%global makemeansRC;
%let makemeansRC=GENERAL FAILURE;
%local dsid;
%local close;
%local dsn;
%local vars;
%local i;
%local found;
%let found=NO;
%let dsn=perm.original;
* test availabiltiy and access;
%let dsid=%sysfunc(open(&dsn));
%if %eval(&dsid<1) %then %do;
   %let dsn=perm2.original;
   %let dsid=%sysfunc(open(&dsn));
   %if %eval(&dsid<1) %then %do;
      %let makemeansRC=data locked or missing;
      %return;
      %end;
   %end;
%let vars=%sysfunc(attrn(&dsid, nvars));
%do i=1 %to &vars;
   %if %upcase(%sysfunc(varname(&dsid,&i)))=VAR1 %then %let found=YES;
   %end;
```

```
%if &found=NO %then %do;
   %let makemeansRC=variable not found;
   %return;
   %end;
data final;
   set &dsn;
   * transformations;
run;
%if &syscc>0 %then %do;
   %let makemeansRC=data step failure;
   %return;
   %end;
proc means data=final;
   var var1;
run;
%if &syscc>0 %then %do;
   %let makemeansRC=proc means failure;
   %return;
   %end;
%else %let makemeansRC=;
%let close=%sysfunc(close(&dsid));
%mend;

%makemeans;
%put RC: &makemeansRC;
```

The revised %MAKEMEANS macro now also produces a return code, &MAKEMEANSRC, that demonstrates program success or failure, which should be validated by subsequent or dependent processes. In this example, a blank return code represents success while other return codes demonstrate specific threats that lead to failure. While the current return code does not include any indication of whether the backup data set is used rather than the primary, this logic could be added to provide additional performance metrics.

The revised program flow now reflects two different exception handling pathways—methods or techniques used to handle specific failure types. If the PERM.Original data set is missing, the process flow switches to a backup data set that can deliver full or nearly complete functionality. When other exceptions or errors are encountered, however, the process flow terminates the macro and writes the exception in the &MAKEMEANSRC return code. The following "Exception Handling Pathways" section describes these and other pathways that SAS developers should keep in their tool belt to ensure maximum value is achieved from software regardless of variability that is encountered.

To be clear, the revised %MAKEMEANS macro is intentionally coded without attention paid to static performance attributes to more clearly demonstrate exception handling logic. This exposes the frequent tradeoff between software readability and other quality characteristics that occurs not only in production software but predominantly throughout software development literature. Thus, the static performance characteristics of this code can be substantially improved through increased focus on maintainability, modularity, stability, testability, and reusability principles.

Exception Handling Pathways

When an exception occurs, software must first detect and subsequently handle it. Most commonly demonstrated in SAS literature, following a runtime error or exception detection, the process produces an exception report (demonstrated and disparaged in the next section), the macro is terminated via the %RETURN statement, or the software is terminated with the %ABORT, ABORT, or ENDSAS statement. The use of these methods in literature is not intended to imply that the exception would be handled similarly in production software, but is intended to improve readability; thus, the *handling* portion of exception handling is rarely demonstrated in literature. This practice is true throughout this text, where %RETURN is commonly implemented to simulate return to some parent process that is not depicted.

Despite the overuse of %RETURN to simulate dynamic exception handling, several exception handling pathways can be harnessed, many of which allow delivery of full or partial functionality despite exceptional conditions. The following exception handling pathways are introduced and demonstrated in detail in a separate text by the author: *Ushering SAS® Emergency Medicine into the 21st Century: Toward Exception Handling Objectives, Actions, Outcomes, and Comms*:[6]

- *Undetected Success*—The user is unaware of the exception, and full functionality is delivered.
- *Rerouted Success*—Program flow is rerouted back to the happy trail, so full functionality is delivered but possibly with some delay.
- *Reattempted Success*—An exception occurs initially, but after waiting some time period, the process is reattempted and succeeds.
- *Prompt User*—More common in applications development, the user is notified of the exception and provides some input that allows the process flow to continue.

- *Partial Success*—Program flow is rerouted around exceptions but some functionality is compromised.
- *Process Termination*—The process must be terminated but program flow can transfer to independent, unrelated processes.
- *Program Termination*—The program must be terminated but exits gracefully.

In many cases, more dynamic (and beneficial) exception handling will require multiple, redundant exception handling pathways. This concept is demonstrated in the %MAKEMEANS macro in the "Happy Trail" section. For example, a backup of the PERM.Original data set may exist as PERM2 .Original. If the first data set is locked or missing, access to the backup data set is attempted. If successful, this second attempt is an example of rerouted success, in which full functionality is delivered, or possibly partial success, in the case that the backup data set represents day-old data.

Regardless of the type and number of exception handling pathways that are implemented within a code module, however, a fail-safe path should always provide a route terminating the process (or program) should all other exception handling strategies fail. For example, because the backup data set could also be missing, robust exception handling would also need to account for this risk. This represents the Hail Mary—the emergency satellite phone tucked away in the glove compartment that you hope never to need.

Exception Handling, Not Exception Reporting!

Exception reporting is a quality control method that detects and reports exceptions that occur but stops short of actually handling software exceptions. The SAS log is an example of an exception report when it includes notes, warning, and runtime errors that occur during execution. Thus, while the log is static and doesn't alter program flow, it represents a historical record of some exception types that occur during execution. Other exception reporting can include additional comments, data, and information conveyed to the log through %PUT or PUT statements, to ODS output through PRINT, REPORT, and other procedures, and to SAS data sets or external files.

Exception handling is a quality assurance method that can improve the functionality and performance of software. Commonly used to detect exceptions, runtime errors, and software defects, exception handling is distinguished from exception reporting because exception handling not only detects but also responds to exceptions by dynamically altering program flow.

Exception reporting, in contrast, is a static quality control method; it only tells you that an exception was encountered, but does nothing to fix or mitigate it. Exception reporting is more commonly used to detect exceptions that occur in data sets. Especially in data analytic languages, exception handling is more closely associated with software quality while exception reporting is associated with data quality.

In SAS literature, exception reporting is often demonstrated in place of exception handling because of ease of implementation and increased readability of code. For example, the following SAS code prints a note (i.e., a very tiny exception report) to the SAS log if the Original data set is missing:

```
%macro reporting();
%let syscc=0;
data final;
   set original;
run;
%if &syscc>0 %then %put data set failed!;
%mend;
```

If Original is missing, however, the exception reporting does little more than the out-of-the-box SAS runtime errors that are printed alongside it. In fact, it's nearly lost in the fray!

```
%reporting;
ERROR: File WORK.ORIGINAL.DATA does not exist.

NOTE: The SAS System stopped processing this step because of errors.
WARNING: The data set WORK.FINAL may be incomplete.  When this
step was stopped there were 0 observations and 0 variables.
WARNING: Data set WORK.FINAL was not replaced because this step
was stopped.
NOTE: DATA statement used (Total process time):
      real time           0.08 seconds
      cpu time            0.01 seconds

data set failed!
```

Not all exceptions create runtime errors and, especially where exceptional data represent outlier, missing, or other invalid data, exception reports are commonly created to record these data mishaps. The following code identifies exceptional data (invalid months) that occur in the Months data set:

```
data months;
   infile datalines delimiter=',';
   length month $10;
```

```
   input month $;
   datalines;
January
Januarry
February
March
march
April
april
May
Mayy
June
;
run;

%macro check();
data final exceptions;
   set months;
   if upcase(month) in
      ('JANUARY','FEBRUARY','MARCH','APRIL','MAY','JUNE','JULY','AUGUST',
      'SEPTEMBER','OCTOBER','NOVEMBER','DECEMBER') then do;
      * do some transformations;
      output final;
      end;
   else output exceptions;
run;
%mend;

%check;
```

Although exception reporting should never be used in isolation to facilitate increased software quality, it can be extremely useful in supporting data quality. As demonstrated in the previous code, especially where quality controls not only demonstrate exceptional data (through exception reporting) but also remove them from data sets, this quality control function can prevent subsequent software failure. For example, by removing invalid months from the data set, a subsequent REPORT procedure might be able to execute with more reliability and accuracy. In this sense, exception reporting that also removes or modifies exceptional data can be considered to be a quality control wrapped within a quality assurance framework.

While often demonstrated as a proxy for exception handling throughout SAS literature, exception reporting should never replace exception handling in production software. Moreover, where exception handling is implemented within a robust system, exceptional data should be reported to a dynamic construct (such as a SAS data set) rather than the SAS log. In saving exceptional data to a data set, subsequent programmatic actions can assess, modify, delete, or further communicate the exceptional data as necessary.

ROBUSTNESS IN THE SDLC

The differentiation between software reliability and robustness is one that should be made early in software planning and design, and that can spur realism within later technical requirements. Perfect reliability hopes to achieve functional and performance objectives all the time, but robustness and fault-tolerance acknowledge that internal and external sources of variability (to software) do exist, will be encountered, and will cause failure. For example, reliability metrics record a customer's desire to have an ETL process complete in under an hour every day while robustness—codified through risk management business rules—tacitly acknowledges some defeat, stating that delayed or partial results (or business value) are better than none at all. Moreover, robustness also demonstrates that preventing invalid results may be as valued as facilitating valid results.

While reliability metrics will be more commonly included in software requirements, the risk management and realism inherent in robust design must be discussed and should also influence requirements. As demonstrated previously in the "Happy Trail" section, achieving robustness capable of diverting program flow to achieve partial business value can be exceedingly complex, but can also endear software products to stakeholders because they will succeed where lesser programs fail. Because of the high cost, a risk assessment should always demonstrate that the increased reliability warrants the effort, whereas a failure log should be maintained to ensure that where identified exceptions or failures occur, they follow business rules prescribed through the risk management framework.

Requiring Robustness

Robustness is often specified through use of a risk register, as depicted in the "Risk Register" section in Chapter 1, "Introduction." For example, Table 6.1 in the "Happy Trail" section demonstrates specific threats to software, such as missing, locked, or invalid data sets. During planning and design, stakeholders must decide which risks to accept and which risks to eliminate to achieve the desired level of robustness and reliability. Risk management and risk resolution are introduced in the "Risk Management" section in Chapter 1, "Introduction."

Some specific threats to software are the focus of later chapters. Portability, for example, describes the robustness of software to function across different environments. Scalability describes the robustness of software to function efficiently when confronted with big data or too many users. These threats also can be recorded in technical requirements or a risk register. For example, stating that software is only intended to operate in the SAS Display Manager 9.4 for Windows defines the landscape in which SAS practitioners must develop software. While software that performs too slowly but completes accurately does not diminish robustness, when software fails to scale to big data and terminates with runtime errors, these functional failures do make software less robust.

One of the simplest ways of facilitating some degree of robustness is to require that software identify general failures by requiring fail-safe post hoc exception handling. Thus, by checking the value of &SYSERR, &SYSCC, and other automatic macro variables at the close of every child process, module, macro, or program, software can detect exceptions that require termination. Implementation of this fail-safe path alone will often not provide additional business value, but it can ensure that software failures don't beget subsequent cascading failures.

Measuring Robustness

It's difficult to measure robustness without measuring reliability, and in fact the primary goal of robustness is inherently to improve software reliability and availability, thus maximizing business value. Unlike reliability, however, robustness can be measured only through analysis of exceptions, warnings, and errors that occur during operation. Thus, if software executes reliably *despite* incurring variability along the road, robustness has succeeded; however, if software executes reliably because no variability is encountered, then no assessment of robustness can be made.

Robustness is most readily assessed by comparing the failure log to the risk register (or other technical requirements that specify risk management business rules that software should follow). If business rules state that software should be robust to missing or locked data sets, then the failure log should not demonstrate failures caused by these exception types. If those failure types are apparent in the failure log, then the required level of robustness has not been achieved, because business rules were not followed. While this analysis may sound like an onerous process, where the risk register and failure log are both maintained in a standardized format (such as a SAS data set), this type of robustness analysis can be achieved through a repeatable, programmatic solution.

WHAT'S NEXT?

Robustness can't guarantee reliability, but it can guard against specific, predictable threats as well as general threats that may be unpredictable or unpreventable. Even robust, reliable software may be useless, however, if it fails to meet other performance objectives, such as execution time thresholds. The next two chapters introduce execution efficiency (i.e., software speed) and efficiency—arguably the most sought-after performance objectives.

NOTES

1. ISO/IEC/IEEE 24765:2010. *Systems and software engineering—Vocabulary*. Geneva, Switzerland: International Organization for Standardization, International Electrotechnical Commission, and Institute of Electrical and Electronics Engineers.
2. ISO/IEC 25010:2011. *Systems and software engineering—Systems and software Quality Requirements and Evaluation (SQuaRE)—System and software quality models*. Geneva, Switzerland: International Organization for Standardization and Institute of Electrical and Electronics Engineers.
3. ISO/IEC/IEEE 24765:2010.
4. Troy Martin Hughes, "Why Aren't Exception Handling Routines Routine? Toward Reliably Robust Code through Increased Quality Standards in Base SAS." Presented at Midwest SAS Users Group (MWSUG), 2014.
5. "Handling Exceptions." *SAS® Component Language 9.4: Reference, Third Edition*. Retrieved from http://support.sas.com/documentation/cdl/en/sclref/67564/HTML/default/viewer.htm#n1xx6zhe41u159n1iixi0diruon2.htm.
6. Troy Martin Hughes, "Ushering SAS® Emergency Medicine into the 21st Century: Toward Exception Handling Objectives, Actions, Outcomes, and Comms." Presented at Midwest SAS Users Group (MWSUG), 2015.

CHAPTER 7

Execution Efficiency

Following my first tour of duty in Afghanistan, I moved to Antigua, Guatemala, to unwind and relax for a few months. One of the most cosmopolitan cities in Central America, Antigua's fabled cobblestone streets are lined with backpacker hostels, quaint cafés, international cuisine, and 17th- and 18th-century cathedrals and colonial architecture, all laid beneath the volcanos Agua, Fuego, and Acatenango.

I'd visited Antigua years before with a friend and, flying into Guatemala City some 26 miles from Antigua, we'd hailed a cab to the bus exchange where we'd caught a chicken bus to Antigua. Aside from the *conductor* (driver) yelling at us at one point to put our heads down (possibly a bad part of town?), it was an uneventful yet laboring hour-long journey winding through the mountains, making stop after stop.

Now living in Antigua, as friends and family would fly in and out of Guatemala City to visit, I'd meet them at the airport, and we'd typically catch a shared shuttle back to Antigua, one of the fastest methods since it bypassed the central terminal and made no extra stops.

But, in first getting from Antigua to the city to pick them up, I had several options. When running extremely late, I called a private cab I knew to be very reliable; if available, he'd race over and ferry me to the city in the fastest possible manner.

Shared shuttles were the next-fastest method, but departed on predetermined schedules primarily from *Parque Central* (the central plaza). Because seats filled up quickly, tickets generally had to be purchased in advance.

For a more leisurely but jostling journey, "express" chicken buses ran from the Antigua terminal to the Guatemala City terminal, from which taxis ran to the airport.

And, for the slowest possible journey, local chicken buses made frequent stops through the 26-mile course, taking nearly 90 minutes. But the stops also provided opportunities for shopping or to take in a movie at the mall on the way.

While speed was typically the deciding factor for choosing one conveyance over another, other factors did exist. Each method had its own price, with taxis being the most expensive and local chicken buses the cheapest. Although I never encountered any threats, shared buses were touted as the safest method. And, while slower and less comfortable, chicken buses always provided the best stories and most memorable encounters with the Guatemalan people—and their livestock!

■ ■ ■

When I made a decision to take one transportation modality over another based solely on the expected duration of travel, I was relying on a single constraint—*execution efficiency*, or speed. Software development stakeholders also typically want the fastest possible software, which can be achieved through both programmatic and nonprogrammatic endeavors.

But in traveling within Guatemala, the ability to make an isolated decision based solely on one factor (speed) was facilitated by the lack of influence of other constraints, including cost, comfort, security, and the adventure factor. Software operation also has competing constraints, with faster software often requiring more memory, input/output (I/O) processing, and CPU cycles. And merely because the constraints aren't considered in a decision doesn't eliminate them.

In assessing transportation cost, a local chicken bus was approximately $1, an "express" chicken bus $2, a shared shuttle $15, and a private taxi $30. To some backpackers who were truly pinching *quetzales* (Guatemalan currency), the price differential was significant enough to sway them toward chicken buses. This attitude is more common in software development environments, in which system resources do limit the speed with which processing can occur. A customer or SAS administrator is likely aware that he could procure more processing power or memory, but might not believe that faster processing justifies the expense.

In justifying their decision to transit solely on chicken buses, backpackers often stated that the chicken buses were more efficient than other methods—effectively assessing the speed of travel respective of the cost of resources, which amounted only to the bus fare. In software development, similarly, decisions often must be made not based solely on speed, but rather on software efficiency—or speed of processing respective of system resource utilization.

This chapter introduces execution efficiency, or throw-caution-to-the-wind speed, irrespective of the associated costs of system resources. The next chapter provides a contrast by focusing on efficiency, which evaluates processing speed respective of system resource utilization.

DEFINING EXECUTION EFFICIENCY

Execution efficiency is "the degree to which a system or component performs its designated functions with minimum consumption of time."[1] *Efficiency*, conversely, is "the degree to which a system or component performs its designated functions with minimum consumption of resources."[2] Execution efficiency (or software speed) examines the ability of software to perform effectively and quickly irrespective of system resource utilization, whereas efficiency does account for resources. This is an important distinction due to the frequent commingling of efficiency and speed in software discussion and literature.

Of all performance characteristics—dynamic and static—speed is most often discussed in SAS literature. Slower SAS software delays results, data products, and data-driven decisions from which business value is ultimately gained. Where time is money, those delays can incur real costs to stakeholders. Central to definitions of both efficiency and execution efficiency is the notion that a system must be functionally effective; therefore, developers must always ensure that software not only is fast but also is accurate.

This chapter aims to disentangle efficiency from execution efficiency, discussing software speed *irrespective* of system resource utilization. Rather than enumerating techniques widely described throughout SAS literature to make code faster, it introduces performance benefits that can be gained through modularized code, software critical path analysis, and parallel processing. For an understanding of software speed *respective* of resource utilization, Chapter 8, "Efficiency," demonstrates the relationship between software execution speed and resource utilization, including its critical role in facilitating software reliability.

FACTORS AFFECTING EXECUTION EFFICIENCY

If software speed is the dependent variable by which software performance so often is measured, then what are the independent variables? Some of the primary factors influencing software speed include:

- Hardware
- Network and infrastructure
- Third-party software
- SAS interface
- SAS version
- SAS options
- Software development best practices
- Data injects

SAS practitioners may have little to no influence over some of these factors, and it's common for developers only to be able to influence software speed programmatically through development best practices. For example, I've worked in research environments in which my computer had been purchased as part of a bulk order by the government. I had no control over our network, I had no influence over the SAS application modules that had been purchased, and I had no ability to upgrade our software to a new version of SAS. While these can be unfavorable development conditions if higher software performance is demanded, SAS practitioners in more restrictive environments are able to focus exclusively on software quality rather than exerting effort to learn and manipulate hardware, network, system, and other resources.

In other environments, however, SAS practitioners may be able to influence more aspects of their infrastructure. If they have the ability to purchase or upgrade hardware or SAS components, they will be able to achieve greater software performance. System resources can be levied to improve performance, such as by throwing additional CPUs, memory, or bandwidth at a problem. Although nonprogrammatic solutions are inherently costlier than programmatic ones, a tradeoff often exists because nonprogrammatic solutions can enable SAS practitioners to spend less time trying to finagle additional speed programmatically from a fixed infrastructure.

Another benefit of considering both programmatic and nonprogrammatic solutions to achieve greater speed is that while programmatic solutions often affect only one software product, an investment in system resources can buttress the infrastructure as a whole with improved hardware, software, or components. Additionally, like reliability and robustness, the fastest performing software will be delivered only by incorporating both programmatic and nonprogrammatic methods.

FALSE DEPENDENCIES

One of the hallmarks of data analytic software is that the code often reads as a novel, from cover to cover. In the beginning chapters, the code ingests some data, in later chapters the data are manipulated or transformed, and in final chapters a series of analyses is performed. Real dependencies exist because later processes rely on the completion of earlier processes; runtime errors that occur during ingestion can cascade into subsequent processes to cause failure, including runtime errors or, more egregiously, invalid data and data products. This phenomenon is discussed in the "Cascading Failures" section in Chapter 6, "Robustness."

As extract-transform-load (ETL) processes gain complexity, however, additional data sources, data volume, and data processes are often incorporated. Each of these aspects individually can contribute to false dependencies that occur when real prerequisites are not required but are nevertheless implemented through serialized code design. For example, a false dependency (of process) exists when a program first performs the MEANS procedure on a data set and subsequently performs the FREQ procedure:

```
data final;
   length char $10 num 8;
run;

proc means data=final;
   var num;
run;

proc freq data=final;
   tables char;
run;
```

In this example, the ordering of the code dictates that the FREQ procedure is performed after the MEANS procedure. However, the MEANS and FREQ procedures are in no way associated—each relies only on the Final data set. Moreover, because both the MEANS and FREQ procedures require only a shared lock on the Final data set, each procedure can in theory be performed in parallel. Thus, by restructuring the program flow to incorporate modular software design and by throwing more resources at the problem, a faster solution can be developed. The solution would not be more efficient—the net sum of resource utilization remains roughly identical—but it would demonstrate execution efficiency because it could be executed faster.

The following sections introduce critical path analysis, an analytic technique that can be used to remove false dependencies during design and development phases. In the later "Parallel Processing" section, theoretical critical path analysis is supplanted by demonstrations of techniques that implement parallel processing to achieve higher performing software. But before learning to run, the first step is crawling toward software modularity.

Toward Modularity

Modular software design is discussed throughout Chapter 14, "Modularity," but must begin with a description of what is *not* modular software. Monolithic—literally, *single stone* (from the Greek *monos* + *lithos*)—represents the common

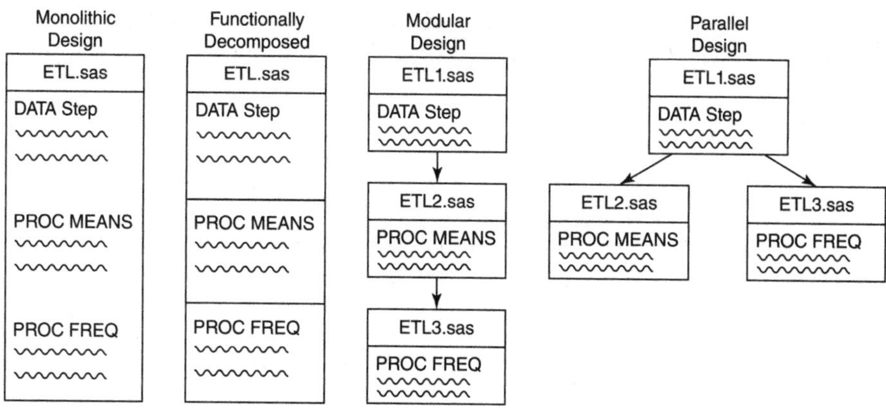

Figure 7.1 From Monolithic to Parallel Program Flow

practice especially in procedural software languages of writing software comprised of single, large programs. The journey toward modularity (and inherently away from monolithic design) facilitates critical path analysis and subsequent parallel processing that can improve processing speed.

The critical path analysis model in Figure 7.1 demonstrates not only parallel MEANS and FREQ procedures, but also parallel Means and Freq modules. Software must first have the capacity to be divided into distinct modules so those modules can be relocated within the flowchart if beneficial. Only then can prerequisites, dependencies, inputs, and outputs be examined to start identifying and analyzing the critical path. Thus, while SAS data analytic software is often first envisioned as serialized processes that can be read like a novel, even novels are broken into discrete chapters. Could you imagine trying to parse this text if it weren't organized in some hierarchy that presented chapter, section, and subsection headings? It would not only be indecipherable to read but also would have been impossible to write!

The benefits of monolithic design in SAS are clear. Global macro variables can be easily passed throughout software if it comprises only one program. Initializations such as SAS library references also must be made only once in monolithic code. Code modules, on the other hand, each must be encapsulated and self-sufficient and, when they represent distinct SAS programs, must be passed parameters through the SYSTASK function, control tables, or other methods since macro variables cannot be passed between separate sessions of SAS. Despite the additional code complexity, modular software is preferred because it is more maintainable, readable, reusable, and testable, and—central to this chapter—faster. Figure 7.1 demonstrates the shift from monolithic to modular design, in which a quagmire of code is transformed from a single SAS program to three separate SAS programs.

Critical Path Analysis

Before exploring false dependencies, it's important to demonstrate a standardized methodology that can capture and display false dependencies. Due to the often serialized nature of data analytic software, flowcharts easily lend themselves to representing data processes. Flowcharts can be simple and temporary; whiteboards and Buffalo Wild Wings napkin sketches often achieve the same objective as enduring solutions like Visio or Microsoft Word SmartArt.

Flowcharting aims to identify inefficiencies in software design that can be corrected to produce faster software. In project management, critical path analysis similarly utilizes flowcharts to represent project activities and determine prerequisites, dependencies, inputs, and outputs. It can demonstrate which tasks can be performed in parallel to save time and which tasks might contribute to bottlenecks and ultimately project delays. More importantly, critical path analysis allows project managers to identify where to place or shift resources to achieve the greatest impact to increase throughput or reduce the project schedule.

To illustrate false dependencies hiding in monolithic software and the performance gains obtainable with modular, parallel program flow, the code in the "False Dependencies" section is demonstrated in Figure 7.2. While the FREQ procedure follows the MEANS procedure in the serialized code, in theory, both procedures could be run in parallel because each only requires completion of

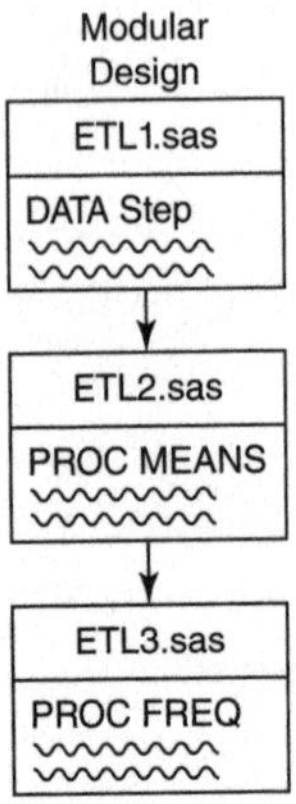

Modular Design	Prerequisites	Inputs	Dependencies	Outputs
ETL1.sas: DATA Step	None	None	PROC MEANS PROC FREQ	Final data set
ETL2.sas: PROC MEANS	DATA Step	Final data set (shared lock)	None	Means output
ETL3.sas: PROC FREQ	DATA Step	Final data set (shared lock)	None	Frequency output

Figure 7.2 Critical Path Analysis

the DATA step (prerequisite) and because each only requires a shared lock on the Final data set (input).

Assuming that the DATA step takes ten minutes to complete and the MEANS and FREQ procedures each take five minutes to complete, the serialized code would complete in 20 minutes, whereas the parallel code would execute in 15 minutes. Hooray, the software is more efficient! Or is it? In actuality, the system resources have not been reduced but only shifted so they are utilized concurrently. Although the software is faster, it is not more efficient.

Critical path analysis can be used to represent true prerequisites and true dependencies, thereby identifying and eliminating false dependencies—at least at the theoretical level. To enable SAS to run the MEANS and FREQ procedures in parallel, their respective code must be disentangled, modularized, and subtly redesigned as demonstrated in the following sections. The example demonstrated in Figure 7.2 is operationalized and automated in the "Synchronicity and Asynchronicity" section in Chapter 12, "Automation."

False Dependencies of Process

Dependencies of process occur when one process is reliant upon another, usually in which one process must complete before a subsequent process can begin. For example, in Figure 7.2, the MEANS procedure must wait until the DATA step ends, thus the MEANS procedure is a dependency of the DATA step. A false dependency, however, occurs when the serialized nature of software directs that two processes execute in series when they could in theory execute in parallel. Also demonstrated in Figure 7.2, the MEANS and FREQ procedures include a false dependency; each could complete in parallel because neither is dependent on the other and neither requires an exclusive lock on the Final data set.

One of the primary factors to consider in assessing process dependencies is the type of file lock that is required by a process. For example, the MEANS and FREQ procedures can execute at the same time because each requires only a shared—not exclusive—file lock on the Final data set. As a rule of thumb, whenever an exclusive lock is required by a module, that module cannot be performed concurrently with any other module requiring access to the same data set. For this reason, while fairly obvious, it's impossible to edit a data set (which requires an exclusive lock) while simultaneously analyzing the data set with the FREQ procedure.

False Dependencies of Data Sets

ETL processes are often efficiently designed to accommodate various types of throughput, such as processing three different transactional data sets on a daily basis. However, efficient software design (i.e., writing less code) can paradoxically contribute to slower software execution due to false dependencies that are created. In the following example, the transactional data sets One, Two, and Three are run through a single ETL process that contains distinct ingestion, transformation, and analysis functionality but that is coded through serialized design. The code prima facie may appear efficient because we've been conditioned to associate the SAS macro language with efficiency. After all, the %ETL macro can process a serialized list of data sets through the DSN parameter!

```
%macro ETL(dsn=);
* ingestion;
data &dsn;
   set &dsn._raw;
run;

* transformation;
data &dsn._trans;
   set &dsn;
run;

* analysis;
proc freq data=&dsn._trans;
   tables char;
run;
%mend;

%macro engine(dsnlist=);
%local i;
%let i=1;
%do %while(%length(%scan(&dsnlist,&i,,S))>1);
   %etl(dsn=%scan(&dsnlist,&i,,S));
   %let i=%eval(&i+1);
   %end;
%mend;

* required to fake ingestion streams;
data one_raw;
   length char $10;
data two_raw;
   length char $10;
```

```
data three_raw;
   length char $10;
run;

%engine(dsnlist=one two three);
```

This code demonstrates real dependencies in that the ingestion module must complete before the transformation module can start, and the transformation module subsequently must complete before the analysis module can start. The code also demonstrates false dependencies, however, in that it requires data set One to be processed before data set Two, and data set Two to be processed before data set Three. In reality, any or all of the modules processing data set One could fail (for example, if data set One had no observations) without negatively affecting the performance of data sets Two or Three. Figure 7.3

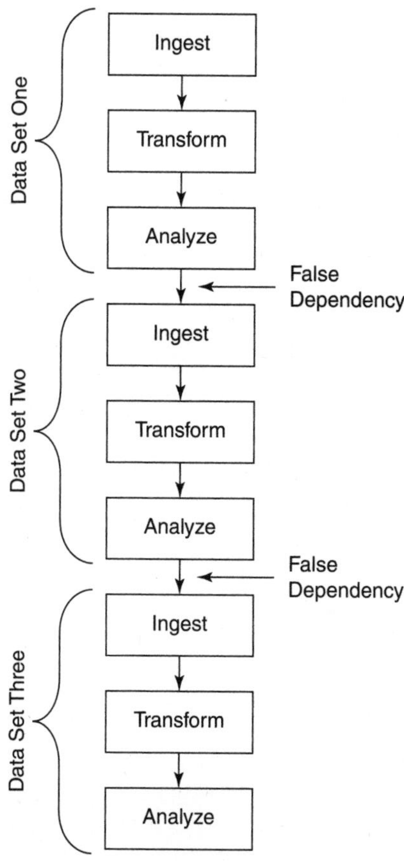

Figure 7.3 Data Set False Dependencies

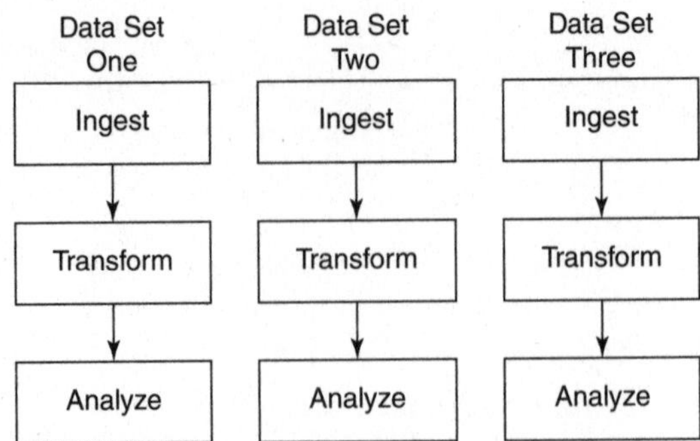

Figure 7.4 Data Set False Dependencies Removed

demonstrates the false dependencies formed when the ingestion of data set Two must wait for data set One to complete, and when the ingestion of data set Three must wait for data sets One and Two to complete.

In theory, a faster solution would simultaneously ingest data sets One, Two, and Three, after which those three process flows would continue with their respective transformations and analyses. Figure 7.4 depicts the revised critical path, implementing parallel processing that removes all false dependencies. Note as before that this solution is not more efficient; however, it completes in approximately one-third the time of the original monolithic process flow.

False Dependencies of Throughput

The previous demonstrations have depicted software that can be modularized into smaller chunks to allow data sets to pass through them more quickly. Sometimes, however, the data set itself can be the source of the bottleneck if it is large and unwieldy. For example, consider a straightforward SAS SORT procedure that sorts 100 observations in less than a second:

```
data reallybig (drop=i);
   length num1 8;
   do i=1 to 100;
      num1=rand('uniform');
      output;
      end;
run;

proc sort data=reallybig;
   by num1;
run;
```

But what happens to execution time when Reallybig actually is really big—say, 100 billion observations or 100 GB? This can present a scalability problem for some environments, causing slow or inefficient performance, or outright functional failure. The effect is similar to what happens when I throw a peeled orange into my koi pond—the koi and turtles quickly battle to get at the orange, completely surrounding it, but the smaller ones are pushed away, and even the larger ones can't nibble through to the center for over an hour. It's tremendously inefficient feeding. Yet when fish flakes are dumped in the pond, they're consumed in minutes because the koi and turtles all can be eating simultaneously. You similarly don't want to wait while your data analytic processes nibble their way to the center of a 100 GB orange!

The same phenomenon typically occurs in data processing, in which the first observation is processed, followed by the second, and so forth. When large data sets are encountered, however, performance can be improved by processing discrete segments of the data sets in parallel and thereafter reconstituting the data or results in a technique referred to as *divide and conquer*. In essence, parallel processing creates a feeding frenzy, allowing everyone to eat simultaneously and efficiently rather than waiting for a large mass to be consumed. The technique is described more fully in the "Load Scalability" section in Chapter 9, "Scalability."

For example, the following code represents the first module (of four) that would be run in parallel to sort the Reallybig data set with 100 observations:

```
proc sort data=reallybig (firstobs=1 obs=25);
   by num1;
run;
```

The second process (running in parallel) would sort observations 26 through 50, and so on. In this manner, all sorting would be done in parallel, after which those sorted data sets could be recombined for a potential performance improvement. This type of dynamic processing can be implemented with the SYSTASK statement, which can spawn batch jobs that execute in parallel. SYSTASK is demonstrated in the "Decomposing SYSTASK" section in Chapter 12, "Automation."

This divide-and-conquer technique can also be applied to DATA steps by processing different segments of observations in parallel. However, as with the SORT procedure, whenever data are divided they must later be recombined, which incurs additional processing time and resources. Because of this extra step, divide-and-conquer methods are always slightly less efficient than processing data as a single unit. However, when sufficiently large and despite this inefficiency, data sets can be processed much faster in parallel than as a single unit. Only through trial and error can thresholds be established to determine when and on what size data sets divide-and-conquer techniques will improve performance.

PARALLEL PROCESSING

By eliminating false dependencies, parallel processing can dramatically increase software speed by performing multiple tasks simultaneously. The previous "False Dependencies" section demonstrates how processes, data sets, and even observations can be decoupled and executed concurrently to increase performance. However, this increased speed typically decreases software efficiency. This occurs because parallel processing requires running multiple SAS sessions, each of which requires overhead resources. For example, when I start the SAS Display Manager, the Windows Task Manager depicts that SAS requires 57 MB of memory just to sit quietly and not even execute code.

A second source of inefficiency in parallel processing is the effort required to coordinate and communicate between SAS sessions. For example, if four SAS sessions are concurrently sorting 100 billion observations, some effort is required to determine which sessions will sort which specific chunks of observations. And each individual session should inform the other sessions once it completes sorting its chunk. A final step—not required for sorts performed on a single data set—is required to reconstitute the four data sets, thus utilizing even more resources. Each of these activities is costly, so while the parallel solution will never be more efficient, it will often be faster given sufficient resources (processors and memory) to perform the tasks.

One way to view parallel processing is that it allows developers to more fully utilize system resources. SAS environments with abundant or unused resources will benefit most fully from parallel processing because the processes effectively won't be in competition with each other. Even SAS practitioners running a single instance of the SAS Display Manager, however, can improve performance by more fully utilizing system resources through parallel processing.

No Multithreading Here!

Parallel processing is not multithreading! The distinction is made in the "Parallel Processing versus Multithreading" section in Chapter 9, "Scalability," but is underscored here. Throughout this text, *parallel processing* depicts processes and programs that run concurrently in separate SAS sessions. Whereas parallel processing typically only increases execution efficiency, multithreading often can also increase efficiency because it operates at the thread—not the process—level.

Hardcoded Concurrency

False dependencies between processes can be overcome through parallel processing that executes processes concurrently rather than in series. In some cases, this can be accomplished relatively easily, without much modification to the software and with little to no communication required between the individual SAS sessions. While hardcoded parallel processing solutions are uncommon in production software, they form the basis for later dynamic parallel design.

The previous "False Dependencies of Data Sets" section demonstrates the false dependency that exists when the data set One must pass through several phases of an ETL process before the data set Two can be processed. The inefficiency is caused by the invocation of the %ENGINE macro, because it requires a parameterized list of data sets that are processed in series.

```
%macro engine(dsnlist=);
%local i;
%let i=1;
%do %while(%length(%scan(&dsnlist,&i,,S))>1);
   %etl(dsn=%scan(&dsnlist,&i,,S));
   %let i=%eval(&i+1);
   %end;
%mend;
```

To overcome this inefficiency, the %ENGINE and %ETL macros don't even need to be modified—modifying only the %ENGINE invocation removes the serialization. For example, a SAS practitioner can open one SAS session and run the following code to process data set One:

```
%engine(dsnlist=one);
```

And, in a separate SAS session, %ENGINE can be invoked only on data set Two:

```
%engine(dsnlist=two);
```

The facile solution works fluidly because no further interaction between the data sets or their derivative processes is required. So long as resource utilization is not taxed by the second SAS session, the software should now process two data sets in the time previously required to process one.

A difficulty in implementing hardcoded solutions like this is that in production software, additional modifications are typically required. This code was substantially advantaged because the data set name was parameterized into the macro variable &DSN and utilized to dynamically create and name every referenced data set, ensuring that data sets from the first session did not conflict with or overwrite data sets from the second session. Data sets were also processed only in WORK libraries, whereas references to shared SAS libraries could have resulted in file access collisions. Finally, no interaction was required between the two sessions as they executed because no dependencies existed between either.

A hardcoded parallel processing solution can be easy to implement but very difficult to maintain because it requires developers or users to maintain separate code bases. For example, two separate macro invocations were required for %ENGINE, one to execute each of two data sets. Because the ultimate goal of production software is often to achieve stable software that can be scheduled, hardcoded parallel processing techniques generally have no place in production software. It is a useful tool to conceptually introduce parallel processing or to test software design, but hardcoded parallel design should typically be improved through dynamic, macro-driven software that does not require maintaining separate code bases.

Executing Software Concurrently

The "Hardcoded Concurrency" section demonstrates separate modules that can be run in parallel to achieve faster performing software. While this requires no communication between the sessions, more advanced parallel processing can execute multiple copies of the same program from separate sessions of SAS. This technique requires execution autonomy, because each software instance must coordinate with all others to ensure that duplication will not result. Moreover, it requires efficient and effective communication between all modules because they must report when they start a process, when they complete a process, and whether the process succeeded or failed.

To achieve execution autonomy, fuzzy business logic should prescribe prerequisites, dependencies, inputs, and outputs for each module. The following code, reprised from the earlier "False Dependencies" section, demonstrates that the FREQ procedure unfortunately must wait for the MEANS procedure to complete:

```
data final;
   length char $10 num 8;
run;

proc means data=final;
   var num;
run;

proc freq data=final;
   tables char;
run;
```

The DATA step process has no prerequisites, so it can execute immediately. The MEANS and FREQ processes each require the Final data set to have been created before they can start. After identifying prerequisites and dependencies and identifying that each process requires only a shared lock of the PERM.Final data set, it's clear (as depicted in Figure 7.2) that MEANS and FREQ can be run in parallel. The two procedures can be modularized through the following code, which is described later throughout the text:

```
%macro means();
%let start=%sysfunc(datetime());
%control_update(process=MEANS, start=&start);
proc means data=perm.final;
   var num;
run;
%lockitdown(lockfile=perm.control, sec=1, max=5, type=W,
   canbemissing=N);
%control_update(process=MEANS, start=&start, stop=%sysfunc(datetime()));
%mend;

%macro freq();
%let start=%sysfunc(datetime());
%control_update(process=FREQ, start=&start);
proc freq data=perm.final;
   tables char;
run;
%lockitdown(lockfile=perm.control, sec=1, max=5, type=W,
   canbemissing=N);
%control_update(process=FREQ, start=&start, stop=%sysfunc(datetime()));
%mend;
```

Communication is the next requirement to transition from monolithic to parallel design. Because global macro variables cannot be used to communicate between SAS sessions, a control table—a SAS data set—can be used to facilitate cross-session communication. In this example, one session will run the DATA step while the second session waits patiently for it to complete. Thereafter, upon detecting DATA step completion, one session will run the Means module while the other session runs the Freq module. Because both sessions must be able to access the Final data set and the control table, both data sets must be moved from the WORK library to a shared library.

```
* saved as c:\perm\parallel.sas;
libname perm 'C:\perm'; * modify to actual location;
%include 'C:\perm\lockitdown.sas'; * modify to LOCKITDOWN location;

%macro control_create();
%if %length(&sysparm)>0 %then %do;
   data _null_;
      call sleep(&sysparm,1);
   run;
   %end;
%if %sysfunc(exist(perm.control))=0 %then %do;
   %put CONTROL_CREATE;
   data perm.control;
      length process $20 start 8 stop 8 jobid 8;
      format process $20. start datetime17. stop datetime17. jobid 8.;
   run;
   %end;
%mend;

%macro control_update(process=, start=, stop=);
data perm.control;
   set perm.control end=eof;
%if %length(&stop)=0 %then %do;
   output;
   if eof then do;
   process="&process";
   start=&start;
   stop=.;
   jobid=&SYSJOBID;
   output;
   end;
   %end;
```

```
%else %do;
   if process="&process" and start=&start then stop=&stop;
   %end;
   if missing(process) then delete;
   run;
&lockclr; * required by LOCKITDOWN macro;
%mend;

%macro ingest();
%let start=%sysfunc(datetime());
%control_update(process=INGEST, start=&start);
data perm.final;
   length char $10 num 8;
   do i=1 to 500000;
      do j=1 to 100;
         char=put(j,$10.);
         num=j;
         output;
         end;
      end;
run;
%lockitdown(lockfile=perm.control, sec=1, max=5, type=W,
   canbemissing=N);
%control_update(process=INGEST, start=&start,
   stop=%sysfunc(datetime()));
%mend;

%macro means();
%let start=%sysfunc(datetime());
%control_update(process=MEANS, start=&start);
proc means data=perm.final;
   var num;
run;
%lockitdown(lockfile=perm.control, sec=1, max=5, type=W,
   canbemissing=N);
%control_update(process=MEANS, start=&start, stop=%sysfunc(datetime()));
%mend;

%macro freq();
%let start=%sysfunc(datetime());
%control_update(process=FREQ, start=&start);
proc freq data=perm.final;
   tables char;
```

```
run;
%lockitdown(lockfile=perm.control, sec=1, max=5, type=W,
   canbemissing=N);
%control_update(process=FREQ, start=&start, stop=%sysfunc(datetime()));
%mend;
```

Because the control table PERM.Control is accessed by multiple sessions of SAS, and because an exclusive lock is required to update the data set, some quality control mechanism must ensure that file collision does not occur. The %LOCKITDOWN macro (or similar functionality) should be implemented before attempting to open the control table to ensure it is not in use. The %LOCKITDOWN macro first assesses the availability of a data set; if it is locked, the macro waits a parameterized amount of time before reattempting access. The %LOCKITDOWN macro is described more fully in Chapter 9, "Scalability," while the full text and code are available in a presentation titled "From a One-Horse to a One-Stoplight Town: A Base SAS Solution to Preventing Data Access Collisions through the Detection and Deployment of Shared and Exclusive File Locks."[3]

The JobID variable in the control table records the unique value of the global macro variable &SYSJOBID. If run in series by a single session of SAS, the JobID will be identical for all processes. When run in parallel, however, the JobID will demonstrate the SAS session that ran each specific module. This information can be useful when trying to troubleshoot parallel processing to ensure that all sessions are executing. Table 7.1 demonstrates use of &SYSJOBID to determine whether parallel processing is running on all cylinders.

The previous code demonstrates processes that create and update the control table, as well as that run the analytic macros MEANS and FREQ. In parallel processing design, however, an engine (AKA controller or driver) often includes the autonomous business logic that makes everything run—in this example, deciding whether to execute a module or sit and wait for other processes to complete. In production software, the engine will often operate as a separate program while spawning (or calling) child processes (batch jobs) saved as separate SAS programs. The following %ENGINE macro should also be included in the previous SAS program; the macro effectively controls and directs execution of separate SAS macros.

```
* continues the program c:\perm\parallel.sas;
%macro engine();
%let ingest=;
%let means=;
%let freq=;
```

```
%control_create();
%do %until(&ingest=COMPLETE and &means=COMPLETE and &freq=COMPLETE);
   %lockitdown(lockfile=perm.control, sec=1, max=5, type=W,
      canbemissing=N);
   data control_temp;
      set perm.control;
      if process='INGEST' and not missing(start) then do;
         if missing(stop) then call symput('ingest','IN PROGRESS');
         else call symput('ingest','COMPLETE');
         end;
      else if process='MEANS' and not missing(start) then do;
         if missing(stop) then call symput('means','IN PROGRESS');
         else call symput('means','COMPLETE');
         end;
      else if process='FREQ' and not missing(start) then do;
         if missing(stop) then call symput('freq','IN PROGRESS');
         else call symput('freq','COMPLETE');
         end;
   run;
   %if %length(&ingest)=0 %then %ingest;
   %else %if &ingest=COMPLETE and %length(&MEANS)=0 %then %means;
   %else %if &ingest=COMPLETE and %length(&FREQ)=0 %then %freq;
   %else %if &ingest^=COMPLETE or &means^=COMPLETE or
      &freq^=COMPLETE %then %do;
      &lockclr;
      data _null_;
         call sleep(1,1);
      run;
      %end;
   %else &lockclr;
   %end;
%mend;

%engine;
```

One hallmark of parallel software is its ability to be initialized from only one session of SAS. Thus, if only one session of the Parallel.sas program is executed, the software will execute in series. However, when multiple instances of the software are executed in separate SAS sessions, the programs will play nicely together because they communicate and coordinate program flow through the %ENGINE macro and the control table. As long as the second instance of the program is executed before the %MEANS macro has completed in the first SAS session, the second instance will access the control table and begin executing the %FREQ macro in parallel. While this example has demonstrated parallel processing that requires manual initialization of multiple SAS sessions and manual execution of multiple instances of the same program, in later sections, examples demonstrate parallel processing in which initialization and execution are automated via the SYSTASK statement.

Starting Concurrent SAS Sessions

The program Parallel.sas in "Executing Software Concurrently" was run manually, which necessitated multiple initializations of SAS sessions. Parallel processing can also be achieved by spawning clone child processes via SYSTASK, by spawning individual child processes via SYSTASK, or by executing batch programs (not batch jobs) from the operating system (OS)—each of which automatically creates concurrent SAS sessions. These methods are discussed more fully in Chapter 12, "Automation," while the "Demand Scalability" section in Chapter 9, "Scalability," discusses considerations and caveats when multiple instances of software are run simultaneously.

The entire code from the "Executing Software Concurrently" section should first be saved as C:\perm\parallel.sas. A *batch program* or *batch file*—as opposed to a batch job—contains OS command line syntax and should be saved with the .bat extension. The following batch file (C:\perm\series.bat) can be created by opening Notebook or another text editor and entering the following syntax *on only one line with no carriage returns*:

```
"c:\program files\sashome\sasfoundation\9.4\sas.exe" -sysin
   C:\perm\parallel.sas
```

The location of the SAS.exe executable program may vary by SAS application and environment, and double quotes are required if its path contains a space, such as the "program files" folder. When the batch file is saved and subsequently executed, the SAS log is automatically saved to a log file—C:\perm\parallel.log—because no log location is specified in the command statement. Before attempting to run the batch file in parallel, view the log to ensure that the program successfully ran in series. Once the log confirms that no runtime errors occurred, delete the PERM.Control and PERM.Final data sets to reset the session. If PERM.Control is not deleted, the program will correctly assess that the Ingestion, Freq, and Means modules all have completed, and thus will exit the software without rerunning any of these modules.

Once the software is confirmed to have run in series from the batch file, it can be tested in parallel. Once again, delete the PERM.Control and PERM.Final data sets to reset the session. The following batch file should be saved to C:\perm\parallel.bat and, as before, the two commands should be on two separate lines without additional carriage returns:

```
start "job1" "c:\program files\sashome\sasfoundation\9.4\sas.exe" -sysin
   "C:\perm\parallel.sas" -log "C:\perm\log1.log" -print
   "C:\perm\output1.lst" -sysparm 0

start "job2" "c:\program files\sashome\sasfoundation\9.4\sas.exe" -sysin
   "C:\perm\parallel.sas" -log "C:\perm\log2.log" -print
   "C:\perm\output2.lst" -sysparm 5
```

The first line of the batch program initiates the first SAS session and runs the first instance of the Parallel.sas program. Without waiting for the first instance of the program to complete, the second line immediately (and asynchronously) initiates a second session of SAS and runs a second instance of the program. Because the default log and output files for each SAS program would have been identically named—as the same SAS program is being executed twice in parallel—unique log and output file names must be specified to avoid file collisions, which would cause errors.

The SYSPARM parameter passes a numeric value to the &SYSPARM automatic macro variable in the %CONTROL_CREATE macro, which waits a parameterized number of seconds before commencing. Without discrepant SYSPARM values—which allow the first instance of the program to build the control table immediately but force the second instance of the program to wait five seconds—each instance of the program would simultaneously assess that the control table did not exist and simultaneously attempt to build it, either overwriting each other or failing with file access errors. Thus, the first SAS session (and instance of the Parallel.sas program) will create the control table and run the DATA step, while the second instance of the program waits five seconds, assesses that the control table exists, and then enters a busy-waiting cycle until the DATA step completes and the control table is updated to reflect this. Thereafter, one session of SAS will execute the %MEANS macro while the other session executes the %FREQ macro.

The benefits of batch program design are numerous. The batch file can be used to launch multiple sessions of SAS automatically and run multiple SAS programs, rather than requiring SAS practitioners to initialize SAS sessions manually to open and run programs. Since batch files are run directly from the OS, they can be automated and scheduled to run through the OS, as discussed in the "Starting Batch from a Batch File" section in Chapter 12, "Automation." Most relevant to increased performance, batch files allow asynchronous and parallel processing that facilitates significantly faster software execution when code is modularized.

Naptime Is Inefficient

I was a huge fan of naptime as a kindergartener—we would each lie down on beach towels, spread across a large area rug, while Mrs. McCarthy read to us. In software development, however, naptime can be extremely inefficient. One of the only negative consequences of parallel software design is the busy-waiting (or spinlock) cycle that is often introduced to put a SAS session to sleep when it has nothing else to do. For example, when the batch file Parallel.bat in the "Starting Concurrent SAS Sessions" section executes, the program running in the second session enters a busy-waiting cycle while it waits for the DATA step to complete in the first session. During this naptime, despite only sleeping, the second SAS session is still consuming overhead system resources as well as additional resources used each time it queries the control table to determine whether the DATA step has completed. Thus, while parallel processing can perform significantly faster, these sleep cycles in part contribute to the lower efficiency demonstrated in parallel processing as compared to serial processing.

Serialized process flows, however, may run slowly, but they are always running—you won't catch a serialized program lying down on the job because it doesn't need to coordinate with or wait for external sessions or programs. While the overhead system resources utilized by a slumbering SAS session are minimal, when multiple SAS sessions are sleeping, or when SAS sessions spend an inordinate percentage of their lives in busy-waiting cycles due to poor management of parallel processes, the cumulative inefficiency can burden the overall SAS system.

While the %LOCKITDOWN macro will prevent failure associated with file access collisions, it will not prevent the associated delays inherent in parallel processing that occur when multiple SAS sessions vie for exclusive access to a single data set. If two SAS sessions simultaneously vie for exclusive access to a control table, the %LOCKITDOWN macro will ensure that one gains access to the data set while the other enters a busy-waiting cycle, rather than terminating with a runtime error. Access to control tables—which typically contain a few dozen to a few hundred observations—only requires a second or two and thus doesn't contribute to inefficiency. When complex processes are run on larger data sets, however, busy-waiting cycles can last hours, during which data accessibility is repeatedly tested at regular intervals.

Thus, in assessing performance gains experienced by parallel processing design, SAS practitioners should also assess potential inefficiency created by sleeping processes. The Parallel.sas program presented in the "Executing Software Concurrently" section, for example, is optimized when run on two SAS sessions because %MEANS can run in one session while %FREQ runs in the other. If run from three concurrent SAS sessions, however, the third program instance will sleep the entire time and add no value, thus further decreasing the software efficiency while not improving performance. Examination of control tables, as demonstrated in the following section, can help demonstrate when concurrent software is being run across an optimal number of SAS sessions.

Control Tables in Parallel Processing

Communication is the most important aspect of parallel processing because it enables individual SAS sessions to speak to each other to facilitate autonomous execution through fuzzy logic business rules. Communication ensures that prerequisites are met before processes begin and further ensures that redundant processing of the same tasks does not occur. Control tables should record when processes start (so that other instances don't accidentally start the same process), when processes complete (so that all instances are aware when prerequisite processes are finished), and when processes fail (so that the process can be restarted or have other actions taken, as well as to ensure that dependent processes do not start following the failure).

Table 7.1 demonstrates a control table from the Parallel.sas program in the "Executing Software Concurrently" section that shows successful completion of the software when run in parallel from two sessions of SAS. The control table demonstrates that the %MEANS and %FREQ macros ran in parallel, which can be assessed by examining start and stop times or by the separate SAS JobIDs, 8040 and 3100. Had only one JobID been listed, this would have indicated either that the software was executed from only

Table 7.1 Control Table Demonstrating Parallel Processing

OBS	Process	Start	Stop	JobID
1	INGEST	01JAN16:12:18:33	01JAN16:12:19:00	8040
2	MEANS	01JAN16:12:19:00	01JAN16:12:19:07	8040
3	FREQ	01JAN16:12:19:01	01JAN16:12:19:15	3100

one session or that other sessions had failed during execution. Had three sessions of SAS run three instances of the Parallel.sas program concurrently, only two JOB IDs would have been demonstrated since the third process would have had no tasks to complete, thus demonstrating the inefficiency of running three sessions.

EXECUTION EFFICIENCY IN THE SDLC

Software speed is a bit different than other performance requirements because it represents a continuous measurement rather than a dichotomous—successful or failed—outcome. For example, software may demonstrate a lack of robustness by terminating with a failure. Software that lacks portability may terminate with a failure when an unexpected (i.e., exceptional) environment is encountered. Thus, while failures associated with many other performance attributes can cause real functional failures, a failure of execution efficiency typically just describes slow execution. This can contrast with an efficiency failure, in which overuse of system resources can cause not only delays but runtime errors and loss of functionality.

Because execution efficiency lies on a continuum rather than existing as a dichotomous outcome, it can be prone to refactoring throughout the software development life cycle (SDLC), even once software is in production. For example, SAS practitioners may initially implement software that provides a functional outcome, but stakeholders may later require an increase in speed. Due to the vast number of methods that can be used to increase software speed—including programmatic and nonprogrammatic best practices—there are often multiple ways to achieve higher performing software over its lifespan. Because of this diversity, however, refactoring for increased speed is one of the most common requests from stakeholders for software already in production.

Requiring Execution Efficiency

This is an easy one—how long should the software take to complete? A typical requirement states only the maximum execution time; however, an average execution time might also be stated. Often in modular software design, in which modules are being independently developed by various SAS practitioners, individual modules will have their own performance requirements that state their respective execution time thresholds.

Because software speed is a continuous rather than discrete measurement, a common objective is to improve the speed of software throughout the SDLC, including once software is in production. I once encountered a customer who was astonished that a SAS process designed to be run hourly could not be compressed into 20 minutes without significant redesign and refactoring. Software quality represents an investment, and like other dimensions of quality, execution efficiency should be planned and implemented throughout the SDLC rather than in an ad hoc fashion once software has been completed. My team was eventually able to expedite the analytic process and run it in less than 20 minutes, but this required starting from scratch with a complete software overhaul. While customers may need to change software requirements drastically during or even after development, accurately conveying realistic software performance requirements during software planning and design will help eliminate later unnecessary redesign and refactoring.

Maximum Execution Time

When a maximum execution time is stated, it represents a standard that should never be exceeded—for example, "The ETL process will *always* complete in 20 minutes or less." However, this simplistic efficiency requirement could allow developers to drain system resources, adversely affecting other processes. For example, to achieve this requirement, developers might design software that spawned 20 concurrent SAS sessions and overwhelmed a SAS server, preventing other unrelated programs from executing. Thus, a more comprehensive requirement might state: "The ETL process will always complete in 20 minutes or less while using no more than two processors concurrently and no more than 2 GB of memory." The latter requirement incorporates both speed and efficiency and might be necessary in environments with many users or constrained system resources. In both examples, however, software that completes in 21 minutes represents a performance failure that should be documented and investigated.

Sometimes a maximum efficiency threshold is stated, but this standard can be exceeded on occasion. For example, a requirement might state that "The ETL process will complete in 20 minutes or less for 95 percent of executions, and 30 minutes or less for 5 percent of executions." After all, requirements are intended both to guide design and development and to regulate function and performance, but should incorporate real-world variability that can exist.

Thus, if SAS practitioners understand that on occasion, larger data sets will be processed that may slow execution speed and efficiency, this variability should be accounted for in software requirements. Another option is to state average execution time in requirements instead of (or in addition to) maximum execution time, thus providing a metric that more readily accommodates variability in run time.

Progressive Execution Time

Especially in Agile and other rapid software development environments, change is espoused and encouraged, and early delivery of functionality and business value is paramount. Thus, expecting customers to state or even know their ultimate performance objectives during early planning will be unrealistic and counter to Agile principles. In these environments, one solution to inevitable change—in which customers eventually will want faster running software—is to implement progressive software requirements. Progressive software requirements for efficiency are demonstrated in the "Velocity as an Indicator" section in Chapter 9, "Scalability."

For example, requirements might state that a SAS process needs to be able to execute in one hour when it is initially released, but additionally specify that the software must execute in 20 minutes or less within three months of software release. In this way, progressive requirements allow developers to focus on the initial release of functionality while also maintaining an awareness of eventual performance objectives, thus enabling more effective refactoring throughout the software lifespan. Because developers are aware from the outset that their software will *someday* be expected to go very fast, they are able to plan for and conceptualize that expectation even if they are not required to implement that speed immediately within software. Maintaining this forward-facing awareness of future requirements is one objective of software extensibility, discussed in the "From Reusability to Extensibility" sections in Chapter 18, "Reusability."

Furthermore, this foresight enables customers to prioritize when additional performance, such as speed, should be implemented into production software during later iterations. If stakeholders determine in two months that the one-hour software execution threshold can remain, they are able to prioritize other functionality or performance rather than requiring developers to make the software faster. Regardless of the software development methodology espoused—Agile or otherwise—progressive software requirements facilitate continuous quality improvement (CQI) that occurs after software is in production.

Outperforming Your Hardware

The reliance on programmatic methods alone to achieve performance objectives, to the exclusion of other nonprogrammatic methods, can cause an unhealthy balance in the use of project and system resources. In many environments, SAS practitioners may only be able to influence software performance programmatically because they have no control over hardware, network, or other IT-related components. However, in more permissive environments that do enable SAS practitioners to influence elements of their environment, all performance-related factors should be considered.

For example, as a customer or sponsor, do you want to pay developers for 80 hours of work spent improving the performance of one SAS program? Yes, the software may run faster after refactoring that incorporates additional software development best practices. Yes, those developers may be able to apply some of the knowledge gained to make future software faster as well. For the price of their labor, however, improved hardware or other infrastructure could have been purchased and implemented, which also would have improved the performance of not only the specific program but also other SAS software running in the environment.

Like many other aspects of performance, gains in software speed typically follow the law of diminishing returns, in which initial programmatic techniques to improve speed may dramatically increase performance, while later techniques may be less influential. Because diminishing returns curves are specific to given hardware and infrastructure, by instead modifying nonprogrammatic aspects of the software operational environment, developers can effectively make a quantum leap from one diminishing returns curve to a higher one, as demonstrated in Figure 7.5.

Figure 7.5 represents that as software continues to be refactored over its lifespan, the performance continues to improve. As the low-hanging fruit are picked, however, speed gains may be reduced with each successive enhancement. And, following several successive performance improvements, the effort (or cost) required to effect further increased performance may be more efficiently invested in nonprogrammatic rather than programmatic improvement. Stakeholders must remain cognizant that often, by instead investing in hardware, infrastructure, or third-party software solutions that interact with

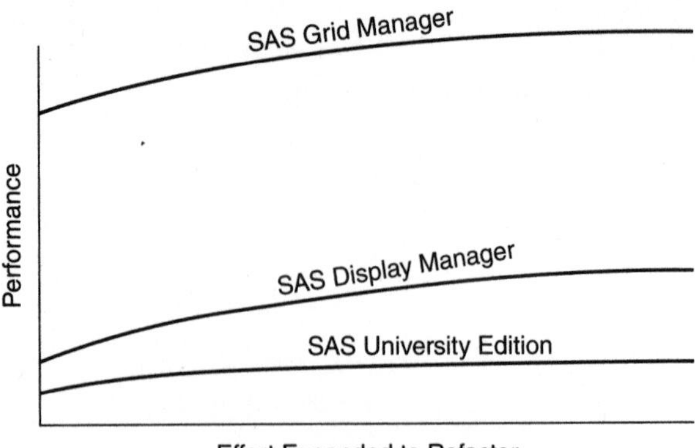

Figure 7.5 Diminishing Return Curves for Software Performance

and improve the performance of SAS software, they can achieve a higher performance curve for a lower price.

Measuring Execution Efficiency

Just as software execution time is a straightforward requirement, it is similarly straightforward to measure. The following code and output define macro variables &DTGSTART and &DTGEND that capture the duration of software execution:

```
%let dtgstart=%sysfunc(datetime());
proc sort data=original out=final;
   by char;
run;

NOTE: There were 1000000 observations read from the data set
WORK.ORIGINAL.
NOTE: The data set WORK.FINAL has 1000000 observations and 2 variables.
NOTE: PROCEDURE SORT used (Total process time):
      real time           0.54 seconds
      cpu time            0.65 seconds

%let dtgstop=%sysfunc(datetime());
%let seconds=%sysevalf(&dtgstop-&dtgstart);

%put TIME: &seconds;
TIME: 0.55299998283386
```

In production software in which run time is specified and required, the macro variable &SECONDS could be saved to a permanent SAS data set to record historical performance metrics. Not only is this necessary for longitudinal analysis to demonstrate that software met performance objectives, but outliers can also be used to investigate potential software exceptions or failures that may have occurred.

The previous method is commonly implemented when an entire program or process needs to be timed. However, when a single DATA step or SAS procedure needs to be timed—as is often required during performance tuning and optimization—a common method is to utilize the SAS FULLSTIMER system option to record execution time and other performance metrics. With the PRINTTO procedure, FULLSTIMER metrics can be directed to a text file in lieu of the log and thus ingested, aggregated, and analyzed in SAS. Another benefit of FULLSTIMER is that it measures three execution time metrics—real time, user CPU time, and system CPU time—which can be used to troubleshoot and optimize software. The FULLSTIMER system option and code that automatically parses FULLSTIMER output are demonstrated in the "Measuring Efficiency" sections in Chapter 8, "Efficiency."

Repeated Measurements

Once software is in production, every run counts—typically, there are no freebies. For this reason, it's important to sufficiently understand and predict future variability during software testing. Software that is run just once in the evening after most of the development crew has gone home for the night might display exemplary performance, but when it is executed the next day and forced to compete against other SAS jobs and users on the server, it might fail miserably. SAS software must be run sufficiently to understand variability patterns to ensure that software will meet performance objectives during operational conditions.

Even under seemingly similar environmental conditions, SAS software execution time from one run to the next can vacillate wildly. Repeated measures—even under conditions that seem to be similar or identical—help demonstrate the true range of performance times for a given process, as well as its variability. For example, the following code creates a 10 million observation (79 MB) data set and sorts it repeatedly (1,000 times) to assess performance metrics:

```
data perm.kindasmall (drop=i);
   length num1 8;
   do i=1 to 10000000;
      num1=rand('uniform');
```

```
      output;
      end;
run;

data historics;
   length secs 8;
run;

%macro repeated_measures();
%local i;
%let i=1;
%do i=1 %to 1000;
   %let dtgstart=%sysfunc(datetime());
   proc sort data=kindasmall out=sorted;
      by num1;
   run;
   %let dtgstop=%sysfunc(datetime());
   %let seconds=%sysevalf(&dtgstop-&dtgstart);
   %put TIME: &seconds;
   data historics;
      set historics end=last;
      if ^missing(secs) then output;
      if last then do;
         secs=&seconds;
         output;
         end;
   run;
   %end;
%mend;

%repeated_measures;
```

Separate code (not shown) parses, collects, and compiles the FULLSTIMER performance metrics, including both real time (actual elapsed time) and total CPU time (time which the CPU spent processing the task). Figure 7.6 demonstrates the tremendous variability in real time when the SORT procedure is performed 1,000 times.

As Figure 7.6 demonstrates, real time measurements are anything but normalized—in fact, the graph represents the typical platykurtic, right-skewed distribution of real time measurements. Because real time is influenced by external factors such as other SAS programs, additional software, or system administrative activities such as virus scanning, it can be incredibly

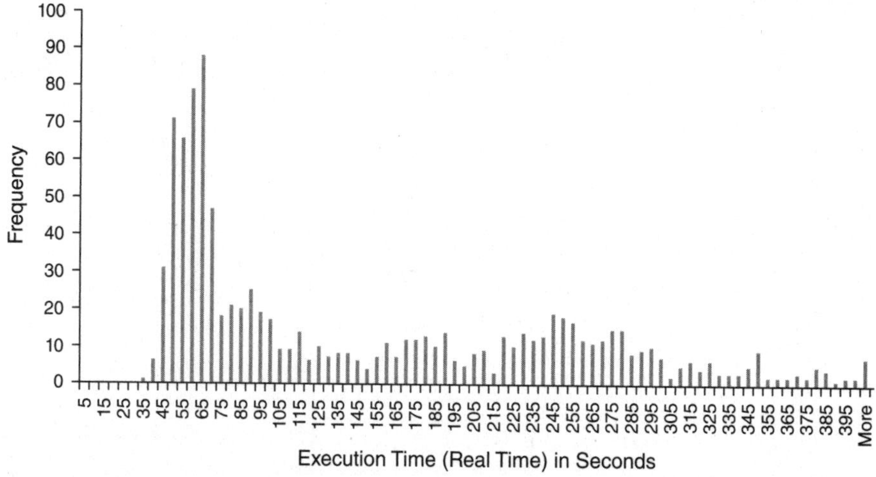

Figure 7.6 Histogram of Real Time for Sorting 10 Million Observations

unpredictable. For example, the SORT procedure took a minimum and maximum of 34.9 and 560.4 seconds, respectively, with a mean of 145.0 seconds and standard deviation of 100.1 seconds! While some of this variability can be eliminated by reducing external processes that are running in the SAS environment, this reduction likely would not represent the typical operational environment.

A more consistent method of calculating performance speed is to utilize CPU time, which is demonstrated in the SAS log when the NOFULLSTIMER option is selected, or which can be calculated by adding user CPU time and system CPU time when the FULLSTIMER option is selected. Figure 7.7 demonstrates the real time measurements (from Figure 7.6) overlaid with CPU time measurements for the same 1,000 iterations of the SORT procedure.

Figure 7.7 puts in perspective the tremendous variation in real time as compared with CPU time. CPU time tends to be very consistent, demonstrating a highly leptokurtic but only slightly right-skewed distribution. For example, the CPU time for the SORT procedure took a minimum and maximum of 13.1 and 27.3 seconds, respectively, with a mean of 16.8 seconds and standard deviation of only 1.5 seconds. The total amount of CPU time—irrespective of its distribution—will also be less than real time because factors such as external processes and the basic I/O functionality required to write the sorted data set incur additional real time. One exception to this rule exists: multithreaded processes can incur a greater CPU time than run time because all parallel threads

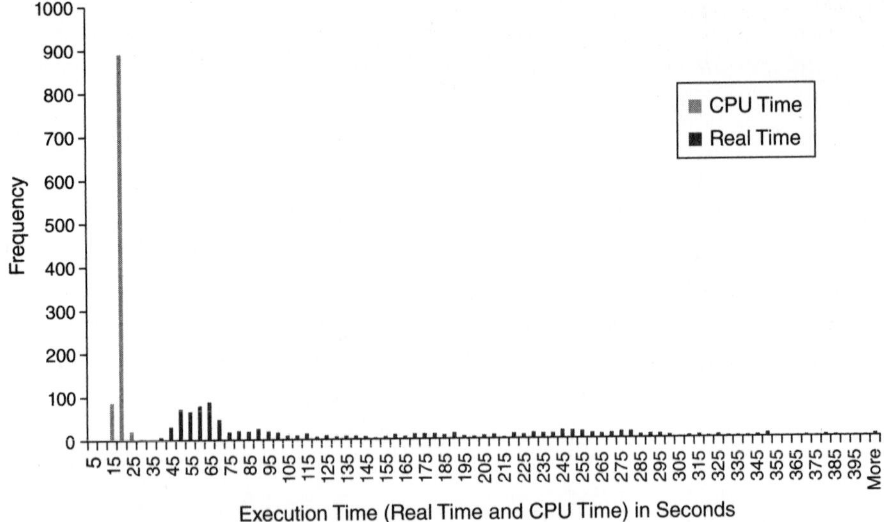

Figure 7.7 Histogram of Real Time and CPU Time for Sorting 10 Million Observations

incur CPU time while real time only measures from the start of the first thread to the completion of the last.

Real time and CPU time each have their respective uses. Customers and most stakeholders will be concerned only with real time because it measures a real construct—how long software took to complete. For this reason, real time is exclusively included within requirements documentation, SLAs, performance logs, and failure logs. For example, if analytic software took 20 minutes too long to complete, stakeholders will not care about its lower CPU time—only the unacceptable real time. During development, optimization, and testing, however, CPU time is invaluable in comparing technical methodologies because it eliminates the variability inherent in real time metrics. In other words, you can run a SAS process a few times and have a decent idea of whether it decreases or increases performance as compared with another method. Once methods have been compared using CPU time, however, testing and validation must ultimately demonstrate that real time—not CPU time—meets established needs and requirements and that its variability will not result in performance failures during software operation. Real time and CPU time are further discussed in the "CPU Processing" section in Chapter 8, "Efficiency."

Inefficiency Elbow

The *inefficiency elbow* is a phenomenon that occurs when software fails to scale to increased data volume, typically causing slowed performance and later functional failure. The SORT procedure most notably displays the inefficiency elbow, which occurs when RAM is exhausted and SAS switches to using less efficient virtual memory. Other SAS processes may hit a wall—at which point they suddenly fail with an out-of-memory error—instead of an elbow. The telltale sequela of the inefficiency elbow is substantially decreased software speed; however, because the cause is rooted squarely in the failure of system resource utilization to scale as data volume increases, this topic is described and demonstrated in the "Memory" section in Chapter 8, "Efficiency," and in the "Inefficiency Elbow" section in Chapter 9, "Scalability."

WHAT'S NEXT?

This chapter has thrown caution to the wind, touting the need for software to go as fast as possible without questioning the utilization (or monopolization) of system resources. As the "Parallel Processing" section demonstrates, however, speed isn't free—it incurs higher memory and CPU consumption, and often makes software less efficient. In Chapter 8, "Efficiency," the real costs of performance are revealed, as system resources—including CPU usage, memory usage, I/O processing, and disk space—are explored.

NOTES

1. ISO/IEC/IEEE 24765:2010. *Systems and software engineering—Vocabulary*. Geneva, Switzerland: International Organization for Standardization, International Electrotechnical Commission, and Institute of Electrical and Electronics Engineers.
2. Id.
3. Troy Martin Hughes. "From a One-Horse to a One-Stoplight Town: A Base SAS Solution to Preventing Data Access Collisions through the Detection and Deployment of Shared and Exclusive File Locks." Western Users of SAS Software, 2014.

CHAPTER 8
Efficiency

As I watched the last bits of my Ivory bath soap float down the Mopan River, I realized that perhaps I hadn't chosen the most effective method to wash my clothes.

San Jose Succotz, Belize. Only a week into the archaeological expedition, my team had returned from our dig site (Minanhá) deep in the highland jungles for sundries, sharper machetes, and naturally to wash our beleaguered, mud-laden clothes.

Standing in the slowly swelling, chest-level current of Succotz's only river, we were aware of the pollution (and occasional body) that floated down the Mopan from neighboring Guatemala, but were comforted by several Maya women washing nearby with their daughters.

In fact, having no idea how to wash clothes in a river—and barely able to even maintain control of our soap and loose clothing—Matt, Catherine, Erica, and I immediately set to watching the Maya women to observe their techniques. But, when they spied our interest, smiled widely, and waved, it was immediately apparent that they were not only washing their clothes—but also themselves.

Embarrassed by our voyeurism, we quickly whirled around and, now facing upstream toward the empty river, were again on our own to formulate a plan.

We had entered the river with a mix of powdered detergent, liquid dish soap, and of course my single bar of Ivory. One failed technique was to place the clothing in a trash bag with the liquid soap and to shake vigorously. But as we realized there was nothing to agitate the clothes (and later added rocks), Catherine was left holding a bag of rocks that was half-sinking, half-drifting downstream.

Undaunted and taking fistfuls of powdered soap and rubbing them directly into the clothing, we tried to forgo the plastic bag, but ended up with plumes of soap dust tearing our eyes as we watched the majority of the crystals float downstream.

And then there was the soap bar that I gingerly stabbed with a stick, confident that I could wield it as a tool to lather the clothing. This worked briefly but, having about a two-minute half-life before the force of the stick would cleave the bar in two, I had had little success before I watched the final discernable soap chunks float down the Mopan.

About this time, and in hysterics at our antics, the Maya women (now fully clothed) waded over and offered their assistance. Apparently I'd been onto something with the bar of soap ... but not stabbing it with a stick. And, half through interpretation and half through gesticulation, we realized they were trying to tell us that we should wash our clothes near the riverbed—not in neck-deep currents into which we'd waded.

Utterly defeated, I rode a horse into Succotz to purchase more soap for my crew as they regrouped near the shore. And, by the time we had finished the endeavor (with questionable results), we'd wasted several hours (and pounds of soap) but had had yet another unforgettable adventure in the jungles of Belize.

■ ■ ■

Our clothes washing was not only terribly slow in occupying the greater part of an afternoon—it was also terribly inefficient. While software performance is typically more predictable than my absurd washing adventures in the Mopan River, it unfortunately can be just as inefficient when resources are neither monitored nor constrained.

When a month's worth of soap products is drifting and foaming down the river, your resources are staring you in the face and it's difficult not to draw the inescapable conclusion that your process is flawed and inefficient. Even if we had effectively and quickly washed our clothes, the loss of those resources would have constituted waste and inefficiency.

But in SAS software development, resources such as CPU cycles or memory consumption aren't always as salient as riverborne detergents. Absent receipt of an out-of-memory error, other general failures, or manual interrogation of FULLSTIMER performance metrics, SAS practitioners are often unaware of how many resources their programs consume.

When resource use is excessive or waste is apparent, stakeholders or bystanders often cry foul, announcing their objection or at least awareness. This had occurred when one of my soap quarters or eighths had split and, escaping my grasp, floated toward the women downstream. Recognizing the intrinsic value in the soap bit, one of the children swam over to us and, grinning, trying not to laugh at our detersive debacle, politely handed me the chunk. She recognized its value.

If you've used up the entire SAS server running some wildly inefficient process, you're likely to get a similar tap on the shoulder, albeit not accompanied by a subtle grin or polite gesture. In other cases, however, the cumulative effect of processes that fail to conserve or efficiently use resources can be detrimental to performance. Achieving efficiency necessitates both an understanding and monitoring of resource utilization.

DEFINING EFFICIENCY

Efficiency is "the degree to which a system or component performs its designated functions with minimum consumption of resources."[1] Efficiency can also be defined as the "relation of the level of effectiveness achieved to the quantity of resources expended."[2] In this definition, Institute of Electrical and Electronic Engineers (IEEE) further notes that "Time-on-task is the main measure of efficiency." This in part explains the frequent commingling of efficiency with execution efficiency (i.e., software speed) in software discussion and literature, and the misconception that software that runs faster is always more efficient. In fact, efficiency metrics should incorporate the utilization of system resources, including memory, CPU cycles, input/output (I/O) functioning, and disk space.

Efficiency is a common objective in software development and literature. Too often, however, software developers attempt to measure efficiency through execution time alone, forsaking system resources that define efficiency. Thus, while describing *efficiency*, a common practice is to instead measure *execution efficiency*—the focus of Chapter 7—because this is a more obtainable and concrete construct. While fast performance is important and an often-stated performance requirement, software functional failures are more likely to be caused by taxing or exhausting system resources than by software execution delays.

System resources play an important role in software execution speed, and software that more appropriately and efficiently utilizes system resources will more likely experience faster execution speeds. For this reason, SAS practitioners interested in increasing performance should look under the hood and investigate how resources interact to improve execution time. But, because SAS software can fail when it runs out of CPU cycles, memory, or disk space, monitoring these resources can also facilitate increased reliability and robustness. This chapter introduces the measurement of SAS system resources to facilitate greater software performance and efficiency.

DISAMBIGUATING EFFICIENCY

If you're confused about efficiency and execution efficiency (i.e., speed), you're not alone—and it's going to get more complicated before it gets better. Even the International Organization for Standardization (ISO) changed what it once called *efficiency* to *performance efficiency* when it updated ISO/IEC 9126-1:2001 to ISO/IEC 25010:2011, citing that it was "renamed to avoid conflicts." In keeping with Project Management Institute and IEEE definitions, however, only *efficiency* is referenced throughout this text—not *performance efficiency*.

When a customer beguilingly compliments a developer, asking "I love your software, but can you make it more efficient?" he is typically asking "Can you make it go faster?" To many nontechnical folk, decreased execution time is synonymous with increased efficiency because run time is a real-world construct measured with no appreciable effort. A customer with no software engineering expertise might casually look at his watch when an analytic process is run, look again later when it finishes, and assess the body of work that was completed in some fixed time period. This measurement, however, depicts *execution efficiency*—the speed with which the software executes—as opposed to *efficiency* because the evaluation occurs irrespective of system resources that were used. The following two sections demonstrate two other terms—*efficiency in use* and *production efficiency*—which are also sometimes confused with efficiency.

Efficiency in Use

The ISO software product quality model distinguishes efficiency from *efficiency in use*, "the degree to which specified users expend appropriate amounts of resources in relation to the effectiveness achieved in a specified context of use."[3] Efficiency in use doesn't describe product or process efficiency, but rather the efficiency with which a product is used or implemented. In the case of software products, efficiency in use can describe how efficiently a user is able to utilize the software, or how efficiently the SAS infrastructure can run the software.

To contradistinguish efficiency and efficiency in use, consider the Guatemalan chicken buses, vividly described for the uninitiated in Chapter 4, "Reliability." The fuel efficiency of a chicken bus is defined by its gas mileage (probably around 1 kilometer per liter), but total efficiency of a chicken bus might additionally incorporate resources such as oil, water, brake pads, and of course chrome polish. Efficiency in use, on the other hand, references the manner in which a product—such as software or a chicken bus—is used.

If a passerby screams, "Hey, that chicken bus is half empty! That's inefficient!" he's describing *efficiency in use*, not the *efficiency* of the bus itself. In essence, the chicken bus is being used inefficiently because it could be carrying an additional 100 people, four goats, and six chickens. Software similarly might perform faster in one infrastructure than another, but this doesn't magically make the code more efficient—only its efficiency in use has improved. Or, conversely, if the intent of SAS software is unclear and code is indecipherable, a SAS practitioner might flounder for an hour trying to execute it. This floundering represents inefficiency in use because the resources—personnel time—are wasted; but the floundering says nothing about the efficiency of the software itself or its use of system resources.

Efficiency in use is relevant when optimizing load or data throughput in software that has a finite capacity, with performance or functional failures occurring when that threshold is exceeded, and inefficiency in use occurring when utilization falls beneath an optimal zone. Some industries operate based on efficiency-in-use principles. For example, chicken buses only leave the terminal when they are full—defined by chicken-bus standards as three persons per seat and 60 or so squatting or standing in the aisle. A bus will not depart until it is bursting at the seams in order to maximize profits per trip, thereby ensuring efficiency in use. Thus, a bus with two few people represents inefficiency in use, while too many people (i.e., passengers on the roof) represents failure.

In data analytic software development, efficiency in use commonly describes the efficiency of data throughput—the quantity of data processed at once. And just like chicken buses, ETL processes can utilize efficiency in use to find the sweet spot for processing bundles of data. If a complex extract-transform-load (ETL) process is run on only 100 observations, this may be an inefficient use of the software because it takes 30 minutes to execute, but it doesn't diminish (or even measure) the efficiency of the software itself. The inefficient use results because the same ETL process could have been run on 10,000 observations and utilized minimally more resources despite processing 100 times more data. But, like overcrowded chicken buses, efficiency in use also often has a ceiling above which failure can occur. Thus, at some point the data load will increase to a level at which the software will begin to slow, perform inefficiently, or possibly terminate with functional failures. Only through optimization efforts can efficiency in use be determined and optimized.

Production Efficiency

Another term often conflated with efficiency in SAS literature is *production efficiency* or *productivity*, "the ratio of work product to work effort."[4] Some SAS literature describes personnel time and costs associated with software development as a software resource, but this is inaccurate. In fact, development time and costs are resources to be considered in the determination of software *development* efficiency (or productivity), but not in software operation efficiency. In other words, software might be produced very inefficiently (by unknowledgeable or dispassionate developers who work slowly), yet the software itself might operate efficiently (in that it does a lot while consuming few system resources). Personnel time and costs should only influence software efficiency insofar as they are required to operate software manually once the software is in production.

The tendency to conflate productivity with efficiency occurs primarily within end-user development environments. Where a clear line delineating software development from software operation does not exist, it may be difficult to distinguish efficiency from productivity. For example, consider the end-user developer who spends eight hours empirically developing and debugging analytic software to produce a multivariate analysis to be used in a report. When asked how long it took to run the software, the end-user developer replies "Eight hours." However, the operational component of the project—the final analytic module that was selected—takes only five minutes to run, while the remainder of the eight hours was spent in development activities. But because the end-user developer doesn't distinguish between development and operational activities, productivity and efficiency are conflated. In reality, the productivity of the developer should be measured as the quantity (or quality) of code produced in 7 hours 55 minutes while efficiency of the code would need to evaluate use of system resources while the code runs for five minutes—two entirely different constructs.

DEFINING RESOURCES

Too many times I have been confronted by a developer who excitedly tells me, "I've made this SAS process more efficient!" My immediate question always is, "How do you know it's more efficient?" Nine times out of ten, the response unfortunately is, "Because it's faster!" Yes, the software is faster, but without examining resource consumption, increased efficiency hasn't yet been shown.

Software execution time is the most salient metric of resource utilization because it's concrete, easily measurable, and readily understood by technical and nontechnical stakeholders. But execution time only indirectly reflects utilization of several related yet distinct software resources. Because efficiency describes the minimization of resource usage—not execution time—it's important first to define relevant system resources. While some resources such as network throughput may not be directly measurable through the SAS application, other resources are discussed, including CPU usage, I/O processing, memory consumption, disk space usage, and personnel required for software execution:

- *CPU Time*—CPU time indicates CPU resource usage, akin to the Microsoft Windows Task Manager that depicts CPU usage. SAS logs demonstrate "cpu time" for all completed processes.

- *I/O Processing*—I/O processing reflects effort expended reading and writing files to disk. UNIX environments include I/O metrics such as page faults, page reclaims, and page swaps under the FULLSTIMER system option, although SAS for Windows fails to provide these metrics.
- *Memory*—Memory includes RAM as well as the significantly less efficient virtual memory. The FULLSTIMER memory metric is printed to the SAS log in both Windows and UNIX.
- *Disk Space*—Disk space use is incurred when SAS writes a data set or other file to disk, as well as when temporary data sets, indexes, or other files are created to facilitate SAS processing. File size can be programmatically assessed by querying the Filesize variable in the DICTIONARY.Tables data set.
- *Personnel*—When SAS practitioners must start SAS software manually, babysit jobs as they run, or manually review execution logs, this investment represents use of personnel as a resource. Note that this does not include effort invested in software development—only that in software operations.

While system resources are defined as discrete constructs, they are often so intertwined that untangling them may be difficult. Despite this complexity, by understanding the relationships between system resources, software efficiency can be better understood and achieved.

CPU Processing

Like many applications, SAS measures CPU usage by *CPU time*—the amount of time processors require to complete software tasks. CPU time differs from execution time, displayed as "real time" in the SAS log and representing the actual elapsed time of software execution. For example, the following code and output demonstrate creation of a 10 million-observation data set:

```
data sortme (drop=i);
   length num1 8;
   do i=1 to 10000000;
      num1=rand('uniform');
      output;
      end;
run;
```

```
NOTE: The data set WORK.SORTME has 10000000 observations and 1 variables.
NOTE: DATA statement used (Total process time):
      real time           0.73 seconds
      cpu time            0.73 seconds
```

By comparing CPU time to execution time, developers can assess the efficiency with which software is performing on a system, but not the efficiency of the software itself. For example, the single-threaded DATA step completes in .73 seconds and also only uses .73 seconds of CPU time, thus representing 100 percent execution efficiency. This is not to say that the code itself is efficient, only that it was executed efficiently by the processor.

The SAS Knowledge Base states that execution time and CPU time should be within 15 percent of each other.[5] Thus, when CPU time falls below this 15 percent threshold, this signals that other processes are slowing the specific software task. As the ratio of CPU time to execution time continues to decrease, this often indicates that SAS software is fighting for resources against competing SAS jobs or external processes. For example, if the previous execution time had been 2 seconds while the CPU time remained .73 seconds, this would have demonstrated inefficient execution of the software, likely due to competing processes running concurrently.

The relevance of CPU time is more straightforward in single-threaded processes but becomes more complicated when multithreading or parallel processing is implemented. When multithreaded processes such as the SORT procedure are executed, CPU time represents the cumulative CPU time used by all threads. For example, the following SAS code and output demonstrate a CPU time that exceeds execution time, making the comparison of the two metrics less intuitive:

```
proc sort data=sortme;
   by num1;
run;
```

```
NOTE: There were 10000000 observations read from the data set
WORK.SORTME.
NOTE: The data set WORK.SORTME has 10000000 observations and
1 variables.
NOTE: PROCEDURE SORT used (Total process time):
      real time           3.60 seconds
      cpu time            6.73 seconds
```

Because the CPU time is roughly half of the execution time, it appears that the multithreaded process ran on two processors. This can be validated by setting the CPUCOUNT system option to 2, which produces similar performance:

```
options cpucount=2;
proc sort data=sortme;
   by num1;
run;

NOTE: There were 10000000 observations read from the data set
WORK.SORTME.
NOTE: The data set WORK.SORTME has 10000000 observations and 1 variables.
NOTE: PROCEDURE SORT used (Total process time):
      real time           3.87 seconds
      cpu time            6.24 seconds
```

By running the SORT procedure with the NOTHREADS option, a single-threaded sort is performed which demonstrates comparable CPU time to the multithreaded sort but significantly increased execution time since the SORT procedure was not performed in parallel:

```
options  cpucount=2;
   proc sort data=sortme nothreads;
by num1;
run;

NOTE: There were 10000000 observations read from the data set
WORK.SORTME.
NOTE: The data set WORK.SORTME has 10000000 observations and 1 variables.
NOTE: PROCEDURE SORT used (Total process time):
      real time           6.06 seconds
      cpu time            6.13 seconds
```

When SAS processes are executed in parallel in separate SAS sessions, a faster (but typically less efficient) solution is achieved because equivalent or slightly greater resources are required. From a project management perspective, parallel processing is equivalent to fast-tracking project tasks to reduce the critical path. Resource use is not minimized—just reorganized. The following code and output demonstrates and compares execution time, CPU time, and memory consumption between single-threaded and multithreaded sorts. While the FULLSTIMER output actually lists separate user CPU time and system

CPU time metrics, these metrics are combined into a single "cpu time" metric for simplicity:

```
proc sort data=sortme out=sorted nothreads;
   by num1;
run;

NOTE: There were 10000000 observations read from the data set
WORK.SORTME.
NOTE: The data set WORK.SORTED has 10000000 observations and 1 variables.
NOTE: PROCEDURE SORT used (Total process time):
      real time           6.18 seconds
      cpu time            6.18 seconds
      memory              317016.00k
      OS Memory           345656.00k

proc sort data=sortme out=sorted;
   by num1;
run;

NOTE: There were 10000000 observations read from the data set
WORK.SORTME.
NOTE: The WORK.SORTED data set has 10000000 observations and 1 variables.
NOTE: PROCEDURE SORT used (Total process time):
      real time           3.67 seconds
      cpu time            6.21 seconds
      memory              474140.00k
      OS Memory           501832.00k
```

As the FULLSTIMER statistics demonstrate, while the multithreaded sort did complete significantly faster, it also used more CPU time and more memory, demonstrating the relative inefficiency of the multithreaded option. However, given modern hardware that offers increasingly more processing power and memory, many environments easily dismiss this decreased efficiency in favor of the higher performance that can be achieved through multithreading. This again underscores the importance of contradistinguishing software speed from software efficiency in SAS literature and technical requirements.

Memory

Memory is required to store data so they can be manipulated by SAS processes. For many SAS procedures and DATA steps, as data volume increases,

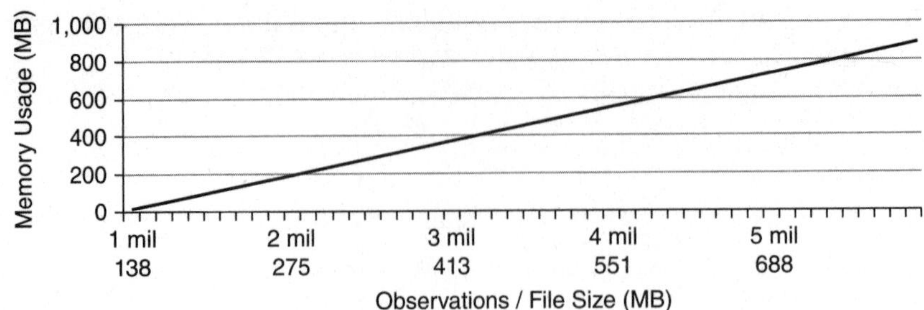

Figure 8.1 Memory Usage by SORT Procedure

memory correspondingly increases in a predictable linear relationship. Figure 8.1 demonstrates the relationship between memory consumption and file size when data sets ranging from 14 MB to 870 MB are sorted with the SORT procedure.

The linear relationship depicted in Figure 8.1 between file size and memory usage will continue until SAS either runs out of memory (producing an out-of-memory error) or begins to utilize virtual memory in place of RAM. Virtual memory temporarily saves data to disk, extending the available memory, but at a tremendous cost of software speed due to additional I/O operations that are required. The switch to virtual memory is apparent due to the dramatic increase in execution time, but it may not be apparent by inspecting memory usage.

For example, while Figure 8.1 depicts a linear relationship of memory consumption, Figure 8.2 demonstrates the corresponding effect that high memory consumption and use of virtual memory play on execution time. At approximately 4.3 million observations, execution time has begun to increase at a higher rate; by 4.5 million observations, the SORT procedure

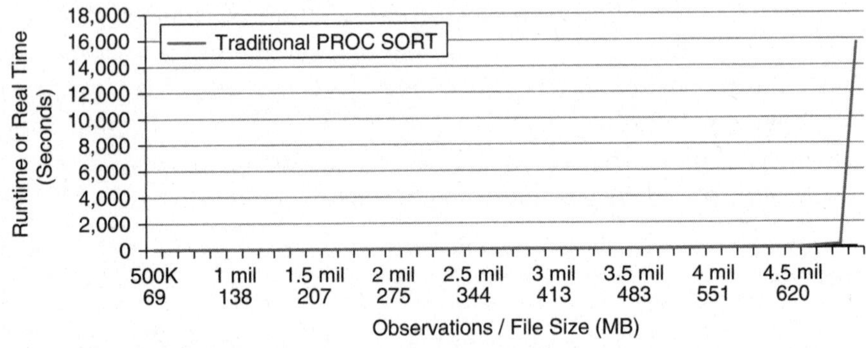

Figure 8.2 Inefficiency Elbow Caused by RAM Exhaustion

requires 4 hours and 20 minutes to complete, more than 200 times slower than would have been predicted by the original linear function!

Figure 8.2 demonstrates an inefficiency elbow typical of decreasing available memory, where executions to the left of the elbow demonstrate efficiency in use and executions to the right of the elbow demonstrate inefficiency in use and a failure of scalability. But the executions to the right of the elbow also demonstrate inefficient software—not because of increased memory utilization, but because of the increase in I/O functions required to accommodate the restricted memory environment. Thus, the number of page faults—demonstrated through FULLSTIMER metrics in UNIX environments—represents I/O usage and increases dramatically beyond the inefficiency elbow. Not all SAS processes will experience an inefficiency elbow since some processes terminate with a failure rather than switching to virtual memory. Only through stress testing, described in the "Stress Testing" section in Chapter 16, "Testability," can SAS practitioners understand both performance failure and functional failure patterns of software.

I/O Processing

I/O processing requires resources to read and write data to disk. I/O processing is often commingled with CPU cycles and memory and may be difficult to disentangle. One clear way to differentiate time spent processing I/O requests versus computational tasks is to utilize the _NULL_ option in a DATA step that has no other input or output. Thus, the DATA step demonstrated in the "CPU Processing" section is reprised here; however, no output is created and the number of observations has been increased to 1 billion:

```
data _null_;
   length num1 8;
   do i=1 to 1000000000;
      num1=rand('uniform');
      output;
      end;
run;
```

```
NOTE: DATA statement used (Total process time):
      real time             36.48 seconds
      user cpu time         34.85 seconds
      system cpu time       0.03 seconds
      memory                750.31k
      OS Memory             32888.00k
```

With no I/O operations required by the _NULL_ option, the computations complete in only 36 seconds, with CPU time taking approximately only 35 seconds, thus representing efficient execution of the code:

```
data sortme (drop=i);
   length num1 8;
   do i=1 to 1000000000;
      num1=rand('uniform');
      output;
      end;
run;

NOTE: DATA statement used (Total process time):
      real time           8:32.74
      user cpu time       1:26.56
      system cpu time     8.25 seconds
      memory              751.59k
      OS Memory           32888.00k
```

However, when I/O operations are required to write the Sortme data set, the CPU time takes four times longer to process, demonstrating that three-quarters of the CPU resources reflect I/O processing (writing the data set), while only 35 seconds (or one-quarter) of CPU resources were used to create the data set. Moreover, note that the efficiency of CPU resources plummeted when I/O processing was incorporated—from 95.6 percent (or 34.9 seconds / 36.5 seconds) to 18.5 percent (or 1 minute 34.8 seconds / 8 minutes 32.7 seconds).

To provide one final example, now that the Sortme data set has been created, it can be used to assess the relative performance of input operations. The following code reads the Sortme data set but writes no output file and performs no other operations:

```
data _null_;
set sortme;
run;

NOTE: There were 1000000000 observations read from the data set
WORK.SORTME.
NOTE: DATA statement used (Total process time):
      real time           1:35.14
      user cpu time       46.03 seconds
      system cpu time     5.95 seconds
      memory              953.21k
      OS Memory           10472.00k
```

CPU time required only 52.0 seconds for the task, but real time required one minute 35.1 seconds, almost twice as long. These performance metrics help demonstrate not only the cost of input operations but also the relative inefficiency. These examples provide a clear indication that software performance (on this system) could be increased through programmatic endeavors to decrease I/O processing, and through nonprogrammatic purchase of hardware to facilitate faster I/O processing. As a general rule, the extent to which reading and writing data sets—especially large ones—can be minimized will significantly improve performance.

Disk Space

Whereas I/O processing represents the speed with which data are written to or read from some location, disk space demonstrates the total storage required by some process. Disk space is differentiated because specific SAS runtime errors occur when SAS runs out of disk space, causing failure. Although some environments may be limited by disk space, due to the decreasing cost of storage, this resource limitation is becoming increasingly uncommon. Still, as the cumulative quantity of data continues to increase in organizations, disk space is a resource that shouldn't be ignored.

When SAS does run out of disk space, it produces a series of runtime errors, as demonstrated in the following output:

```
proc sort data=perm.bigdata out=perm.sorted;
   by char1;
run;

ERROR: No disk space is available for the write operation.  Filename =
       C:\Users\me\AppData\Local\Temp\SAS Temporary
       Files\SAS_util0001000003CC_comp\ut03CC000029.utl.
ERROR: Failure while attempting to write page 1352 of sorted run 15.
ERROR: Failure while attempting to write page 44934 to utility file 1.
ERROR: Failure encountered while creating initial set of sorted runs.
ERROR: Failure encountered during external sort.
ERROR: Sort execution failure.
NOTE: The SAS System stopped processing this step because of errors.
NOTE: There were 80493346 observations read from the data set
SAFESORT.BIGDATA.
WARNING: The data set SAFESORT.SORTED may be incomplete.  When this
step was stopped there were 0 observations and 10 variables.
```

```
NOTE: PROCEDURE SORT used (Total process time):
      real time           9:55.38
      cpu time            2:15.97
```

In this example, SAS has created utility files—temporary files that the SORT procedure uses to perform multithreaded sorting—but the cumulative size of those files (which are saved by default to the WORK library) has exceeded the allowable size of the WORK library. A full WORK library is one of the most common sources of disk space errors and can occur for several reasons:

- Failure to delete contents of WORK between SAS programs
- Extremely large data sets saved to WORK
- Iterative DATA steps saving incrementally named data sets to WORK
- Utility files, the behind-the-scenes temporary files created by SAS

I grew up in a house with a septic tank; you typically don't realize it's full until it's too late, there's a telltale stench emanating from the yard, and the lawnmower starts sinking into brown puddles. The SAS WORK library operates similar to a septic tank and won't give any indication it's approaching its capacity—only when it's full. But despite the lack of warning, there are steps that can be taken to lessen the likelihood of an overflowing disk. One of the simplest solutions is to delete the entire WORK library before or after usage with either the DATASETS or DELETE procedure.

When large data sets are saved to the WORK library, WORK will need to be deleted more often—sometimes multiple times within a single program. This can also occur when software iteratively creates numerous successive data sets in the WORK library in the process of creating some permanent data set. For example, the following development pattern is common (yet not always advisable) within SAS software:

```
data temp1;
   * read/create data;
run;

data temp2;
   set temp1;
   * transform data;
run;

data temp3;
   set temp2;
   * transform data;
run;
```

The development pattern is common because it allows a SAS practitioner to inspect data sets throughout successive processes and because it facilitates manual recreation of later data sets from earlier data sets if errors necessitate that data must be recreated. At the very least, when the ultimate data set in the sequence finally is created (and validated), the prerequisite data sets should be deleted so they don't clutter the WORK library.

Another best practice that prevents the WORK library from filling up and that supports recoverability is to save temporary data sets instead to a shared library so that these can be used as checkpoints in the event of software failure. This is discussed and demonstrated in the "Rerun Program" section in Chapter 5, "Recoverability." For example, data could be saved to PERM.Temp2 (rather than WORK.Temp2), enabling automatic recovery from checkpoints given implementation of appropriate business rules. And, as before, once the final data set is created (and validated), the incrementally named "Temp" data sets can be deleted from the PERM library to minimize disk space usage.

If these commonsense solutions to reducing disk space usage in the WORK library still fail to produce desired results, the WORK library can be permanently associated to a new disk volume. And for disk space errors occurring outside the WORK library, purchasing additional storage is the obvious solution.

Personnel

The time that SAS practitioners spend developing software is a software development resource that should be associated with project or software development costs, but should not be considered an operational resource. However, when personnel are responsible for manually executing software, validating SAS logs, or performing other manual operational functions, this time does represent resource utilization and can be considered when assessing the overall efficiency of a software product. This underscores the importance of software automation as the final step in building production software. Removing the human element from operational software can tremendously improve performance and efficiency, freeing developers to pursue developmental activities rather than repetitive, operational tasks.

EFFICIENCY IN THE SDLC

The most important decision about the role of efficiency in the SDLC is whether to require or monitor efficiency at all. As demonstrated, efficiency is more complex than assessing execution time or CPU time. Moreover, because CPU time, memory, I/O processes, disk usage, and personnel use different units of measurement, it's difficult to provide a multivariate solution that weighs components against each other, let alone that includes multiple resources. For example, if a SAS process is modified so that it uses less memory but now requires more CPU cycles, has it been made more or less efficient? That will ultimately depend on how individual resources are valued by stakeholders.

The value of resources can vary between organizations and stakeholders; over time, as hardware or infrastructure is updated, resource valuation can vary even within an organization. For example, large SAS grid environments may exemplify processing power that feels limitless and that represents a threshold that would be difficult for SAS processes to exceed. In such an environment, SAS practitioners might largely ignore CPU time in efficiency calculations or considerations because of its seemingly limitless capacity. Thus, efficiency might be defined uniquely by each organization or team based on relative resource value, real constraints, imposed limitations, and the overall effect on other performance attributes such as execution speed.

Given the complexity of the multivariate relationships inherent in resource utilization and optimization, the most common method to tackle the efficiency conundrum is to use software execution time as the primary performance metric for efficiency. However, in assessing software speed, it's important to understand the resources that collectively can make software faster or slower and to interrogate these resources individually to drive higher performing software.

Requiring Efficiency

For better or for worse, efficiency requirements are commonly referenced in terms of execution time alone with no mention of resource utilization. An awareness of resource utilization, however, is required—otherwise you're measuring only speed, not efficiency. Thus, requirements for efficiency can reference execution time but must also reference system resource utilization. Where this misnomer exists, requirements documentation should be amended to represent the true quality characteristic that is being discussed, required, or measured—whether speed or efficiency.

For example, consider the customer who specifies only that a SORT procedure must complete in less than ten minutes, but fails to specify any actual efficiency requirements or constraints. The savvy SAS practitioner could sneak out, buy a faster processor, and deliver software that meets this solitary requirement. The process still might consume identical resources, but could complete much faster, thus tremendously improving execution efficiency but not efficiency. Another SAS practitioner might instead perform six simultaneous sorts in parallel, thus implementing a divide-and-conquer technique to achieve several sorted data sets that are subsequently merged into a composite sorted data set. This solution would also be wicked fast, but would consume more resources in a limited time period, thus possibly monopolizing system resources. Again, this solution would also improve execution efficiency, but would actually decrease efficiency due to the increased use of resources.

In each case, the SAS practitioner would have delivered the required speed requirement, but may also have incurred unintended consequences related to resource utilization. In the first example, the cost of new hardware represents additional project (not operational) resources that the customer might not have intended to purchase. In the second example, the monopolization of resources could cause other processes to run slowly or fail if the additional processing power and memory were not available when requested. Thus, efficiency metrics can be required to specify resource utilization thresholds that software should not exceed.

Resource Exception Handling

It can be difficult to estimate CPU, memory, I/O, or disk space usage in planning or design phases, and often these resources can be reliably assessed only once software has been developed. Thus, rather than trying to stab in the dark and risk describing inaccurate resource utilization metrics in requirements documentation, a better approach is to briefly state a commitment to resource monitoring and governance to support robustness and fault-tolerance against resource overuse or exhaustion.

For example, if load and stress testing have demonstrated that a SORT procedure begins to perform inefficiently as the file size approaches 1 GB, a useful requirement might state that file size must not exceed 900 MB, that file size will be dynamically monitored by software, and that exception handling will proactively and automatically split larger files and sort them using divide-and-conquer techniques to prevent inefficient performance. While the 1 GB limitation most likely would not be known during software planning, a proxy value could be used to demonstrate the intent to develop software that will be robust to diminished resources. As development teams become more aware of their specific environments and respective resource utilization and limitations, they can become eerily accurate at predicting resource utilization even in planning and design phases. Data governors that can automatically detect and handle exceptionally large data are discussed in the "Data Governor" section in Chapter 9, "Scalability."

When SAS processes are executed, both FULLSTIMER and NOFULLSTIMER SAS logs demonstrate real time and CPU time, thus enabling a rough estimate of process efficiency to be gauged by dividing the latter into the former. Monitoring this quotient will not demonstrate programmatic efficiency—whether the software is efficiently coded—but can provide an indication of how efficiently the process is executing in the environment. Thus, monitoring these performance metrics over time (by automatically collecting, saving, and analyzing SAS logs) can be used to isolate processes that more often execute inefficiently. Armed with this information, SAS practitioners can highlight processes that are relatively less efficient and can implement either programmatic solutions to improve the performance of those specific processes, or nonprogrammatic solutions to improve the infrastructure as a whole.

Limit Parallel Processing

I once introduced the concept of parallel processing to a SAS team, and in about a month the server was in need of a serious laxative—nothing could get through it no matter how hard we pushed. Even with a small team, we had so many processes running simultaneously that we had effectively used the server's entire bandwidth. While parallel processing, described in Chapter 7, "Execution Efficiency," and Chapter 12, "Automation," should be implemented where software requires expedited processing, SAS practitioners must be cautioned that this concentrates resource utilization and can overwhelm the infrastructure.

When we met to discuss how and why the server was suddenly failing, we discovered that nearly everyone had converted SORT procedures and other processes to run in parallel across sometimes six or more SAS sessions! And, when these parallel processes were running concurrently with other parallel processes, it appeared to the SAS server as though our team size had suddenly quadrupled or sextupled overnight. Although parallel processing is one of the most effective methods to boost execution efficiency, it does not increase efficiency and, if implemented injudiciously, can stall or crash a SAS server. Thus, in some instances, requirements documentation may actually need to explicitly prohibit or limit parallel processing.

For example, in SAS environments that suffer from very real resource constraints, I've encountered teams that limit parallel processing during core business hours (i.e., when the server is more likely to be hit with manually executed jobs) by restricting SAS practitioners to one, two, or three sessions. Thus, you could run a three-pronged divide-and-conquer sort, or you could execute three separate analytic routines, but don't you dare run them all at once. A more exact yet less tenable approach might be to describe the maximum amount of resources that a SAS practitioner should be using at any one time. But, since a valid and useful requirement must be measurable, this more exact solution is often a better theoretical construct to be used for anticipating server load than for actually limiting resource utilization. Moreover, as production software is stabilized, automated, and scheduled, these jobs—which may over time account for the majority of SAS server bandwidth—are run administratively, rather than from user accounts.

Measuring Efficiency

System resources are most effectively measured programmatically with the FULLSTIMER SAS system option, which provides more performance metrics than the default NOFULLSTIMER option. The following example demonstrates FULLSTIMER invoked within the SAS Display Manager for Windows, depicting execution time (real time), CPU time, and memory usage but not I/O resource utilization:

```
options fullstimer;
data perm.sortme (drop=i);
   length num1 8;
   do i=1 to 1000000000;
      num1=rand('uniform');
      output;
      end;
run;
```

```
NOTE: The data set PERM.SORTME has 1000000000 observations and
1 variables.
NOTE: DATA statement used (Total process time):
      real time           2:32.63
      user cpu time       1:29.57
      system cpu time     6.81 seconds
      memory              751.59k
      OS Memory           32888.00k
      Timestamp           04/18/2016 10:40:42 PM
      Step Count                        132337  Switch Count  0
```

A benefit of SAS performance metrics written to the log is their robustness to software failure. If a process terminates prematurely with an error, FULLSTIMER metrics are still produced that demonstrate resource utilization up to the point of failure. This is beneficial when failures occur due to resource-related runtime errors, such as the infamous out-of-memory error. By saving FULLSTIMER metrics to an external file and monitoring that file for not only resource utilization but also runtime errors, SAS practitioners can pinpoint resource utilization levels when failures occur. Awareness of critical resource utilization thresholds is necessary when implementing data governors, as discussed in the "Data Governors" section in Chapter 9, "Scalability."

FULLSTIMER System Option

FULLSTIMER metrics vary by SAS environment, with Windows environments providing far fewer metrics than UNIX environments. Because the SAS University Edition runs on LINUX, when the previous code is executed (on the same box with identical system resources, albeit from this different SAS interface), it produces the following output:

```
options fullstimer;
data perm.sortme (drop=i);
   length num1 8;
   do i=1 to 1000000000;
      num1=rand('uniform');
      output;
      end;
run;

NOTE: The data set PERM.SORTME has 1000000000 observations and
1 variables.
NOTE: DATA statement used (Total process time):
      real time           3:48.17
      user cpu time       2:00.22
      system cpu time     17.82 seconds
      memory              786.28k
      OS Memory           26780.00k
      Timestamp           04/20/2016 06:36:33 AM
      Step Count                        9  Switch Count  46
      Page Faults                       0
      Page Reclaims                     386
      Page Swaps                        0
      Voluntary Context Switches        489961
```

```
Involuntary Context Switches        737
Block Input Operations              0
Block Output Operations             8
```

Not only are additional metrics available, but also note the slower performance and significantly increased CPU utilization within SAS University Edition as compared with the SAS Display Manager for Windows. In most cases, software development and optimization occurs on a single system, so FULLSTIMER metrics will be consistently represented. However, as demonstrated in the "Contrasting the SAS University Edition" section in Chapter 10, "Portability," significant differences exist between the performance of the SAS University Edition and other SAS interfaces, which can be demonstrated through FULLSTIMER metrics.

In the next example, a subset of the 1 billion-observation Sortme data set is created, but the WHERE option is mistakenly placed in the DATA step statement rather than the SET statement. This requires the SET statement to read all observations, after which only approximately 10 percent are output to the Subset data set:

```
data perm.subset (where=(num1<.1));
   set perm.sortme;
run;
```

```
NOTE: There were 1000000000 observations read from the data set
PERM.SORTME.
NOTE: The data set PERM.SUBSET has 99994212 observations and 1 variables.
NOTE: DATA statement used (Total process time):
      real time           9:46.37
      user cpu time       2:38.26
      system cpu time     6.33 seconds
      memory              961.18k
      OS Memory           32888.00k
```

The revised code produces the same functionality, but more efficiently places the WHERE option in the SET statement, thus only reading approximately 10 percent of the Subset data set. Efficiency is improved due to the reduced CPU cycles required, but because the variable Num1 still has to be interrogated to determine if each observation should be included, memory consumption remains identical between the two methods:

```
data perm.subset;
   set perm.sortme (where=(num1<.1));
run;
```

```
NOTE: There were 99994212 observations read from the data set
PERM.SORTME.
      WHERE num1<0.1;
NOTE: The data set PERM.SUBSET has 99994212 observations and 1 variables.
NOTE: DATA statement used (Total process time):
      real time           6:57.14
      user cpu time       56.58 seconds
      system cpu time     9.93 seconds
      memory              961.59k
      OS Memory           32888.00k
```

Although not demonstrated in the SAS Display Manager output, when the two methods are compared in the SAS University Edition, the more efficient solution also clearly demonstrates the reduced I/O resources utilized, given that only 10 percent of the data were ingested into the data set. The inefficient code requires almost twice as many page reclaims. However, if a SAS practitioner were not aware of the best practice of including the WHERE option in the SET statement (as opposed to the DATA step statement), he might have considered the first solution efficient. Thus, when optimizing SAS software, a solution often isn't considered inefficient until it is outperformed either in theory or practice.

This underscores the relativity of measuring and assessing efficiency. Solutions are rarely decried as "the most efficient way ever" but rather as "more efficient," "less efficient," or "efficient enough." Thus, if the first (inefficient) solution were in use in SAS software, and stakeholders were pleased with the software's performance and efficiency, it probably wouldn't be judged as inefficient because it met the business need. However, when compared to the second (optimized) code, it is evident that the first solution is relatively inefficient compared to the second because it fails to incorporate the SAS software development best practice of WHERE placement.

FULLSTIMER Automated

A common goal of production software is to maximize operational efficiency by removing the human element and automating SAS jobs. When software is automated, log parsing (and validation) also should be automated to eliminate tedious, error-prone, manual review of the SAS log. Numerous texts demonstrate automated methods to parse SAS logs, thus freeing SAS practitioners to pursue more creative and meaningful endeavors. Most examples in literature, however, use the PRINTTO procedure to write the entire log to a text file, after which it is parsed when software terminates. For many purposes, this is sufficient; however, it does not enable dynamic processing based on log results.

A second goal of production software is often to embed an exception handling framework that aims to increase the robustness and fault tolerance of software through dynamic, data-driven, fuzzy logic. For this reason, FULLSTIMER metrics might need to be accessed immediately after a process has completed so they can dynamically alter program flow (if needed). For example, if FULLSTIMER detects that a process used too much memory or that the ratio of CPU time to execution is below some established threshold, program flow can be altered due to inefficient or overutilization of system resources, by sending an alert email to stakeholders, by automatically terminating lower-priority jobs, or through other methods.

The following %READFULLSTIMER macro ingests a text file that includes FULLSTIMER log output and subsequently assigns global macro variables to values found in the FULLSTIMER metrics. For example, when "real time" is detected in the file, the global macro variable &REALTIME is assigned to the value. Because some UNIX FULLSTIMER metrics are not available in the Windows environment, missing metrics are assigned a value of –9:

```
%macro readfullstimer(textfile=);
* converts all times to seconds SS.xx format from HH:MM:SS.ss format;
%let syscc=0;
%global fullstimerRC;
%global realtime;
%global usercputime;
%global systemcputime;
%global memory;
%global osmemory;
%global stepcount;
%global switchcount;
%global pagefaults;
%global pagereclaims;
%global volcontextswitches;
%global involcontextswitches;
%global blockinops;
%global blockoutops;
%global errtextlong;
%let fullstimerRC=;
%let realtime=;
%let usercputime=;
%let systemcputime=;
```

```
%let memory=;
%let osmemory=;
%let stepcount=;
%let switchcount=;
%let pagefaults=;
%let pagereclaims=;
%let volcontextswitches=;
%let involcontextswitches=;
%let blockinops=;
%let blockoutops=;
%let errtextlong=;
data _null_;
   length tab $100 errtextlong $2000 erryes 8;
   infile "&textfile" truncover;
   input tab $100.;
   if _n_=1 then errtextlong='';
   if _n_>=5 then do; *** needs to be made dynamic based on which
         version of FULLSTIMER is used;
      if find(tab,"ERROR")>0 or find(tab,"WARNING")>0 then erryes=1;
      else if find(tab,"real time")>0 then erryes=0;
      if erryes=1 then errtextlong=strip(errtextlong) || ' *** ' ||
         strip(tab);
      call symput('errtextlong',strip(errtextlong));
      if lowcase(substr(tab,1,9))='real time' then do;
         if count(scan(substr(tab,10),1,' '),':')=0 then call
            symput('realtime',scan(substr(tab,10),1,' '));
         else if count(scan(substr(tab,10),1,' '),':')=1 then call
            symput('realtime',(input(strip(scan(substr(tab,10),
            1,':')),8.0) * 60) +
            input(strip(scan(substr(tab,10),2,':')),8.2));
         else if count(scan(substr(tab,10),1,' '),':')=2 then call
            symput('realtime',(input(strip(scan(substr(tab,10),
            1,':')),8.0) * 3600) +
            (input(strip(scan(substr(tab,10),2,':')),8.0) * 60) +
            input(strip(scan(substr(tab,10),3,':')),8.2));
         end;
      else if lowcase(substr(tab,1,13))='user cpu time' then do;
         if count(scan(substr(tab,14),1,' '),':')=0 then call
            symput('usercputime',scan(substr(tab,14),1,' '));
         else if count(scan(substr(tab,14),1,' '),':')=1 then call
            symput('usercputime',(input(strip(scan(substr(tab,14),
            1,':')),8.0) * 60) +
            input(strip(scan(substr(tab,14),2,':')),8.2));
```

```
      else if count(scan(substr(tab,14),1,' '),':')=2 then call
         symput('usercputime',(input(strip(scan(substr(tab,14),
         1,':')),8.0) * 3600) +
         (input(strip(scan(substr(tab,14),
         2,':')),8.0) * 60) +
         input(strip(scan(substr(tab,14),3,':')),8.2));
      end;
   else if lowcase(substr(tab,1,15))='system cpu time' then do;
      if count(scan(substr(tab,16),1,' '),':')=0 then call
         symput('systemcputime',scan(substr(tab,16),1,' '));
      else if count(scan(substr(tab,16),1,' '),':')=1 then call
         symput('systemcputime',(input(strip(scan(substr(tab,16),
         1,':')),8.0) * 60) +
         input(strip(scan(substr(tab,16),2,':')),8.2));
      else if count(scan(substr(tab,16),1,' '),':')=2 then call
         symput('systemcputime',(input(strip(scan(substr(tab,16),
         1,':')),8.0) * 3600) +
         (input(strip(scan(substr(tab,16),2,':')),8.0) * 60) +
         input(strip(scan(substr(tab,16),3,':')),8.2));
      end;
   else if lowcase(substr(tab,1,6))='memory' then call
      symput('memory',scan(substr(tab,7),1,' k')/1024); *convert
      KB to MB;
   else if lowcase(substr(tab,1,9))='os memory' then call
      symput('osmemory',scan(substr(tab,10),1,' k')/1024);
      *convert KB to MB;
   else if lowcase(substr(tab,1,10))='step count' then do;
      call symput('stepcount',scan(substr(tab,11),1,' '));
      call symput('switchcount',scan(substr(tab,11),4,' '));
      end;
   else if lowcase(substr(tab,1,11))='page faults' then call
      symput('pagefaults',scan(substr(tab,12),1,' '));
   else if lowcase(substr(tab,1,13))='page reclaims' then call
      symput('pagereclaims',scan(substr(tab,14),1,' '));
   else if lowcase(substr(tab,1,26))='voluntary context switches'
      then call symput('volcontextswitches',scan(substr(tab,27),
         1,' '));
   else if lowcase(substr(tab,1,28))=
      'involuntary context switches' then call
      symput('involcontextswitches',scan(substr(tab,29),1,' '));
   else if lowcase(substr(tab,1,24))='block input operations' then
      call symput('blockinops',scan(substr(tab,25),1,' '));
```

```
      else if lowcase(substr(tab,1,25))='block output operations'
         then call symput('blockoutops',scan(substr(tab,26),1,' '));
      end;
   retain erryes errtextlong;
run;
* set all missing values to -9;
%let macrolist=realtime usercputime systemcputime memory osmemory
   stepcount switchcount pagefaults pagereclaims volcontextswitches
   involcontextswitches blockinops blockoutops;
%let i=1;
%do %while(%length(%scan(&macrolist,&i,,S))>1);
   %let mac=%scan(&macrolist,&i,,S);
   %put MAC: &mac &&&mac;
   %if %length(&&&mac)=0 %then %let &mac=-9;
   %let i=%eval(&i+1);
   %end;
%let errtextlong=%sysfunc(compress("&errtextlong",,kns));
%if &syscc>0 %then %do;
   %let fullstimerRC=FAILURE;
   %return;
   %end;
%mend;
```

For example, to implement the %READFULLSTIMER macro on the DATA step that creates PERM.Subset, the following code is executed.

```
libname perm 'c:\perm';
%let path=%sysfunc(pathname(perm));
options fullstimer;
proc printto log="&path/out.txt" new;
run;
data perm.subset;
   set perm.sortme (where=(num1<.1));
run;
proc printto;
run;
%readfullstimer(textfile=&path/out.txt);
```

Because the SAS log is saved to an external file Out.txt, and because that file may be repetitively overwritten by successive invocations of %READFULLSTIMER, exception handling inside the %READFULLSTIMER macro captures runtime errors that may occur. Given this framework, software can respond dynamically to real-time resource utilization by assessing the global macro variables that are created. These global macro variables can also be saved in a historical data set for longitudinal analysis of software performance.

WHAT'S NEXT?

Execution efficiency and *efficiency* describe rapid performance and appropriate utilization of system resources. But in data analytic environments, often the greatest source of variability in production software are the data themselves. As data volume or velocity increase, the performance and efficiency of software can wane, demonstrating a lack of scalability. Chapter 9, "Scalability," demonstrates scalable software that supports increased data volume, demand, and hardware infrastructure.

NOTES

1. ISO/IEC/IEEE 24765:2010. *Systems and software engineering—Vocabulary*. Geneva, Switzerland: International Organization for Standardization, International Electrotechnical Commission, and Institute of Electrical and Electronics Engineers.
2. IEEE Std 26513-2013. *IEEE standard for adoption of ISO/IEC 26513:2009 systems and software engineering—Requirements for testers and reviewers of documentation*. Geneva, Switzerland: Institute of Electrical and Electronics Engineers.
3. ISO/IEC 25010:2011. *Systems and software engineering — Systems and software Quality Requirements and Evaluation (SQuaRE)—System and software quality models*. Geneva, Switzerland: International Organization for Standardization and Institute of Electrical and Electronics Engineers.
4. ISO/IEC/IEEE 24765:2010.
5. "FULLSTIMER SAS Option." *SAS® Knowledge Base/Focus Areas*. Retrieved from http://support.sas.com/rnd/scalability/tools/fullstim/.

CHAPTER 9
Scalability

Standing in line at the Hedman Alas office in Tegucigalpa, Honduras, I was anxious to leave city life and head to colonial, cosmopolitan Granada, Nicaragua. Only 275 miles away, the "direct" route would be one of my lighter travel days, but I'd booked a *primera clase* (first class) bus with reclining seats, snacks, moist towelettes, and even a bathroom. Most importantly, Hedman Alas doesn't allow chickens!

I had hoped to catch a bus to Leon, Nicaragua, but there were none (as well as no one who spoke English at the office). I understood we'd be departing to Granada in the morning—and very few other details.

While other Hedman Alas buses had played Spanish-speaking movies with English subtitles, within seconds of pulling out of the terminal, the *conductor* (bus driver) blasted an all-ABBA soundtrack that persisted throughout the voyage. Painful as it was, we were making good time, and an afternoon arrival in Granada seemed imminent.

But at the intersection of CA1 (Central American Highway 1)—or the Pan-American Highway—in Jicaro Galan, the Mamma Mia tour bus first slowed then crawled into a dusty lot. No, we hadn't broken down, but it was announced we'd be stopping for a few hours. *¿¡Que!?*

Busing to Tegucigalpa a few days earlier we'd passed the same remote, rundown, rusted-out hotel—obviously closed for decades—devoid of any signs of life. Despite the bleak façade, however, the Hotel Oasis Colonial was magnificent inside, and definitely aptly named.

A full-service bar, parrots squawking about, and palm trees all framing a large, pristine pool—it took me all of two seconds to order the first round of *rum y Coca-Colas*, request fresh towels, and strip down to something remotely swimmable.

But after a relaxing couple of hours, someone fortunately noticed the lone gringo was missing from the bus as it was departing and, sopping wet, I jumped through the open door as it pulled away—which reminds me, I owe the Oasis a towel…

Already substantially delayed, we passed through Nicaraguan customs as dusk was falling, yet the pace continued to plummet. Despite the so-called "direct" route, we seemed to be dropping off passengers every few miles at rural, unmarked destinations. And because everyone had stowed luggage beneath the bus, we were delayed at each stop while they opened the compartment and rummaged about.

As we pulled into Leon, which I hadn't realized was on the route and where I'd originally wanted to stop, my GPS now predicted us reaching Granada sometime around midnight—a far cry from my hopeful afternoon arrival. I made the split-second decision to disembark and, in a headlong dash, backpack in tow, sprinted toward the front of the bus and escaped just as narrowly as I had entered it hours earlier. Far better to be searching for a hostel at 9 PM in Leon than midnight in Granada!

■ ■ ■

Despite various scalability challenges during my trip, the outcome was positive. But scalability issues can be far more pernicious to data analytic development if left unchecked.

The first scalability challenge arose from my inability to process Honduran Spanish—significantly faster than the Guatemalan Spanish—at the Hedman Alas ticket counter. Because of the rapidity, I failed to understand our route, stops, trip duration, and the fact that I could easily disembark in Leon or anywhere along the way. Data velocity can also pose a challenge to data analytic software and, as the pace of a data stream picks up, software must scale effectively to that rate.

At each seemingly unplanned roadside stop, scalability again reared its head as the *conductor* was forced to dig through a mountain of luggage to locate a traveler's bag. As the night wore on and the busgoers slowly evaporated, the speed of locating luggage gradually increased. A host of data operations such as data set joins or lookup tables will also vary in performance based on the size of the data being accessed or searched.

A final scalability issue that the trip presented was the overall size of our traveling party—an entire busload. It slowed us at the Oasis because we had to wait hours to exchange a few passengers with another bus. It slowed us at the Honduran–Nicaraguan border because we had to get in line behind other buses. And it slowed us in Nicaragua because dozens of people expected to be dropped off at nondescript patches of highway. A car, taxi, or any smaller vehicle could have made the journey in one-fourth the time.

Data processes too can experience decreased performance as data sets increase in size, resulting in slowed processing, inefficiency, or even functional failures. Arriving in Leon hours behind schedule only represented a performance failure—arriving late—which was easily remedied when I immediately found a room and Aussie *compañeros* (companions) in the first hostel. But had I arrived in Granada at midnight and had to sleep in a park, I definitely would have considered this to have been a functional failure.

In my trek, I avoided this potential functional failure by getting off the bus earlier than anticipated, and by hitching and busing to Granada via Managua a couple days later. Sometimes in data analytic environments, big data also need to be chunked into pieces—like my trek—to improve performance or reduce the possibility of failure. I would ultimately look back at my split-second decision to dart off the bus in Leon as one of the best decisions of my entire adventure.

DEFINING SCALABILITY

Scalability is "the degree to which assets can be adapted to support application engineering products for various defined measures."[1] It describes the ability of a system or software to effectively and efficiently adapt to and incorporate additional resources, demand, or load (i.e., throughput). Common system resources include memory, processing power, and disk space, and are often scaled through distributed processing environments such as grid architecture. Demand must scale when software is utilized by multiple users, either running the same software concurrently or distributed copies of software in a shared environment. And, especially relevant to SAS data analytic development, scaling load reflects software that can process data of increasing volume, velocity, or variability—the so-called 3*V*s of big data.

Scalability is important because it demonstrates the adaptability of software effectiveness, efficiency, and other performance to environmental variability. Software may be effective in a test environment with smaller sample data sets, but through scalability, software can be demonstrated to still perform efficiently with higher actual data volumes. And, as data may vary or increase throughout the software lifespan, scalable software will continue to demonstrate efficient performance. Demand scalability also facilitates parallel processing because it enables multiple versions of software to run concurrently.

This chapter introduces the concept of scalability as it relates to system resources, demand, and load. Resource scalability is briefly introduced but, because it is largely configuration-dependent, is not described in detail. Demand scalability concepts describe the encapsulation of software to enable multiple copies to be run concurrently without interference, as well as the alternative goal of facilitating communication between concurrently running software through intentional interaction. The majority of the chapter, however, focuses on facilitating scalable data analytic processes that can competently tackle high-velocity and high-volume data.

THE SCALABILITY TRIAD

Scalability can refer to three distinct concepts, each posing respective opportunities and challenges. A customer discussing only a "scalable software solution" might be referencing scalable hardware in a distributed network environment that will support the software. But *scalability* just as easily could reference software scaling to increased data volume. For this reason, the chapter is divided into resource scalability, demand scalability, and load scalability.

Some of the confusion surrounding scalability erupts because it describes both programmatic and nonprogrammatic methods. For example, if load scalability is required to facilitate SAS extract-transform-load (ETL) software that can process larger data sets, this can often be achieved through a hardware solution that increases SAS processing power in a distributed network environment. In many organizations, however, SAS practitioners may be powerless to upgrade hardware and infrastructure, thus only programmatic solutions can be implemented. While scalability principles in software development can increase software's effectiveness to process larger data loads, maximum scalability can only be achieved through solutions that incorporate both programmatic and nonprogrammatic aspects.

Because this text focuses on programmatic methods and best practices, resource scalability is only briefly introduced. Myriad hardware-centric solutions do exist that enable SAS to more efficiently leverage available resources, as well as to incorporate additional resources to support faster processing, greater throughput, or increased users. Some of these solutions are SAS-centric, such as the SAS Grid Manager, while others represent hardware hybrids that leverage third-party infrastructure and software, such as Teradata or Hadoop.

The remaining two dimensions of scalability—demand and load—are discussed more thoroughly in the chapter. Demand scalability can enable SAS software to run concurrently over several SAS sessions, maximizing the use of available system resources. In some cases, the intent is to maintain distance between users running separate software instances so that they don't interfere with each other. In other cases, the intent conversely is to enable all software instances to collaborate in parallel to accomplish some shared task. In all cases, the intent of demand scaling is to enable SAS software to operate in more diverse environments while maintaining or improving performance.

Load scalability is often a challenge first bridged as software transitions from using test data to the increased volume and complexities of production data. While *load testing* describes testing data analytic software with the expected data volume, *stress testing* aims to determine when software performance will begin to decrease or fail due to increased load. Thus, if tested thoroughly, the scaling of software to increased throughput should occur in a predictable pattern demonstrated through load and stress testing. Load and stress testing are described in the "Load Testing" and "Stress Testing" sections, respectively, in Chapter 16, "Testability."

RESOURCE SCALABILITY

Software performance is typically governed by available system resources such as processors, memory, and disk storage. Resource scalability aims to increase performance through the addition of system resources, but ineffective implementation can limit the benefits of system and hardware upgrades. Thus, a critical component of resource scalability is directing how those resources will interact to achieve maximized benefit. For example, an additional SAS server and accompanying processors could promote high availability and system failover to facilitate a more reliable system, or those same resources could instead be used to extend processing power through parallel processing and load balancing. Configuration management and optimization will ultimately determine to what extent additional resources will support scalability objectives, which sometimes can compete against reliability objectives for resources.

Distributed SAS Solutions

The SAS Grid Manager supports software scalability by facilitating high availability through software failover. For example, if SAS software running on one machine fails due to a hardware fault, it will automatically restart on another machine from the last successful checkpoint, thus improving recoverability. Service level agreements (SLAs) for SAS software supporting critical infrastructure will often state availability requirements that must be maintained above some threshold. Because even robust software cannot overcome hardware failure, a hardware solution is necessary when high availability is required. The SAS Grid Manager is the software platform that can organize and optimize those hardware components.

In addition to supporting high availability, SAS Grid Manager also supports load balancing of distributed processing. Large SAS jobs can be deconstructed and run in parallel across multiple processors for increased performance. This is similar to the parallel processing demonstrated in the "Executing Software Concurrently" section in Chapter 7, "Execution Efficiency," but is optimized to ensure that individual jobs are efficiently allocated across machines.

With the release of SAS 9.4M3, the SAS Grid Manager for Hadoop was also made available for environments that leverage Hadoop, an open-source framework that runs across clusters of similarly configured commodity computers.

Other third-party software solutions include incorporation of in-database products such as Teradata, Netezza, Oracle, and Greenplum. While tremendous performance improvement can be gained through these systems, this chapter focuses on programmatic methods to scale Base SAS software.

DEMAND SCALABILITY

Demand scalability describes the ability of software to run concurrently in the same or across different environments. With some software, especially web applications, the goal of scalability is to enable as many users as possible to use the application without affecting each other's function or performance. For example, millions of users can run simultaneous Google searches without one user's search affecting another's search. A scalable solution would also dictate that a high volume of searches not slow the processing of individual searches. In the SAS world, software tools might be similarly designed so that multiple users could run separate instances of the same software simultaneously while achieving dynamic results based on varying input or customizable settings. This version of demand scalability is described later in the section "The Web App Mentality."

In other cases, a distributed solution is sought in which SAS software is coordinated and executed in parallel across multiple processors. This can be accomplished through solutions that scale and manage resources, such as the SAS Grid Manager, that can enable one SAS job to be distributed across processors. But SAS software can also be written so that multiple instances can be run simultaneously to achieve a collective solution. For example, several instances of ETL software might be run in parallel with each instance processing a different data stream or performing different tasks to achieve maximum performance and resource utilization. This type of distributed processing requires significant communication and coordination between all sessions to ensure that tasks are neither skipped nor duplicated, and is demonstrated in the "False Dependencies of Data Sets" section in Chapter 7, "Execution Efficiency."

The Web App Mentality

The majority of SAS software is intended to be run by a single user or scheduled to be run as a single batch job. This common usage is largely fed by the notion that software should perform some function or series of tasks that reliably produces identical results each time it is run. By incorporating principles of software flexibility and reusability, however, software can be made to be much more dynamic, enabling functionality to be altered based on parameterized input and user customization. However, with software reuse comes the possibility that two instances of a program will be simultaneously executed by separate users. Thus, where software modules or programs are designed for reuse, they should be encapsulated sufficiently to prevent cross-contamination if multiple instances are run. This perspective is found commonly in other software development, and typifies web application development in which software is intended to be distributed and run by thousands or millions of users concurrently.

Often in data analytic development, various data models and statistical models are empirically tested in succession. Software is executed with one model, after which refinements are made and the software is rerun. The following code simulates this type of analytic development with a malleable data transformation module Transform.sas that is being refined:

```
* data ingestion module ingest.sas;
libname perm 'c:\perm';
data perm.mydata (drop=i);
   length num1 8;
   do i=1 to 1000;
      num1=round(rand('uniform')*10);
      output;
      end;
run;

* data transformation module transform.sas;
data perm.transformed;
   set perm.mydata;
   length num1_recode $10;
   if num1>=7 then num1_recode='high';
   else num1_recode='low';
run;
```

Another analyst may want to reuse the transformation module but want to apply a different data model—for example, by recoding NUM1 values of 5 (rather than 7) or greater as "high." If only one central instance of the software exists, then this modification would overwrite the current logic, making it difficult for multiple analysts to utilize the software for their individual needs. But even if a copy of the software is made and only that single line of business logic is modified, the PERM.Transformed data set references a shared data set that would be overwritten. To overcome this limitation and enable parallel versions of the software to be run simultaneously without conflict, the Transformed.sas program must dynamically name the created data set and must dynamically transform the data.

The following business logic can be separated from the code and saved to the SAS file C:\perm\translogic1.sas. The %TRANSLOGIC1 macro can be maintained and modified by one SAS practitioner while other users maintain and customize their own data models:

```
%macro translogic1;
length num1_recode $10;
if num1>=7 then num1_recode='high';
else num1_recode='low';
%mend;
```

The following code, saved as C:\perm\transform.sas, can be used to call the %TRANSLOGIC1 macro or other business logic macros dynamically, as specified in the %TRANSFORM macro invocation:

```
* data transformation module;
%macro transform(dsnin=, dsnout=, transmacro=);
%include "C:\perm\&transmacro..sas";
data &dsnout;
   set &dsnin;
   %&transmacro;
run;
%mend;
```

The %TRANSFORM invocation specifies that the %TRANSLOGIC1 business logic should be used. In a production environment, rather than specifying parameters in a macro invocation, the SYSPARM option could be utilized to pass parameters to the &SYSPARM automatic macro variable in spawned batch jobs:

```
%transform(dsnin=perm.mydata, dsnout=perm.transformed,
   transmacro=translogic1);
```

With these improvements, multiple instances of the software can now be run in parallel because static aspects that would have caused interference have been parameterized. Moreover, the software is more reusable because it can produce varying results based on business rules that are dynamically applied. And, because essentially only one instance of the software is being utilized, maintainability is substantially increased.

To continue this example, the %MEANS macro is saved as C:\perm\means .sas and simulates an analytic process that would be more complex:

```
* saved as means.sas;
%macro means;
ods listing close;
ods noproctitle;
ods html path="c:\perm" file="means.htm" style=sketch;
proc means data=perm.mydata;
   var num1;
run;
ods html close;
ods listing;
%mend;
```

If multiple instances of the software are executed in parallel, however, the HTML files will still overwrite each other. A more robust and reusable solution would be to implement dynamism into software so that the software can be reused with modification only to customizable parameters, but not to the code itself. This concept is discussed in Chapter 17, "Stability," in which more flexible software supports code stability and diverse reuse. The updated code now requires three parameters in the macro invocation, again simulating parameters that could be passed via SYSPARM in a batch job:

```
%macro means(dsnin=, htmldirout=, htmlout=);
ods listing close;
ods noproctitle;
ods html path="&htmldirout" file="&htmlout" style=sketch;
proc means data=&dsnin;
   var num1;
run;
ods html close;
ods listing;
%mend;

%means(dsnin=perm.mydata, htmldirout=c:\perm, htmlout=means.htm);
```

Several instances of the Means.sas code can now be executed in parallel without interference. This method is useful when competing data models are being examined simultaneously. For example, analysts can run the same software—albeit with different customizations—to produce varying output while developers can maintain a single version of the software independently.

This design paradigm is discussed and demonstrated in the "Custom Configuration Files" section in Chapter 17, "Stability." Use of the SYSPARM option and corresponding &SYSPARM automatic macro variable are described in the "Passing Parameters" section in Chapter 12, "Automation."

Distributed Software

A more common implementation of demand scalability is used to run SAS software across multiple processors through load distribution. The SAS Grid Manager, discussed in the previous "Distributed SAS Solutions" section, accomplishes this by splitting code among multiple processors. However, if a SAS environment has not paid to license SAS Grid Manager, SAS practitioners can still implement distributed solutions by initiating multiple instances of one program in parallel. While these instances will not have the added benefits of optimized load balancing and high availability that SAS Grid Manager offers, they will benefit from increased resource utilization and faster performance.

Because this use of demand scalability effectively breaks software into modules that can be run concurrently, a tremendous amount of communication and coordination is required to ensure that tasks are neither skipped nor duplicated. Some of the threats to this type of parallel processing are discussed in the "Complexities of Concurrency" section in Chapter 11, "Security." For example, each module must be validated to demonstrate that it completed successfully, and to ensure that other instances of the software will not rerun and duplicate the same task. These validations can also serve as checkpoints that demonstrate the last process to execute successfully. This type of distributed processing typically also increases software recoverability because software can be restarted from checkpoints automatically rather than from the beginning. This functionality is demonstrated in the "Recovering with Checkpoints" section in Chapter 5, "Recoverability."

Parallel Processing versus Multithreading

Parallel processing broadly represents processes that run simultaneously across multiple threads, cores, or processors. Multithreading is a specific type of parallel processing in which multiple threads are executed simultaneously to perform discrete tasks concurrently or divide and conquer a single task. Base SAS includes multithreading capabilities in several procedures to maximize performance and resource utilization. The "Support for Parallel Processing" chapter of *SAS 9.4 Language Reference Concepts: Third Edition*, lists the following SAS procedures as supporting multithreading, starting with SAS versions 9:[2]

- SORT
- SQL
- MEANS
- REPORT
- SUMMARY
- TABULATE

SAS practitioners have limited control over threading on these multithreaded procedures by specifying the THREADS system option, by specifying the THREADS procedure option, or by increasing the value of the CPUCOUNT system option. SAS practitioners unfortunately cannot harness multithreading when building their own SAS software, unlike object-oriented programming (OOP) languages. Notwithstanding this limitation of Base SAS software, SAS practitioners can benefit by understanding how multithreading increases performance and by replicating aspects of these techniques using parallel processing.

For example, when the SORT procedure runs with multithreading enabled, SAS divides the initial data set into numerous smaller data sets, sorts these in parallel, and finally joins the sorted subsets into a composite, sorted data set. Figure 9.1 depicts the modest performance improvement that the SORT procedure achieves through multithreading on two CPUs. On systems that were tested, increasing CPUCOUNT beyond two CPUs failed to increase performance further.

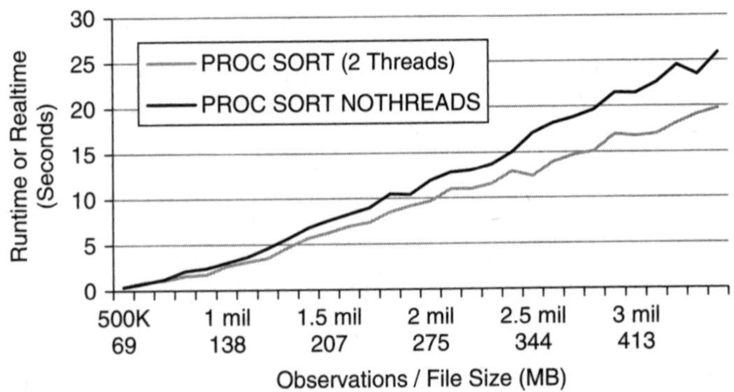

Figure 9.1 Multithreaded versus Single-Threaded SORT Procedure

Multithreading facilitates the optimization of threads by the SAS application by evenly distributing threads across processors. Moreover, because only one instance of the SAS application is running, system resource overhead is minimized. Notwithstanding, SAS practitioners can improve performance of the SORT procedure by running multiple procedures in parallel. For example, the following code simulates parallel processing of the SORT procedure for the PERM.Mydata data set, which has 1,000 observations:

```
* first sort module to run in parallel;
proc sort data=perm.mydata (firstobs=1 obs=500) out=sort1;
   by num1;
run;

* second sort module to run in parallel;
proc sort data=perm.mydata (firstobs=501 obs=1000) out=sort2;
   by num1;
run;

data sorted;
   set sort1 sort2;
   by num1;
run;
```

An initial process (not shown) would determine how many parallel processes to run (based on file size or observation count), after which two (or more) SAS sessions could be spawned with the SYSTASK statement to perform the sorting. After the batch jobs had completed, a final DATA step in the original SAS program would rejoin the two sorted halves for a faster solution than the native SORT procedure.

Two primary methods exist for implementating parallel processing in SAS software. The first, as described previously, uses an engine (AKA controller or driver) to spawn multiple SAS sessions to execute external code in parallel. For example, the values for FIRSTOBS, OBS, and the output data set name for each SORT procedure would be dynamically encoded so that external code (not shown) could be called with SYSTASK and executed. The second method (demonstrated in the following section) uses software that is actually designed to have two or more instances run simultaneously so that they can work together to accomplish some shared task (like sorting) faster.

Distributed SORT-SORT-MERGE

The SORT-SORT-MERGE represents the common task of joining two data sets by some unique key or keys. The process can be accomplished using the

SQL procedure or hash object but is often performed by two successive SORT procedures followed by a DATA step that joins the two data sets:

```
proc sort data=temp1;
   by num1;
run;

proc sort data=temp2;
   by num1;
run;

data sorted;
   merge temp1 temp2;
   by num1;
run;
```

This code creates a false dependency, described in the "False Dependencies of Process" section in Chapter 7, "Execution Efficiency," because the sorts are not dependent on each other yet are performed in series rather than parallel. Especially where the two data sets are equivalent in size and take roughly the same amount of time to sort, the speed of the SORT-SORT-MERGE can be increased by performing the two sorts in parallel. To coordinate this, SAS software should be written that is dynamically capable of sorting either Temp1 or Temp2 based on whichever data set has not been sorted. Thus, when two instances of the software are executed simultaneously, one will sort Temp1 and the other—detecting that Temp1 has already commenced its sort—will sort Temp2. When both SORT procedures complete, one of the software instances will perform the merge while the other patiently waits or exits. A similar technique is demonstrated in the "Executing Software in Parallel" section in Chapter 7, "Execution Efficiency," to perform independent MEANS and FREQ procedures in parallel.

The following code initializes the PERM.Data1 and PERM.Data2 data sets—each of which contains 100 million observations—that will be individually sorted and subsequently merged:

```
libname perm 'C:\perm';
data perm.data1 (drop=i);
   length num1 8 char1 $10;
   do i=1 to 100000000;
      num1=round(rand('uniform')*1000);
      char1='data1';
```

```
      output;
      end;
run;

data perm.data2 (drop=i);
   length num1 8 char2 $10;
   do i=1 to 100000000;
      num1=round(rand('uniform')*1000);
      char2='data2';
      output;
      end;
run;
```

The following code should be saved as C:\perm\distributed.sas. The program can be run in series in a single SAS session, which will perform the typical SORT-SORT-MERGE in succession. For faster performance, however, the software can be opened in two different SAS sessions and run simultaneously. The two SORT procedures will perform in parallel, after which one of the sessions will merge the two data sets into the composite PERM.Merged:

```
libname perm 'C:\perm';
%include 'C:\perm\lockitdown.sas';

%macro control_create();
%if %length(&sysparm)>0 %then %do;
   data _null_;
      call sleep(&sysparm,1);
   run;
   %end;
%if %sysfunc(exist(perm.control))=0 %then %do;
   %put CONTROL_CREATE;
   data perm.control;
      length process $20 start 8 stop 8 jobid 8;
      format process $20. start datetime17. stop datetime17. jobid 8.;
   run;
   %end;
%mend;

%macro control_update(process=, start=, stop=);
data perm.control;
   set perm.control end=eof;
%if %length(&stop)=0 %then %do;
```

```
   output;
   if eof then do;
   process="&process";
   start=&start;
   stop=.;
   jobid=&SYSJOBID;
   output;
   end;
   %end;
%else %do;
   if process="&process" and start=&start then stop=&stop;
   %end;
   if missing(process) then delete;
   run;
&lockclr;
%mend;

%macro sort(dsn=);
%let start=%sysfunc(datetime());
%control_update(process=SORT &dsn, start=&start);
proc sort data=&dsn;
   by num1;
run;
%lockitdown(lockfile=perm.control, sec=1, max=5, type=W,
   canbemissing=N);
%control_update(process=SORT &dsn, start=&start,
   stop=%sysfunc(datetime()));
%mend;

%macro merge();
%let start=%sysfunc(datetime());
%control_update(process=MERGE, start=&start);
data perm.merged;
   merge perm.data1 perm.data2;
   by num1;
run;
%lockitdown(lockfile=perm.control, sec=1, max=5, type=W,
   canbemissing=N);
%control_update(process=MERGE, start=&start, stop=%sysfunc(datetime()));
%mend;

%macro engine();
%let sort1=;
```

```
%let sort2=;
%let merge=;
%control_create();
%do %until(&sort1=COMPLETE and &sort2=COMPLETE and &merge=COMPLETE);
   %lockitdown(lockfile=perm.control, sec=1, max=5, type=W,
      canbemissing=N);
   data control_temp;
      set perm.control;
      if process='SORT PERM.DATA1' and not missing(start) then do;
         if missing(stop) then call symput('sort1','IN PROGRESS');
         else call symput('sort1','COMPLETE');
         end;
      else if process='SORT PERM.DATA2' and not missing(start) then do;
         if missing(stop) then call symput('sort2','IN PROGRESS');
         else call symput('sort2','COMPLETE');
         end;
      else if process='MERGE' and not missing(start) then do;
         if missing(stop) then call symput('merge','IN PROGRESS');
         else call symput('merge','COMPLETE');
         end;
   run;
   %if %length(&sort1)=0 %then %sort(dsn=PERM.DATA1);
   %else %if %length(&sort2)=0 %then %sort(dsn=PERM.DATA2);
   %else %if &sort1=COMPLETE and &sort2=COMPLETE and %length(&merge)=0
      %then %merge;
   %else %if &sort1^=COMPLETE or &sort2^=COMPLETE or
      &merge^=COMPLETE %then %do;
      &lockclr;
      data _null_;
         call sleep(5,1);
         put 'Waiting 5 seconds';
      run;
      %end;
   %else &lockclr;
   %end;
%mend;

%engine;
```

The control table (PERM.Control) depicted in Table 9.1 demonstrates the performance improvement: each data set was sorted in approximately four minutes in parallel, after which the %MERGE macro merged the two data sets. Note that the JOBID variable indicates that two separate jobs were running.

Table 9.1 Control Table Demonstrating Parallel Processing

Process	Start	Stop	JobID
SORT PERM.DATA1	01APR16:13:16:23	01APR16:13:20:15	14776
SORT PERM.DATA2	01APR16:13:16:26	01APR16:13:21:01	13368
MERGE	01APR16:13:21:01	01APR16:13:21:56	13368

Another example of distributed processing is found in the "Executing Software in Parallel" section in Chapter 7, "Execution Efficiency." Whether the SYSTASK statement is used to spawn separate SAS sessions or a control table is used to broker and coordinate separate SAS sessions that are operating concurrently, parallel processing can provide significantly faster processing.

LOAD SCALABILITY

Load scalability describes the ability of a system or software to accommodate data throughput of increasing volume or velocity. Variability—the third *V* in the big data triad—represents less of a scalability challenge and more of a robustness challenge when unexpected or invalid data are encountered that are not handled through appropriate quality controls. Load scalability is arguably the most relevant scalability facet to data analytic development because of the inherent accretionary nature of data sets, databases, data streams, and other data sources. In fact, scaling resources (by increasing hardware) and scaling demand (by increasing the instances of SAS software executing in parallel) are often necessitated to enable SAS software to scale successfully as load increases.

The expectations of software scalability vary intensely with software intent. User-focused applications typically demand instantaneous service without delay, regardless of data volume. For example, a Google search today should be just as quick and effective as a Google search in a year when the quantity of searchable data may have doubled. In the applications world, scalability generally denotes no change in performance as data volume scales. And, in recent years, SAS has promulgated several in-memory solutions, such as SAS In-Memory Statistics and SAS Visual Analytics, which provide instantaneous processing, scaling effectively without diminished performance.

Base SAS, however, traditionally supports data analytic endeavors rather than applications development. In data analytic environments, users are typically not waiting for an instantaneous response as is common in software and web applications. Thus, in many SAS environments, as data load increases,

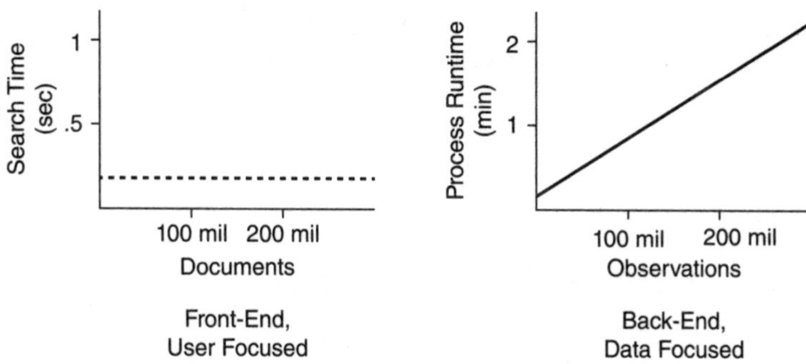

Figure 9.2 Contrasting Scalability Performance

execution time is not only allowed to but also expected to commensurately increase. For example, if a data set doubles in size within a year, for many purposes it's acceptable that the time to sort those data equivalently doubles, or at least follows a predictable algorithm. In data analytic environments, a scalable solution is more commonly defined as one that continues to operate efficiently despite greater throughput while software applications more commonly define scalability as equivalent performance, especially software execution time.

This disparity is demonstrated in Figure 9.2, which contrasts the scalability required by user-focused applications such as a Google search with the scalability demonstrated by a process in data analytic software.

For example, in data analytic development, as long as the correlation of file size and execution time for the SORT procedure remains linear, scalability is demonstrated and the efficiency of the procedure remains stable. Thus, while increased resources are often utilized as the software scales to respond to larger data sets, the resources are consumed at the same rate. Efficiency can begin to decrease as increasingly more resources are utilized, however, and as system resources (such as CPU cycles or memory) are depleted, a threshold may be reached after which the process performs inefficiently and thereafter fails. This type of failure is described later in the "Inefficiency Elbow" section.

In other cases, however, data analytic processes may have fixed performance thresholds that cannot be surpassed. For example, an hourly financial fraud program may be required by organizational policy to run every hour; despite increasing data volume (or velocity), this threshold represents a steadfast performance requirement. Given these requirements, and likely having exhausted programmatic methods to improve software speed, stakeholders would need to achieve scalability through nonprogrammatic methods that either scale resources or scale demand. In all cases, a shared understanding of load scalability by all stakeholders must be achieved given the variability in defining what constitutes and does not constitute scalable performance.

Data Volume

Data volume describes the size of a data set or data product, commonly measured and referenced as file size or observation count. The number of observations can be useful in tracking incremental data sources as well as validating processes in which the number of observations is expected to increase, decrease, or remain constant. Observation count is often anecdotally used in SAS literature to describe or make claims to "big data," although this is a poor metric when used in isolation. Blustering about a "10 billion-observation data set"—without additional context—conveys little more than hype. In most circumstances, file size represents a more useful, quantifiable, comparable metric for conveying the enormity of data.

Big data became a focus of the data analytic community because scalability challenges were encountered and overcome, but the challenges almost exclusively arise from file size—not observation count. File size is critical to understanding data volume because it influences the performance of input/output (I/O)-bound processes such as reading or writing data. The %HOWBIG macro in the "Data Governors" section depicts a reusable macro that programmatically assesses file size and which can be used in calculating actual and predicted data volume to avoid and overcome scalability challenges.

Diminished Performance

As demonstrated in Figures 9.1 and 9.2, when data volume increases, performance generally will slow as additional system resources are consumed to handle the additional load. Performance typically decreases in a predictable, often linear path correlated most significantly with file size. This increased execution time can cause software to fall short of performance requirements. For example, if an ETL process is required by a customer to complete in 55 minutes, as data volume increases over the software lifespan, execution time may increase until it fails to meet this performance objective.

In many cases, performance requirements are predicated upon temporal or other limitations. For example, the requirement that software execute in less than 55 minutes may exist because the ETL program is intended to be run every hour. Thus, failure to complete in a timely manner might cause dependent processes to fail or might delay or hamper the start of the subsequent hourly execution. In a robust, fault-tolerant system, while runtime errors will not be produced by these failures, business value can be diminished or eliminated when dependent processes cannot execute. This phenomenon is discussed more in the "Functional Failures" section.

As data volume increases, or when it's expected to increase in the future, it's important for developers and other stakeholders to understand customer intent regarding software execution time limits. For example, requiring that software must complete in 55 minutes or less is very different from requiring that software must produce output every 55 minutes. In the first example, the customer clearly specifies that from the moment data are received, they should be extracted, transformed, and loaded and ready for use in 55 minutes. However, if the customer only wants data updated once per hour, it may be acceptable to create a program flow that takes 90 minutes, but that is run in parallel and staggered to still achieve output once per hour.

Figure 9.3 demonstrates two interpretations of the 55-minute requirement. On the left, the phase-gate approach is shown across a four-hour time period, demonstrating the requirement that the software complete in 55 minutes or less without the ability to overlap any processes. On the right, an overlapping schedule is demonstrated in which two copies of the software are able to run for 90 minutes, but because one starts on even hours and one starts on odd hours, they collectively produce results once an hour. In this example, the extract module is able to overlap with the load module to facilitate parallel processing and a control table or other mechanism would be required to coordinate the concurrent activities.

Parallel execution and other distributed processing models are a primary method to reduce execution time of software as data volume increases. Another benefit of the overlapping execution method noted in Figure 9.3 is the tendency

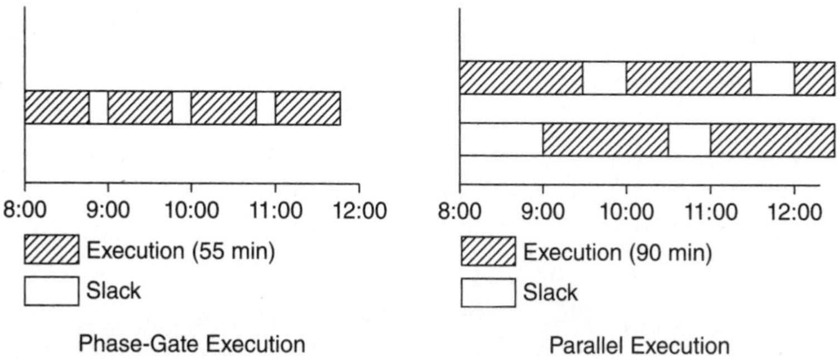

Figure 9.3 Phase-Gate and Parallel Software Models

to process smaller data sets, which reduces execution time. For example, an ordinary serialized ETL process that completed in 90 minutes would only ingest incoming data every 90 minutes. However, because the parallel version ingests data every 60 minutes, smaller chunks are being processed each time, thus reducing overall execution time.

Inefficiency Elbow

The *inefficiency elbow* represents a specific type of performance failure that occurs as resources are depleted due to increased data volume. Because execution time is the most commonly assessed performance metric, the elbow is typically first noticed when software performance drastically slows due to a relatively nominal increase in data volume. Figure 9.4 demonstrates the inefficiency elbow within the SAS University Edition, observed when execution times for the SORT procedure are measured in 100,000 observation increments from 100,000 to 4.5 million observations. Execution time takes only 36 seconds at 4.2 million observations, 98 seconds at 4.3 million observations, and 4 hours 21 minutes by 4.5 million observations!

The inefficiency elbow depicts the relative efficiency of a process as it scales to increasing data volume, indicating relatively efficient performance to the left of the elbow and relatively inefficient performance to the right. This is to say that efficient performance does not necessarily denote that the code in the underlying process is coded as efficiently as possible, only that as data volume scales, the software less efficiently utilizes resources until it fails outright.

To better understand what occurs as the efficiency elbow is reached, Table 9.2 demonstrates abridged UNIX FULLSTIMER performance metrics,

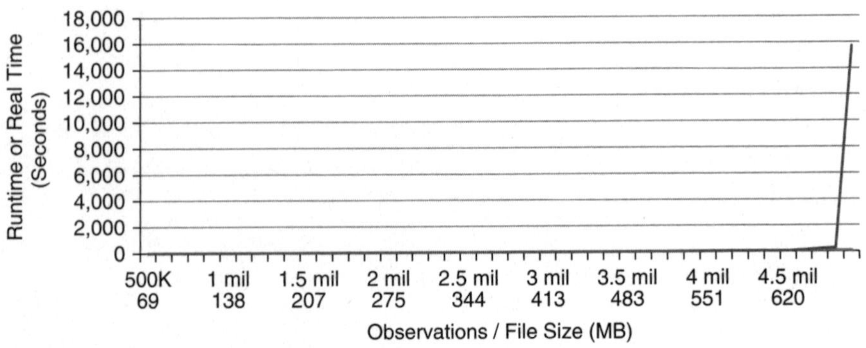

Figure 9.4 Inefficiency Elbow on SORT Procedure

Table 9.2 FULLSTIMER Metrics for SORT Procedure at Inefficiency Elbow

OBS	File Size (MB)	Seconds	User CPU Time	System CPU Time	Memory	OS Memory	Switch Count	Page Faults	Invol Switches	Block In Ops
3,500,000	482	29	17.96	5.59	675.36	706.1	41	0	666	0
3,600,000	496	30	18.68	5.86	694.56	725.3	41	0	602	0
3,700,000	509	31	19.27	5.76	713.76	744.51	49	1	712	32
3,800,000	523	33	19.37	6.34	732.96	763.71	49	9	553	352
3,900,000	537	39	20.11	6.98	747.37	778.11	49	200	717	6552
4,000,000	551	35	20.29	7.96	766.57	797.32	49	47	987	2640
4,100,000	564	36	21.4	7.41	785.77	816.52	49	39	962	2128
4,200,000	578	98	20.36	14.87	806.54	837.29	97	5663	2546	338608
4,300,000	592	190	13.03	31.83	825.74	856.49	249	35156	3744	1213088
4,400,000	606	264	9.91	20.46	844.94	875.69	185	21417	3077	836704
4,500,000	620	15679	0.02	521.79	864.14	894.89	1083	893270	15978	34001456

including real time, CPU time, memory, OS memory, switch counts, page faults, involuntary context switches, and block input operations. SAS documentation contains more information about these specific metrics.[3] Metrics surrounding the inefficiency elbow (from 3.5 to 4.5 million observations) demonstrate the substantial increase in I/O processing as SAS shifts from RAM to virtual memory, and finally the later jump in CPU usage.

As Figure 9.2 demonstrates, although the inefficiency elbow is not reached until around 4.2 million observations (as defined by the change in execution time), I/O metrics presaged defeat of the SORT procedure several iterations earlier. For example, both page faults and block input operations had consistently been 0 until 3.7 million observations, when both began increasing. Given this information, a quality assurance process monitoring these metrics could have detected levels above some threshold and subsequently terminated the process. This technique is described in the following section, "Data Governors."

Without understanding the contributing factor that increased data volume can play in inefficiency, stakeholders may mistakenly attribute process slowness to network or other infrastructure issues. Worse yet, stakeholders may accept reduced performance as the status quo and their penance for big data, rather than investigating the cause of performance deficiencies. In reality, several courses of action can either move or remove the inefficiency elbow, the first of which begins with stress testing to determine what volume of data will induce inefficiency or failure in specific DATA steps, SAS procedures, and other processes. Without proper stress testing, discussed in the "Stress Testing" section in Chapter 16, "Testability," the inefficiency elbow will be a very unwelcome surprise when it's encountered.

To reduce the effects of the inefficiency elbow, stakeholders have several programmatic and nonprogrammatic solutions that entail more efficient use of resources, more distributed use of resources, or procurement of additional system resources:

- *More efficient software*—The inefficiency elbow denotes a change in *relative* efficiency; thus, it's possible that software is not truly operating efficiently even to the left of the elbow. For example, in Figure 9.4, it's possible that a WHERE statement in the SORT procedure—if appropriate—would reduce the number of observations read, thus reducing I/O consumption and shifting the elbow to the right.
- *Increased infrastructure*—Increased infrastructure can include increasing the hardware components as well as installing distributed SAS solutions such as SAS Grid Manager or third-party solutions like Teradata or Hadoop. By either increasing system resources or increasing the efficiency with which those resources can be utilized, the inefficiency elbow is shifted to the right, allowing more voluminous data to be processed with efficiency.
- *Distributed software*—As demonstrated in Figure 9.3, when processes can overlap or be run in parallel, the resource burden exhibited by a single SAS session can be distributed across multiple sessions and processors. Moreover, when this parallel design distributes data volume across multiple sessions, this doesn't alter the inefficiency elbow, but does decrease the position of execution time on the graph since fewer observations are processed through each independent SAS session.
- *Data governors*—A data governor is a quality control device that restricts data injects based on some criteria such as file size. For example, if a certain process is stress tested and demonstrates relative inefficiency when processing data sets larger than 1 GB, a data governor placed at 1 GB would block or subset files (for later or parallel processing) to prevent inefficiency.

Figure 9.5 demonstrates the effect of these methods on software performance and the inefficiency elbow.

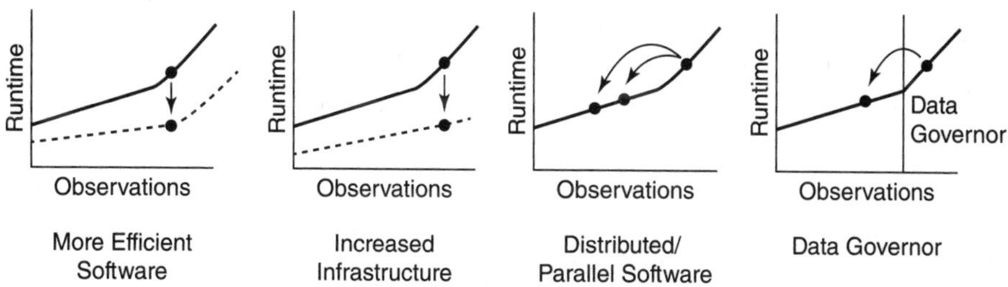

Figure 9.5 Inefficiency Elbow Solutions

SAS literature is replete with examples of how to design SAS software that is faster and more efficient, so these best practices are not discussed in this text. Infrastructure is discussed briefly in the "Distributed SAS Solutions" section, but this also lies outside of the focus of this text. Parallel processing and distributed solutions are discussed throughout Chapter 7, "Execution Efficiency," and data governors are discussed in the following section.

Data Governors

Just as a governor in a vehicle is installed to limit speed beyond some established threshold, a data governor can limit the velocity or, more typically, the volume of data for specific processes. Data governors can be implemented immediately before the inefficiency elbow, as demonstrated in Figure 9.5, or anywhere else to ensure a process does not begin to function inefficiently or fail. While data volume is the primary factor affecting performance, other factors can also lead to more substantial variability. For example, if other memory-hogging processes are already running, performance or functional failure might be reached at a significantly lower data volume because resources are more constrained. The "Contrasting the SAS University Edition" section in Chapter 10, "Portability," also demonstrates the influence that nonprogrammatic factors such as SAS interface have on the inefficiency elbow.

A data governor must first determine the file size of the data set. The following %HOWBIG macro obtains the file size in megabytes from the DICTIONARY.Tables data set. Note that a generic return code—the global macro variable &HOWBIGRC—is generated to facilitate future exception handling. However, robust production software would need to implement additional controls—for example, to validate that a data set exists and is not exclusively locked by another process, either of which would cause the SQL procedure to fail:

```
%macro howbig(lib=, dsn=);
%let syscc=0;
%global howbigRC;
%global filesize;
%global nobs;
%let howbigRC=GENERAL FAILURE;
proc sql noprint;
   select filesize format=15.
      into :filesize
      from dictionary.tables
      where libname=%upcase("&LIB") and memname=%upcase("&DSN");
   select nobs format=15.
      into :nobs
      from dictionary.tables
      where libname=%upcase("&LIB") and memname=%upcase("&DSN");
   quit;
run;
%if %eval(&syscc>0) %then %do;
   %let howbigRC=&syscc;
   %let filesize=;
   %let nobs=;
   %end;
%else %do;
   %let filesize=%sysevalf(&filesize/1048576); *convert bytes to MB;
   %let nobs=%sysevalf(&nobs);
   %end;
%if &syscc=0 %then %let howbigRC=;
%put filesize: &filesize;
%put observations: &nobs;
%mend;
```

In this example, the inefficiency elbow was shown to be around 1GB, so to account for some variability, the data governor will be set at 0.9 GB through conditional logic similar to the following code:

```
%let governor=%sysevalf(1024*.9); * in MB;
%if %sysevalf(&filesize>&governor) %then %do;
   * subset the data here or perform other busines logic;
   %end;
* run process on full data set or data subset;
```

If a 1.25 GB data set is encountered by the governor, business logic could specify that the first 0.9 GB of the data set be processed initially and the last 0.35 GB be processed either in series—that is, after the first batch completes—or immediately as a parallel process. One facile way to implement a data governor is through FIRSTOBS and OBS statements that can be used to subset the data before it is even read, efficiently eliminating unnecessary I/O processing. Because FIRSTOBS and OBS are dependent on observation number while data governors and other performance quality controls should be driven by file size, a conversion to observation count is necessary.

In a separate example, the following code demonstrates the creation of the 1 million-observation PERM.Mydata data set which, when interrogated through the %HOWBIG macro, is shown to be 7.875 MB:

```
libname perm 'c:\perm';
data perm.mydata (drop=i);
   length num1 8;
   do i=1 to 1000000;
      num1=round(rand('uniform')*10);
      output;
      end;
run;
```

Suppose that stress testing has indicated that efficiency begins to decrease shortly after 6 MB (apparently this process has time-traveled back to 1984), indicating that the governor should be installed at 6 MB. The following macro %SETGOVERNOR calls the %HOWBIG macro to determine the file size and observation count and, if the file size exceeds the 6 MB limit, then the value to be used in the OBS option (to ensure a file size equal to or less than 6 MB) will be initialized in the &GOV macro variable.

```
* determines file size and provides the observation number for OBS;
%macro setgovernor(lib=, dsn=, govMB=);
%let syscc=0;
%global setgovernorRC;
%global governor;
%local gov;
%let setgovernorRC=GENERAL FAILURE;
%let governor=;
%howbig(lib=&lib, dsn=&dsn);
%if %length(&howbigRC)>0 %then %do;
   %let setgovernorRC=HOWBIG failure: &howbigRC;
   %return;
   %end;
```

```
%if %sysevalf(&filesize>&govMB) %then %do;
   %let gov=%sysfunc(round(%sysevalf(&nobs * &govMB / &filesize)));
   %let governor=(firstobs=1 obs=&gov);
   %end;
%if &syscc=0 %then %let setgovernorRC=;
%mend;
```

The following code invokes the %SETGOVERNOR macro to determine how many observations should be read into the Test data set. Because the file size 7.875 MB is greater than the 6 MB governor, the OBS statement (codified in the &GOVERNOR macro variable) specifies that only 761,905 observations (or 1,000,000 × 6 / 7.875) should be read into the Test data set.

```
%setgovernor(lib=perm, dsn=mydata, govMB=6);
data test;
   set perm.mydata &governor;
run;
```

To test the accuracy of this math, the %HOWBIG macro can subsequently be run on the Test data set, which produces the following output, indicating that the data set is indeed now only 6 MB and equal to the established governor.

```
%howbig(lib=work, dsn=test);
NOTE: PROCEDURE SQL used (Total process time):
      real time           0.00 seconds
      cpu time            0.00 seconds


filesize: 6
observations: 761905
```

In this example, accuracy of the subset data set size is guaranteed because the data set was created through a standardized process, thus all observations are of equal size. In more variable data sets or especially in heteroscedastic data sets in which the file size of each observation is correlated with the observation number itself (e.g., values are more likely to be blank in the first half of the data set than the second half), this method will be less accurate. In many situations, however, this type of data governance provides an effective means to limit input into some process so that it does not incur performance or functional failures due to high data volume. And, depending on business logic, the remaining observations that were excluded could be executed in parallel or in series following the completion of the first process.

Data governors are most effective when inter-observation comparisons are not required. For example, if the DATA step had included by-level processing, utilized the RETAIN or LAG functions, or otherwise required interaction between observations, the data set could not have been split so easily. In some instances, it's possible to take the last observation from the first file and replicate this in the second file, thus providing continuity between the data set chunks. In other cases, such as a data governor placed on the FREQ procedure, the output from each respective FREQ procedure would need to be saved via the OUT statement and subsequently combined to produce a composite frequency table. Notwithstanding these complexities, where resources are limited and inefficiency overshadows big data processing, data governors can be a cost-effective, programmatic solution to scaling to big data.

As discussed, the %HOWBIG macro does not detect specific exceptions that could occur, such as a missing or exclusively locked data set that would cause the SQL procedure to fail. However, it includes basic exception handling to detect general failures via the &SYSCC global macro variable. The return code &HOWBIGRC is initialized to GENERAL FAILURE and is only set to a missing value (i.e., no failure present) if no warnings or runtime errors are detected. The %SETGOVERNOR macro follows an identical paradigm but, because it calls %HOWBIG as a child process, %SETGOVERNOR must dynamically handle failures that may have occurred in %HOWBIG through exception inheritance. Inheritance is required to achieve robustness in modular software design and is introduced in the "Exception Inheritance" section in Chapter 6, "Robustness."

Functional Failures

As data volume continues to increase and as system resources continue to be depleted, functional failure can become more likely; SAS software can terminate with a runtime error because it runs out of CPU cycles or memory. In some instances, the inefficiency elbow will have first been reached, indicating that performance will have slowed substantially as the process became less efficient. In other cases, the inefficiency elbow may not exist but a performance failure will occur because the execution time exceeds thresholds established in requirements documentation. Still, in other cases, some processes fail with runtime errors with no prior performance indicators (aside from FULLSTIMER metrics that may not have been monitored).

Figure 9.6 demonstrates performance of the SORT procedure when the volume of data is varied from 100,000 to 5 million observations. In a development environment that espouses software development life cycle (SDLC) principles, load testing would have demonstrated performance of the procedure

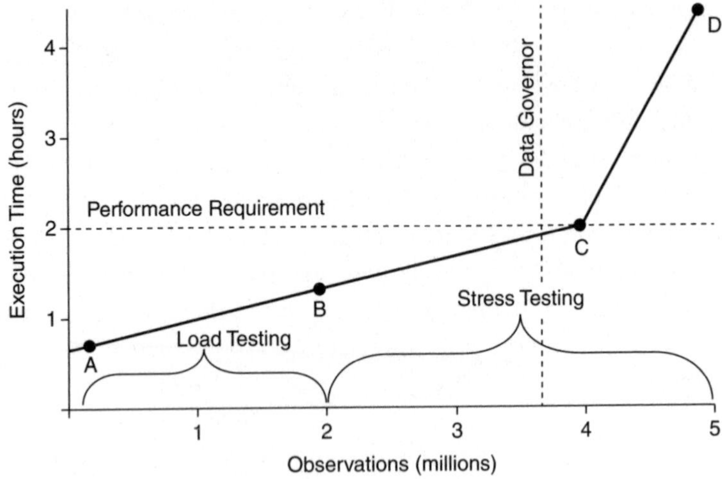

Figure 9.6 Scaling with Efficiency

at the predicted load of 2 million observations. However, realizing that data volume might increase over time, additional stress testing would have tested increasingly higher volumes of data to determine where performance and functional failures begin to occur.

If an inefficiency elbow is demonstrated through stress testing, performance requirements for execution time and resource consumption should be established below that point. To ensure that this threshold is never crossed, a data governor can be implemented to prevent both performance failures and functional failures from occurring when exceptionally large data sets are encountered. Figure 9.6 demonstrates the placement of these elements to support efficient performance.

Even with adequate technical requirements, robust quality control measures such as data governors, and attentive monitoring of system resources, SAS processes can still fail due to resource consumption while processing big data. For this reason, production software should espouse fault-tolerant programming within an exception handling framework that monitors all completed processes for warnings and runtime errors that can signal functional failures due to resource constraints. The use of return codes and global macro variable error codes such as &SYSCC and &SYSERR that facilitate failing safe are discussed extensively in Chapter 3, "Communication," and Chapter 6, "Robustness."

Data Velocity

Data velocity typically describes the rate at which data are processed or accumulate. In this usage, when velocity increases, it will often also increase data volume. For this reason, the challenges of increased data velocity will only be discussed briefly in this section. But velocity can also describe a change in requirements in which data are required to be processed faster. Both data-driven and requirements-driven velocity increases can cause performance and functional failures.

Data-driven velocity increases result from data pouring into a system at a faster rate. Because data are often processed in batches at set intervals, such as every 15 minutes, this increased velocity can cause these chunks to grow in size, at which point they also represent increased data volume. As demonstrated in the prior "Data Volume" section, data volume can pose serious challenges to SAS processes. The second failure pattern occurs not when the actual pace of data increases but when the requirements for some process are increased. For example, an ETL process initially was mandated to run hourly and a customer has now required that it be run every 15 minutes. While this will mean fewer data must be processed at every 15-minute interval, the processing must now occur in one-fourth the time. While the cause is requirements-driven, the challenge simulates other scalability issues in which too many data exist to be processed within some time-boxed period.

Velocity as an Indicator

During planning and design, at least a rough conception of eventual data volume should be obtainable. In many data analytic development environments, the data are already held so velocity is zero. For example, clinical trials software might be developed to analyze a study sample that has already been collected and to which data are not being added. In other cases, however, data accumulate at some velocity, which can be useful in predicting future data volume. For example, clinical trials software might be developed to analyze a longitudinal study that is expected to continue to grow in size for five years. Knowledge of data velocity can be used to establish load testing parameters to determine if performance requirements will be adequate throughout the life of the software. If an initial data set has 10,000 observations and is expected to grow by 10,000 observations each year thereafter, performance requirements would need to expect file sizes of 50,000 to 60,000 observations.

Modest file sizes such as this likely would not incur any noticeable performance decreases, so neither data volume nor velocity would be of significant concern in this example. But in environments where data volume can be staggering, data velocity should be used to facilitate accurate load and stress testing at accurate data volume levels. This information can be critical to stakeholders, helping them predict when software or hardware upgrades will be needed to maintain equivalent performance.

To continue the example demonstrated in Figure 9.6, the current data volume of some SAS processes is 2 million observations. If stakeholders further determine that the volume will grow by approximately 1 million observations per year, then they can predict that the process will still be operating efficiently and within the stated performance requirement of two hours' execution time. However, if data velocity instead accelerates, performance or functional failures could result as data volume increases and the software takes longer and longer to run. This underscores the importance of establishing expected software lifespan during planning and design phases of the SDLC.

The ability to predict future software performance is beneficial to Agile software development environments in which stakeholders aim to release software incrementally through rapid development that delivers small chunks of business value. An initial SAS process that meets performance requirements could be developed quickly and released, with the understanding that subsequent incremental releases will improve this performance to ensure that performance requirements will be met in the future as data volume continues to increase. Progressive software requirements to facilitate increasing the performance of software over time are discussed in the "Progressive Execution Time" section in Chapter 7, "Execution Efficiency."

The role of data volume and velocity in supporting effective Agile development is demonstrated in Figure 9.7. In this example, an initial solution is released in the first iteration that meets the two-hour performance requirement but that would begin to fail to meet this requirement when the data volume increased to approximately 2.5 million observations. The software is refactored in the second iteration to produce faster performance, which shifts the predicted performance function (as estimated through load and stress testing). After refactoring a second time in the third iteration, the software performance increases to a level that will meet the performance requirements to 5 million observations, giving stakeholders ample time to decide how to

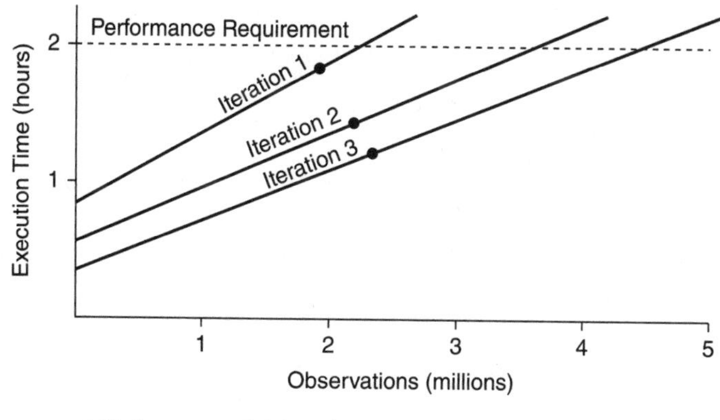

Figure 9.7 Incremental Refactoring to Support Increased Software Performance

handle that scalability challenge when it is encountered. Thus, Agile development enables SAS practitioners to produce a solution quickly that meets immediate requirements while allowing developers to keep an eye toward the future and eventual performance requirements that software will need to meet to scale to larger data. This forward-facing awareness is the objective of software refactoring and extensibility, discussed in the "From Reusability to Extensibility" sections in Chapter 18, "Extensibility."

Scalability Idiosyncrasies

Load scalability is typically construed as causing performance failures (due to increased execution time) or functional failures (due to depleted system resources). However, increased data volume can also indirectly cause idiosyncratic results, runtime errors, and other failures. This further underscores the necessity of stress testing all production software to determine where syntax might need to be made more flexible to adapt to shifts in business logic or Base SAS system thresholds. These two issues are discussed in the following sections.

Invalid Business Logic

Invalid business logic essentially amounts to errors that SAS practitioners make in software design when they fail to predict implications of higher data volume. The failure patterns are endless but one example occurs when observation counters exceed predicted limits. For example, the following code simulates an ingestion process (with increasing data volume) followed by a later process:

```
libname perm 'c:\perm';

* simulates data ingestion;
data perm.mydata;
   length num1 8;
   do i=1 to 1000;
      num1=round(rand('uniform')*10);
      output;
      end;
run;

* simulates later data process;
filename outfile 'c:\perm\mydata_output.txt';
data write;
   set perm.mydata;
   file outfile;
   put @1 i @4 num1;
run;
```

The output demonstrates that the second process will operate effectively when the data set contains fewer than 100 observations, but when this threshold is crossed, the data no longer are delimited by a space and are indistinguishable. At 1,000 observations, a functional failure occurs when the data are overwritten:

```
1  4
2  1
3  4
...
9  1
10 7
11 6
...
99 0
1008
1013
```

Logic errors such as this are easily remedied but can be difficult to detect. However, if load testing and stress testing are not performed or fail to detect underlying software defects, they can go unnoticed indefinitely.

SAS Application Thresholds

Programmatic limitations exist within the SAS Base language just as in any software language. For example, variable naming conventions specify acceptable alphanumeric characters that can appear in a SAS variable name, and maximum name lengths are specified for SAS variables, macro variables, libraries, and other elements. Under the SAS 9.4 variable naming standard (V7), both data set variable names and macro variable names are limited to 32 characters in length. Thus, failures could result if variable names were being dynamically created through macro language, if the name included an incrementing number, and if that number caused the length of the variable to exceed 32 characters. Many of these SAS thresholds are detailed in *SAS® 9.4 Language Reference: Concepts: Fifth Edition.*[4]

Other SAS language limitations, however, are published nowhere and are discovered solely through trial and error—that is, when software has been performing admirably, but suddenly and inexplicably fails. For example, a commonly cited method to determine the number of observations in a data set is to utilize the COUNT statement in the SQL procedure to create a macro variable. The following output demonstrates that the PERM.Mydata data set has 1,000 observations:

```
proc sql noprint;
   select count(*)
   into :nobs
   from perm.mydata;
quit;

%put OBS: &nobs;
OBS: 1000
```

While this method lacks robustness (because the SQL procedure fails with a runtime error if the data set does not exist, is exclusively locked, or contains no variables), it can be a valuable tool when these exceptions are detected and controlled through exception handling. But as data volume increases, another exception can occur when the default SAS variable format shifts from standard to scientific notation. Thus, 10 million is represented in standard notation as 10000000, while 100 million is represented in scientific notation as 1E8:

```
%put OBS: &nobs; * for 10 million observations;
OBS: 10000000
```

```
%put OBS: &nobs; * for 100 million observations;
OBS: 1E8
```

This notational shift can destroy SAS conditional logic and other processes, creating aberrant results. For example, if %SYSEVALF is not used to compare the resultant value, then operational comparisons are character—not numeric—which causes scientific notation to yield incorrect results:

```
* equivalent values in standard and scientific notation;
%let nobs_stand=100000000;
%let nobs_sci=1E8;

%macro test(val=);
%if &nobs_stand<9 %then %put standard notation: 100000000 is less
   than &val;
%else %put standard notation: 100000000 is NOT less than &val;
%if &nobs_sci<9 %then %put scientific notation: 1E8 is less &val;
%else %put scientific notation: 1E8 is NOT less than &val;
%mend;

%test(val=9);
standard notation: 100000000 is NOT less than 9
scientific notation: 1E8 is less 9
```

1E8 is not less than 9, but when a character comparison is made, the fallacy is demonstrated because SAS only assesses that the first character, 1, is less than 9. The %EVAL function is used to force numeric comparisons in the macro language and typically works—that is, until scientific notation is encountered:

```
%macro test(val=);
%if %eval(&nobs_stand<9) %then %put standard notation: 100000000 is less
   than &val;
%else %put standard notation: 100000000 is NOT less than &val;
%if %eval(&nobs_sci<9) %then %put scientific notation: 1E8 is less &val;
%else %put scientific notation: 1E8 is NOT less than &val;
%mend;

%test(val=9);
standard notation: 100000000 is NOT less than 9
scientific notation: 1E8 is less 9
```

The scientific notation unfortunately fails again, but now for a different reason. The %EVAL function does not recognize scientific notation and thus views this as an attempt to convert a character value to a number. This is demonstrated in the following output that results when the %EVAL function is used to print standard and scientific notation:

```
%put standard: %eval(&nobs_stand);
standard: 100000000
%put scientific: %eval(&nobs_sci);
ERROR: A character operand was found in the %EVAL function or %IF
condition where a numeric operand is required. The condition was: 1E8
scientific:
```

This failure is overcome by substituting the %EVAL function with the %SYSEVALF function, which does interpret scientific notation. Now, as expected, both standard and scientific notation are accurately read and can be operationally evaluated:

```
%macro test(val=);
%if %sysevalf(&nobs_stand<9) %then %put standard notation: 100000000
   is less than &val;
%else %put standard notation: 100000000 is NOT less than &val;
%if %sysevalf(&nobs_sci<9) %then %put scientific notation: 1E8 is
   less &val;
%else %put scientific notation: 1E8 is NOT less than &val;
%mend;

%test(val=9);
standard notation: 100000000 is NOT less than 9
scientific notation: 1E8 is NOT less than 9
```

While %SYSEVALF is not always required and can be replaced by %EVAL for many operations, as data sets increase in volume, %SYSEVALF may be required to ensure that scientific notation is accurately assessed. Another possible solution is to add FORMAT=15. to the SELECT statement in the SQL procedure, as demonstrated previously in the "Data Governors" section. The duality of these two solutions is discussed further in the "Risk versus Failure" section in Chapter 1, "Introduction."

SCALABILITY IN THE SDLC

The role that scalability plays throughout the SDLC is emphasized in the "Load Scalability" section, demonstrating that scalability should be something you plan for in software, not something that happens to software unexpectedly. Through load testing and stress testing, SAS practitioners should be able to demonstrate with relative certainty the performance and efficiency that software will exhibit as it scales. With robust design that avoids inefficiency through data governors and other controls, and fault-tolerant design that detects and responds to runtime errors experienced when software fails to scale, software reliability should not be diminished as it encounters big data.

One of the most important reasons to understand predicted load scalability as early as possible in the SDLC is that it can alert stakeholders to future actions that they may need to take to ensure software performance objectives continue to be met. For example, when SAS software demonstrates that it has limited ability to scale to higher volumes of data that are expected in the future, stakeholders can begin discussing whether a programmatic solution (like increasing the demand scalability of the software through parallel processing) or a nonprogrammatic solution (such as purchasing additional hardware and a Teradata license) is warranted and when the solution will be necessary.

Requiring Scalability

While this chapter has demonstrated three different facets of scalability, requirements documentation won't necessarily state *how* scalability is to be achieved, but rather *what* performance must be achieved. A requirement might state that a back-end SAS process must be able to complete in 10 minutes if processing 10 million observations and in 15 minutes if processing 12 million observations, demonstrating the scalability of performance as data volume increases. These objectives might be met through programmatic enhancements, through nonprogrammatic enhancements, or through a combination thereof. Where progressive requirements are implemented, the requirements might state that the software will not need to accommodate the higher velocity and volume of data for a number of months or years or until the increased data level has been reached.

The necessity for scalability is most apparent in front-end processes for which users are explicitly waiting for some task to complete. Performance requirements for front-end processes often state that performance must remain static while data volume scales to higher levels. For example, a requirement might state that a process should complete in two seconds or less as long as fewer than 100 million observations are processed. When scalability requirements effectively confer that execution time cannot appreciably increase—despite increasing load—solutions that include both scalable resources and scalable demand must typically be implemented to achieve the objective.

Note that in both requirements examples, a maximum threshold was established, effectively limiting the scope of the software project with data volume or velocity constraints. Without this threshold, SAS practitioners will have no idea how to plan for or assess the scalability of software, including where to establish bounds for load and stress testing. Especially when a SAS process is designed to process data streams or other third-party data sources, it's prudent to establish data volume and velocity boundaries in requirements documentation so SAS practitioners understand the scope of their work and the ultimate requirements of their software.

Requirements documentation should also specify courses of action that should be followed when exceptional data—those exceeding volume or velocity restrictions—are encountered. For example, if data sets exceeding 1 GB are received, the first gigabyte should be subset and processed, after which additional observations from the original data set can be processed thereafter in series. This specification clarifies to developers that they won't be responsible for designing a system scalable beyond data thresholds. An alternate requirement, however, might instead state that when data sets that exceed the stated threshold are received, they should be processed through a complex, distributed, divide-and-conquer routine capable of processing all data in parallel. These two vastly different solutions would require tremendously different levels of development (and testing) to implement, underscoring the need to state explicitly not only how software should handle big data, but also how it should handle exceptional data that are too big.

Measuring Scalability

For the purposes of this chapter, scalability has been demonstrated as a scalar value, represented by the execution time. The graphical representations—including Figures 9.1, 9.4, and 9.6—each depict software scalability demonstrated through load and stress testing, thus facilitating precise determination of when performance and functional failures occur. While this is the most common measurement for scalability, in environments for which these vast data points are not available, scalability may be represented instead as a dichotomous outcome—for example, a process succeeded at 4 million observations but ran out of memory and failed at 5 million observations. The major weakness of dichotomous assessments is their lack of predictive ability. For example, with only two data points collected—one success and one failure—it's impossible to establish a trend that might have predicted performance or functional failure as data volume increased.

Therefore, when planning with software scalability in mind, load and stress testing should assess software execution times as the data load is systematically increased. This testing will provide SAS practitioners with confidence that the software will meet performance requirements now and in the future, as well as elucidate bizarre failures such as those described previously in the "Scalability Idiosyncrasies" section.

While load scalability can often be assessed by measuring execution time alone, a fuller picture of software performance can be gained by interrogating system resources recorded via FULLSTIMER metrics. These metrics can provide invaluable information that reflect whether software could benefit from programmatic solutions, such as distributed processing, or whether nonprogrammatic solutions such as increased hardware would be recommended. Poor performance in the face of big data can be identified through execution time alone; however, its cause—and, equivalently, its solution—can be obtained only through an understanding of system resource utilization and constraints.

WHAT'S NEXT?

The last two chapters have distinguished software efficiency from software speed, while this chapter demonstrated how these quality characteristics can be used to predict and measure software performance as it scales to increased load. In the next chapter, software portability is examined, which aims to achieve equivalent functionality and performance across disparate SAS environments, interfaces, versions, and processing modes. Portability can ensure that SAS dynamically responds to environmental cues and facilitates software execution across diverse landscapes, and is key to maximizing the generalizability of software.

NOTES

1. IEEE Std. 1517-2010. *IEEE standard for information technology—System and software life cycle processes—Reuse processes*. Geneva, Switzerland: Institute of Electrical and Electronics Engineers.
2. "Support for Parallel Processing." *SAS® 9.4 Language Reference Concepts, Third Edition*, pp. 189–190.
3. "FULLSTIMER SAS Option." *SAS® Knowledge Base/Focus Areas*. Retrieved from http://support.sas.com/rnd/scalability/tools/fullstim/.
4. "Names in the SAS Language." *SAS® 9.4 Language Reference: Concepts, Fifth Edition*. Retrieved from http://support.sas.com/documentation/cdl/en/lrcon/68089/HTML/default/viewer.htm#p18cdcs4v5wd2dn1q0x296d3qek6.htm.

CHAPTER 10

Portability

"Why is there a machete in this ambulance?!"

Not "Do you guys carry epinephrine?" or "Where is the bag valve mask?" No, my first question after jumping in the back of the ambulance as we raced to a call for suspected anaphylaxis was to question the legitimacy of the machete tucked precariously inside the sliding door.

If you're a firefighter, one of your first stops in any city—especially overseas—is the local firehouse. While it's not uncommon for a drop-in to become a half-day of riding calls, somehow I found myself volunteering at the *Bomberos Voluntarios* (volunteer fire department) in Antigua, Guatemala, not even a week after moving to the city. So much for relaxing after Afghanistan ...

The predominantly Spanish-speaking crews welcomed me, given that Antigua is rife with tourists, and Americans were especially elated when an English speaker showed up after a stabbing or tuc-tuc accident. While I didn't always expect to understand my crew or patients, I did expect emergency medical training to be relatively universal in tools, techniques, and protocols. This wasn't always the case.

Climbing into the back of the minivan—oh, yeah, the ambulance was a converted Chrysler minivan—the glint of the machete blade first caught my eye, but the amazement didn't end there.

The hatchback had neither latch nor lock, so a seatbelt had been tied to the metal hasp on the floor (where the door originally would have locked) and also the hatchback—such that the rear door was secured only by buckling the seatbelt.

Yet the hatchback hydraulics still somehow functioned, so while the door had to be closed from the outside, the seatbelt could only be fastened (to secure the door) from the inside, which made for some interesting acrobatics on a two-man crew.

Of course the belt left a gap of a couple inches, so the door incessantly bounced and rattled as we raced through the cobblestone streets of Antigua. Because the litter was not secured to the ambulance, half my job seemed to entail reassuring patients that I would prevent them from sliding out the back—and the other half was actually preventing them from sliding out the back!

Start an IV—sure thing, that's a transferable skill in any language. Start an IV while bouncing down cobblestone streets—no, thanks, please pull this thing over.

There was oxygen in the ambulance, but we were advised not to use it (in nearly all cases) because the bottles could only be refilled in Guatemala City, an hour's drive away. The *Bomberos Voluntarios* didn't have the money for that commute.

Most supplies were haphazardly garnered through donations from the United States and Europe, so we took nothing for granted and made every effort to conserve tape, gauze, bandages, and everything else. But gas was the scarcest resource of all, with the gas gauge of every apparatus I ever rode registering on empty—just enough to triangulate to town, the hospital, and back to the firehouse. There were absolutely no supermarket runs, joyriding, or "staging" in interesting locations, as are common back in the States.

While the vast majority of my skills and training were transferable, the environment, calls, and equipment were at times so different from anything I'd ever experienced that I simply had to take a knee and wait for instruction.

Despite the many differences, as I stood in a receiving line one night helping distribute firefighter patches and badges to a graduating class of *bomberos,* I was certain that the pride I felt was no different from their own.

And what of the machete? Although I asked every firefighter in the station, the most common responses were "*¡Ten cuidado!*" (Be careful) or "*¿Por qué no?*" (Why wouldn't we have a machete in the ambulance?!)

■ ■ ■

Something I often take for granted in the United States is the level of care provided through emergency medical services. Despite subtle protocol and equipment differences, treatment is largely standardized, and patients may not realize discrepancies in level of care that exist elsewhere in the world. SAS practitioners who spend their career working within a single environment also may never encounter system and environmental variability discussed in this chapter.

The majority of my medical training transferred directly from the United States to Guatemala. After all, there are only so many ways to splint an arm or bandage wounds after a bike accident. The Base SAS language is also tremendously portable between environments and, in the vast majority of cases, software written in one environment will perform equivalently in another.

In other cases, however, protocols differed or equipment was unavailable, but we made do. We once transported a patient to an ambulance in a wheelbarrow, following his collapse during the *Carrera de Charolas* (race of the trays)—a sprint through the cobblestone streets in which waiters, waitresses, and bartenders each must single-handedly carry a tray of beverages. But the solution worked.

A similar aim of software portability is to deliver equivalent functionality despite environmental variability, and this can be achieved when SAS software can detect its environment and respond dynamically.

Sometimes we couldn't replicate equipment; for example, we didn't carry pulse oximeters in the ambulance, so oxygen saturation vitals were never collected. SAS software, too, can encounter environments in which some function or task cannot be completed, but this should be detected and handled as an exception, rather than allowing software to fail aimlessly.

In all cases, my goal while working with the *Bomberos Voluntarios* remained to provide the highest standard of care possible, regardless of what environmental obstacles we faced. Software portability strives for the same objective, often overcoming environmental variability to achieve equivalent results that deliver full or partial business value.

DEFINING PORTABILITY

Portability is "the ease with which a system or component can be transferred from one hardware or software environment to another."[1] As a subcomponent of transferability within the International Organization for Standardization (ISO) software product quality model, software portability describes the ability of software to operate equivalently across variable environments. Variability can be manifested in the hardware, operating system (OS), infrastructure (e.g., network, folder structure), SAS application (e.g., interface, version, available components, compatibility options), or SAS processing mode (i.e., interactive, batch).

With few exceptions, Base SAS is naturally portable between environments without code modification, and this flexibility is expected in fourth-generation languages (4GLs) that largely aim to remove environmental-specific code that must be customized by developers. The compounding effects of the numerous dimensions of environmental variability can cause rare but unwanted results in SAS software function or performance. Therefore, implementing portability in software effectively validates that the software has been tested across specific environments and that it can be safely transferred to and executed in environments identical to those in which it was tested.

This chapter introduces several dimensions of portability that can effectively expand the intended audience of SAS software, making it viable in a greater range of environments. Even within development environments that are relatively stable over time, portability is sought because it can facilitate software that executes without modification in separate development, test, and production environments, militating against the inefficient practice of maintaining separate code bases. To the extent that development, test, and production environments can execute identical software, code maintainability will be substantially improved.

DISAMBIGUATING PORTABILITY

Transferability is "the degree to which the software product can be transferred from one environment to another."[2] Within the ISO software product quality model, the transferability characteristic includes four sub-characteristics: portability, installability, adaptability, and compliance. Portability is the focus of this chapter and typically reflects the ease, efficiency, and effectiveness with which software operates across two or more environments. Portability is often defined dichotomously as software that either functions or does not. However, as demonstrated throughout this chapter, gradations of portability can exist, especially where software functionality is portable between systems but performance varies substantially.

Installability is "the degree to which the software product can be successfully installed and uninstalled in a specified environment."[3] Installability can facilitate software that "unpacks" its environment, including required infrastructure, folders, and files, and often enables users to customize software adequately during initialization. Often, in literature, both portability and installability are defined as static performance attributes (and thus characteristics of internal software quality) because examination of code can demonstrate the environments to which software is intended to be portable. However, because portability and installability failures are demonstrated through software execution (not inspection), they are included as dynamic performance requirements within this text.

Adaptability is defined as synonymous to *flexibility*, "the degree to which the software product can be adapted for different specified environments without applying actions or means other than those provided for this purpose for the software considered."[4] Flexibility is referenced throughout this text, principally in the roles that it plays in supporting software stability, reusability, and extensibility. *Transferability compliance* is not discussed in this text but is "the degree to which the software product adheres to standards or conventions relating to portability."[5]

Although subsumed under the compatibility quality characteristic in the ISO software product quality model, *interoperability* can be closely tied to portability and is defined as "the degree to which the software product can be cooperatively operable with one or more other software products."[6] The SAS application demonstrates interoperability in its ability to interface with third-party software products that can include drivers, APIs, and databases. Interoperability is not discussed further in this text.

3GL VERSUS 4GL PORTABILITY

Software portability is so ubiquitous that it is widely overlooked and taken for granted—and that is the intent. I wake up in the morning, and before even climbing out of bed, I perform a Google search on my phone to check the weather and news. Minutes later, I flip open a laptop and check traffic on Google Maps before driving to work. At work, it's a different computer and possibly a different Internet browser, but the same Google interface to conduct countless searches. Millions of people are using the same web application around the world on various devices and interfaces, all having similar experiences due to software portability. Without this portability, user experience on the application could vary tremendously by Internet browser, hardware, device, and other environmental characteristics.

Data analytic software often differs from traditional software applications because solutions are developed to meet niche analytical requirements rather than for broad dissemination. SAS practitioners don't typically buy SAS Enterprise Guide to develop software and then attempt to market that software beyond its original use or environment. SAS software may be shared and reused within an organization, and sometimes more broadly disseminated through literature and technical white papers, but it inherently differs from applications development because en masse distribution and use are never goals.

4GLs such as Base SAS are inherently more portable than third-generation languages (3GLs). 4GLs focus less on resource management and more on functionality. Thus, Base SAS is designed to offer similar service across the breadth of hardware on which it operates, so that running the general linear model or GLM procedure in one environment will perform equivalently to other environments. When stakeholders purchase the SAS application, they expect the SAS software they author to function identically regardless of the environment in which it's run.

Thus, SAS practitioners have far fewer portability obstacles than can plague 3GL developers, and in many cases can ignore the environment in which they are operating. Most SAS software written on one system will function on another. Because portability can often be overlooked in SAS development, however, it's often not contemplated until functional or performance failures are investigated and portability is determined to be the culprit. The following "Facets of Portability" sections define the overarching dimensions of SAS environments that can pose portability challenges to SAS practitioners.

FACETS OF PORTABILITY

Portability aims to overcome functional and performance discrepancies that exist among systems due to the myriad environmental attributes that can vary in SAS environments. These attributes can include the following:

- *OS*—for example, Windows, UNIX
- *SAS interface*—for example, SAS Display Manager, SAS Enterprise Guide, SAS Studio
- *SAS processing mode*—for example, batch or interactive
- *SAS version*—for example, 9.3, 9.4
- *Licensed SAS components*—for example, SAS/GRAPH, SAS/STAT, SAS/ACCESS
- *SAS environment portability*—the local installation of SAS, including folder and file organization, SAS options, and customization of configuration files
- *SAS data*—data set file formats
- *Software development life cycle (SDLC)*—for example, independent environments that comprise the development, testing, and production phases

The following sections don't aim to give exhaustive lists of all portability challenges that may exist. Rather, they demonstrate aspects of the environment that may pose portability challenges and of which SAS practitioners should be aware.

OS Portability

The vast majority of the Base SAS language operates equivalently across OS environments, such as Windows, UNIX, and z/OS. In fact, most Base SAS technical documentation is contained in OS-agnostic literature, such as the *SAS® 9.4 Macro Language Reference*,[7] *Base SAS® 9.4 Procedures Guide*,[8] or the *SAS® 9.4 Functions and CALL Routines: Reference*.[9] Within those compendia, the beauty of Base SAS portability is revealed in that SAS practitioners rarely have to make programmatic decisions based on the OS on which they are developing. To a large extent and in much of data analytic development, the OS can be ignored.

In some cases, idiosyncrasies do exist in SAS architecture, implementation, syntax, or how code executes in different environments. Thus, in addition to the broad SAS software documentation, SAS also produces OS-specific documentation such as the *SAS® 9.4 Companion for UNIX Environments*[10] or the *SAS® 9.4 Companion for Windows.*[11] Because Windows is a popular environment for the installation of SAS client software, and because LINUX/UNIX is common for server environments and undergirds the SAS University Edition, SAS practitioners may occasionally perceive lack of portability between these OSs.

Capitalization in Windows and UNIX

Regardless of the OS environment, the SAS application is not case sensitive in its interpretation of SAS code. Even reserved words such as procedures, functions, and SAS global macro variables don't require capitalization as in some languages. For example, while the SAS procedures SORT and MEANS are capitalized within the text of this book, they have not been capitalized in code examples. The Windows environment operates similarly to the SAS application and is not case sensitive. UNIX, however, does enforce capitalization. Thus, when SAS software must interact with its environment through I/O functions or environmental variables, this discrepancy can cause SAS software to fail.

For example, the following code retrieves the SASROOT operating environment variable that demonstrates the location of the SAS.exe executable file. The code executes in both Windows and UNIX environments because SASROOT has been capitalized:

```
%put %sysget(SASROOT);
C:\Program Files\SASHome\SASFoundation\9.4
```

However, when SASROOT is not capitalized (as might be common practice in some SAS Windows environments) the code fails when executed in UNIX because "sasroot" is not recognized:

```
WARNING: The argument to macro function %SYSGET is not defined as
a system variable.
%put %sysget(sasroot);
```

A straightforward development best practice that overcomes this limitation is to capitalize environmental variables to ensure portability. However, capitalization restrictions extend beyond environmental variables to references of SAS path names. For example, the SAS University Edition uses the default folder /folders/myfolders/ (all lowercase) as the root directory for the SAS environment. The following LIBNAME statement successfully assigns the library LOWER to this location:

```
libname lower '/folders/myfolders/';
NOTE: Libref LOWER was successfully assigned as follows:
      Engine:        V9
      Physical Name: /folders/myfolders
```

However, the following library assignment fails in the SAS University Edition because its capitalization is inconsistent:

```
libname lower '/folders/MyFolders/';
NOTE: Library LOWER does not exist.
```

Thus, in some cases, Windows users may experience runtime errors when porting software from Windows to UNIX.

SAS Interface Portability

The SAS Display Manager, SAS Enterprise Guide, and SAS Studio represent individual SAS interfaces that compile and execute Base SAS software. While collectively referred to as the "SAS application" throughout this text, these unique interfaces can demonstrate subtle differences in functionality and performance. For example, SAS environmental macro variables may be assigned uniquely within an interface and not be available in all interfaces. Performance characteristics such as software speed can also vary greatly between SAS interfaces, and because SAS practitioners most commonly operate on a single SAS interface, these performance differences may be widely unrecognized.

Environmental Variables

SAS environmental variables can vary by SAS interface. For example, when you open a new SAS Display Manager session, no global macro variables are initially defined. This is demonstrated by submitting the following code, which produces no output:

```
%put _global_;
```

However, within Enterprise Guide, the output is very different:

```
%put _global_;
GLOBAL SASWORKLOCATION "C:\Users\stud\AppData\Local\Temp\SEG17308\
   SAS Temporary Files\_TD15952_COMPUTADOR_\Prc2/"
```

```
GLOBAL _CLIENTAPP 'SAS Enterprise Guide'
GLOBAL _CLIENTAPPABREV EG
GLOBAL _CLIENTMACHINE 'STUD'
GLOBAL _CLIENTPROJECTNAME ''
GLOBAL _CLIENTPROJECTPATH ''
GLOBAL _CLIENTTASKLABEL 'Program'
GLOBAL _CLIENTUSERID 'stud1'
GLOBAL _CLIENTUSERNAME ''
GLOBAL _CLIENTVERSION '7.100.1.2805'
GLOBAL _EG_WORKSPACEINIT 1
GLOBAL _SASHOSTNAME 'stud'
GLOBAL _SASPROGRAMFILE
GLOBAL _SASSERVERNAME 'Local'
```

When submitted from the SAS University Edition, the code again produces an entirely different set of SAS environmental variables:

```
%put _global_;
GLOBAL BASEDIR /folders/myfolders/
GLOBAL GRAPHINIT
GLOBAL GRAPHTERM
GLOBAL OLDPREFS /folders/myfolders/.wepreferences
GLOBAL OLDSNIPPETS /folders/myfolders/.mysnippets
GLOBAL OLDTASKS /folders/myfolders/.mytasks
GLOBAL STUDIODIR /folders/myfolders/.sasstudio
GLOBAL STUDIODIRNAME .sasstudio
GLOBAL STUDIOPARENTDIR /folders/myfolders
GLOBAL USERDIR /folders/myfolders
GLOBAL _BASEURL http:localhost:10080SASStudio
GLOBAL _CLIENTAPP SAS Studio
GLOBAL _CLIENTAPPVERSION 3.4
GLOBAL _EXECENV SASStudio
GLOBAL _SASPROGRAMFILE
GLOBAL _SASWSTEMP_ foldersmyfolders.imagescda3fc30dc864cf9a17bbd7afcad3d83
GLOBAL _SASWS_ foldersmyfolders
```

These differences may seem insignificant until you're writing software that is designed to be ported across different SAS interfaces that must rely on one or more of these variables. Consider the seemingly straightforward task of obtaining the current folder in which a SAS program is saved, which is complicated because the SAS Display Manager, SAS Enterprise Guide, and SAS University Edition each utilize different methods for obtaining this environmental variable. To execute, the %GETCURRENTPATH macro must be saved to a named SAS program file:

```
* saved as /folders/myfolders/test/mycode.sas;
%macro getcurrentpath();
%let syscc=0;
%global path;
%let path=;
%global getcurrentpathRC;
%let getcurrentpathRC=GENERAL FAILURE;
%if %symexist(_clientprojectpath) %then %do;
   %let path=%sysfunc(dequote(&_clientprojectpath));
   %let path=%substr(&path,1,%length(&path)-%length(%scan(&path,-1,\)));
   %end;
%else %if &SYSSCP=WIN %then %do;
   %let path=%sysget(SAS_EXECFILEPATH);
   %let path=%substr(&path,1,%length(&path)-%length(%scan(&path,-1,\)));
   %end;
%else %if &_CLIENTAPP=SAS Studio %then %do;
   %let pathfil=&_SASPROGRAMFILE;
   %let pathno=%index(%sysfunc(reverse("&pathfil")),/);
   %let path=%substr(&pathfil,1,%eval(%length(&pathfil)-&pathno+1));
   %end;
%else %do;
   %let getcurrentpathRC= Environment Not Defined!;
   %put &getcurrentpathRC;
   %end;
%if &syscc=0 and %length(&path)>0 %then %let getcurrentpathRC=;%mend;
```

When executed from within the SAS University Edition, the global macro variable &_CLIENTAPP is detected and the following output is produced:

```
%getcurrentpath;
%put PATH: &path;
PATH: /folders/myfolders/test
```

If executed from an unnamed SAS program, the %GETCURRENTPATH macro will not execute and produces the following warning:

```
%getcurrentpath;
WARNING: Argument 2 to macro function %SUBSTR is out of range.
WARNING: Argument 3 to macro function %SUBSTR is out of range.
%put PATH: &path;
PATH:
```

The exception handling in the %GETCURRENTPATH macro detects general failures by validating the value of &SYSCC and detects the specific failure that occurs when an unknown interface is encountered. While this fault-tolerance attempts to prevent errors that would occur if the code were run in interfaces other than the three specified, the degree of portability still is unknown until tested in those specific interfaces. For example, because SAS Display Manager was only tested on a Windows platform, it's unclear whether SAS Display Manager on a UNIX platform would behave identically. This underscores the complexities of portability that can occur when dimensions of portability interact, as well as the need to test software in the specific environment in which it will be operated to validate its portability.

Contrasting the SAS University Edition

The SAS University Edition is the gateway drug to SAS, offered free to students, to schools, and for general noncommercial use. When Dr. Goodnight announced the release of the SAS University Edition at the 2014 SAS Global Forum, I flashed a Cheshire-Cat grin as thousands of attendees wildly applauded. The cost-prohibitive nature of SAS (as proprietary software) had limited its accessibility, while the surging popularity of Python, R, and other open-source and free software languages within the analytics and data science communities had provided fierce competition. Thus, the introduction of SAS University Edition allows neophytes to get their feet wet for free, just as they do with other popular data analytic languages.

Despite its license limitations for noncommercial use, the SAS University Edition has other technical limitations that may be less well known, such as the limit of running only two CPUs, which can produce decreased performance compared with other SAS interfaces. For example, Figure 10.1 demonstrates the substantially higher performance of the SORT procedure in the SAS Display Manager as compared with the SAS University Edition when run on the same computer.

Because of the decreased resources available to the SAS University Edition, the software encounters the inefficiency elbow (discussed in the "Inefficiency Elbow" section in Chapter 9, "Scalability") at significantly smaller file sizes than the SAS Display Manager. This difference results in functional failures (out-of-memory errors) that also occur on much smaller data sets in SAS University Edition than in either the SAS Display Manager or SAS Enterprise Guide.

These performance disparities are important to understand, especially for readers whose sole experience with SAS may be the SAS University Edition. Especially because the SAS University Edition is intended to be the

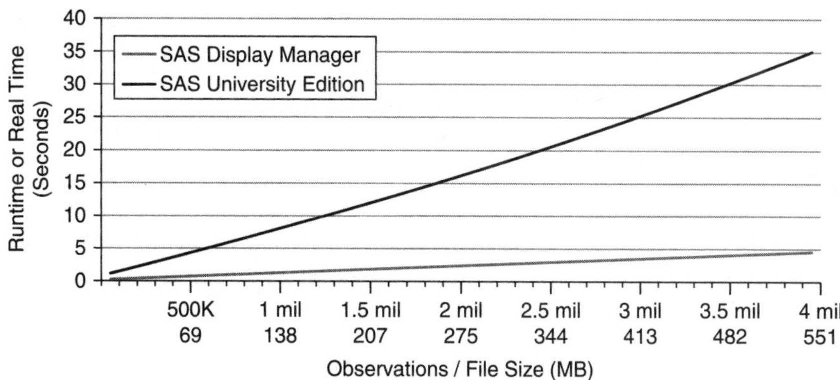

Figure 10.1 SAS University Edition Performance Limitations

first face of SAS that many SAS practitioners will encounter, it's important to understand that other pay-to-play SAS interfaces (like SAS Display Manager and SAS Enterprise Guide) do provide higher performance than the SAS University Edition.

Users experienced with other SAS interfaces may recognize other specific functional limitations in the SAS University Edition. For example, the following code attempts to shell to the OS and list files in the current directory. However, it fails because the interface for University Edition by default does not enable SYSTASK:

```
systask command "dir";
ERROR: Insufficient authorization to access SYSTASK COMMAND.
```

With SYSTASK disabled in SAS Studio, this tremendously limits parallel processing because independent SAS sessions cannot be asynchronously spawned, as demonstrated in the "Decomposing SYSTASK" section in Chapter 12, "Automation." Moreover, while batch jobs were introduced within the SAS University Edition in 2016, they must be run from within a web browser from inside the virtual machine, further preventing scheduled production jobs from being run in this interface. Notwithstanding these subtle limitations, the SAS University Edition offers substantial SAS functionality and performance for free.

SAS Processing Mode Portability

The processing mode represents whether SAS is running interactively, in which code is submitted and executed manually, or in batch mode, in which SAS programs are executed from the SYSTASK command or the OS directly via a command line statement or batch program. SAS software is typically developed in the interactive mode because it enables SAS practitioners to execute code, immediately view results in the log, output, and data windows, and continue to modify the code as necessary. However, because a common goal is to automate and schedule SAS production software to execute with little to no human interaction, the transition to SAS batch jobs is often the final step in development.

Few differences exist between how software executes in interactive mode versus batch; however, because batch opens the SAS application—not just a specific program—system options and other options (such as an Autoexec.sas or configuration file) can be specified at initialization, allowing tremendous diversity in software execution. At times, it may be necessary to determine programmatically whether software is executing interactively or in batch so that program flow can be dynamically altered. To demonstrate the distinction between the two modes, the following code can be saved to C:\perm\batchtest.sas.

```
%put SYSENV: &SYSENV;
%put SYSPROCESSMODE: &SYSPROCESSMODE;
```

When the code is executed from the SAS Display Manager (in interactive mode), the following output is produced. Note that the SAS automatic macro variable &SYSENV is set to FORE to indicate that SAS is running interactively in the foreground.

```
%put SYSENV: &SYSENV;
SYSENV: FORE
%put SYSPROCESSMODE: &SYSPROCESSMODE;
SYSPROCESSMODE: SAS DMS Session
```

To execute Batchtest.sas in batch mode from Windows, the following command should be entered (in a single line) at the command prompt.

```
"C:\program files\SASHome\SASFoundation\9.4\sas.exe" -noterminal
   -sysin c:\perm\batchtest.sas -log c:\perm\batchtest.log
```

When executed, the log file Batchtest.log indicates that the &SYSENV macro variable now reflects that the software was executed in the BACK (background) rather than foreground while the &SYSPROCESSMODE macro variable now also indicates batch operation.

```
%put SYSENV: &SYSENV;
SYSENV: BACK
%put SYSPROCESSMODE: &SYSPROCESSMODE;
SYSPROCESSMODE: SAS Batch Mode
```

Note that the NOTERMINAL command line option must be specified to enable batch mode; without NOTERMINAL specified, the &SYSENV variable will always reflect FORE, even when run from batch mode.

Because SAS system options can be passed via batch invocation, and customized parameters can be passed with the SYSPARM option, batch jobs can function very differently than their interactive equivalents. In fact, if software depends on SAS options or parameters passed through batch invocation, batch software may lack backward compatibility, in that it no longer can be executed directly from the interactive mode. The "Batch Backward Compatibility" section in Chapter 12, "Automation," demonstrates the importance of facilitating software that can still be run interactively to support necessary maintenance and testing.

SAS Version Portability

As each SAS version is released, additional functionality is included, and for the most part, backward compatibility is assured so that code that functioned in the past will continue to do so in the future. Performance can also be improved, such as SORT and SQL, which were upgraded to embrace multithreading with the SAS 9 release. As a result, sorts performed in SAS 9 should perform faster than their equivalent SAS 8 sorts despite identical code. The FULLSTIMER performance metrics comparing multithreaded and single-threaded sorting are demonstrated in the "CPU Processing" section in Chapter 8, "Efficiency."

Version portability is typically not a problem for organizations; after all, why would a team using a newer version of SAS suddenly switch to an older version? On occasion, however, multiple versions may be simultaneously maintained in an environment, especially when the SAS applications are being upgraded. For example, it's not uncommon for a team to upgrade its development server to get the kinks out before upgrading test and production servers. Thus, for periods of time, SAS practitioners may need to become cognizant of functionality or performance differences between the environments.

Version compatibility also becomes an issue when code is ported from SAS white papers or other literature using newer versions of SAS. For example, with the release of SAS 9.4, I've started using the DELETE procedure that debuted. However, because many organizations still run SAS 9.3, if I were publishing code, it might be beneficial to utilize the older DATASETS procedure to delete a file to ensure portability to prior SAS versions. In rare cases—and I won't shame a certain federal agency that in 2016 still uses SAS 8—a significant amount of work may be required to translate new functionality into antediluvian SAS versions and environments.

When software is developed without regard for backward compatibility and utilizes newer functionality, one method to prevent catastrophic failure (in the event that the software later finds itself being run by an older version of SAS) is to implement exception handling that validates SAS software version. By testing the automatic macro variable &SYSVER, software can fail safe or dynamically perform equivalent functionality when encountering an older version of SAS. For example, the following code ensures SAS 9.4 is running before executing the remainder of the %TESTVER macro, which requires the DELETE procedure:

```
%macro testver();
%if &sysver^=9.4 %then %do;
   %put Requires SAS v9.4;
   %return;
   %end;
proc delete data=perm.mydata;
run;
%mend;

%testver;
```

However, when SAS 9.5 (or 10?) is released, this code will unintentionally fail because it uses the not equal operator. A more robust and reusable solution tests a parameterized version number and provides a return code to demonstrate success or failure if an equivalent or newer version of SAS is detected:

```
* tests whether SAS version is suffiviently high enough;
%macro testver(ver= /* version number in 9 or 9.4 format--no V */);
%global testverRC;
%let testverRC=FAILED;
%let realver=9.3;
%if %sysevalf(&sysver<&ver) %then %do;
   %put Sad Panda Alert: Requires SAS &ver but you only have &sysver;
   %return;
   %end;
%else %let testverRC=;
%mend;
```

Running the code on SAS 9.4 now produces the expected exception and reroutes program flow when SAS 9.5 is required through the invocation:

```
%testver(ver=9.5);
Sad Panda Alert: Requires SAS 9.5 but you only have 9.4
%put RC: &testverRC;
RC: FAILED
```

In production software, rather than printing a message about the exception in the SAS log, the return code &TESTVERRC would be utilized to drive program flow dynamically within the parent process that had called the %TESTVER macro.

SAS Component Portability

SAS software is licensed as separate components or modules, the most common of which is Base SAS, which provides foundational software development and SAS macro development capabilities. Other components, such as SAS/GRAPH, SAS/STAT, or SAS/ACCESS, can be purchased individually or are sometimes bundled together in licensing packages. Even if different organizations have roughly equivalent hardware, infrastructure, and system resources, SAS functionality and performance can still vary wildly based on add-on components that have been purchased.

For example, I recently reviewed the code for the first SAS white paper I ever presented: *Winning the War on Terror with Waffles: Maximizing GINSIDE Efficiency for Blue Force Tracking Big Data*.[12] I had received an email referencing the paper and wanted to answer a question posed by a reader. But when I ran the code, I received a spate of runtime errors; the GINSIDE procedure—at the heart of the paper—did not function because I no longer have a license for SAS/GRAPH. In the same light, organizations should carefully deliberate the decision to eliminate SAS licenses and first ensure that production software does not rely on SAS procedures or other functionality that specific licenses support.

The SETINIT procedure lists all SAS licensed components and licensing periods to the log. The following example demonstrates the format of output in the SAS Display Manager, although licenses and dates will differ by installation, and format may differ by SAS interface and OS:

```
proc setinit;
run;
```

```
NOTE: PROCEDURE SETINIT used (Total process time):
      real time           0.05 seconds
      cpu time            0.00 seconds

Original site validation data
Current version: 9.04.01M3P062415
Site name:     'SASSY McSASSY'.
Site number:   12345678.
Expiration:    14OCT2019.
Grace Period:  45 days (ending 28NOV2019).
Warning Period: 45 days (ending 12JAN2020).
System birthday:    12SEP2013.
Operating System:   WX64_WKS.
Product expiration dates:
---Base SAS Software
      14OCT2016
---SAS/Secure 168-bit
      14OCT2016
---SAS/Secure Windows
      14OCT2016
---SAS Enterprise Guide
      14OCT2016
---SAS Workspace Server for Local Access
      14OCT2016
```

The SAS log is useful to someone interested in confirming license availability but unfortunately offers no automatic macro variables that can be used to retrieve license information programmatically. For example, in more robust code, it would be useful to validate SAS/GRAPH component availability before attempting the GINSIDE procedure to ensure the procedure is licensed.

The following code saves the SETINIT output to a text file, parses the file, and generates an asterisk-delimited macro variable &COMPONENTS that contains a list of all licensed SAS components:

```
%macro get_components();
%global components;
%let components=;
%local out;
%let out=c:\perm\components.txt;
proc printto log="&out" new;
run;
proc setinit;
```

```
run;
proc printto;
run;
data _null_;
   length text $100 components $1000;
   infile "&out" truncover end=eof;
   input text $100.;
   if _n_=1 then components='';
   if substr(text,1,3)='---' then components=strip(components)
      || strip(substr(text,4,90)) || '*';
   if eof then call symput('components',substr(strip(components),1,
      length(strip(components))-1));
   retain components;
run;
%mend;
```

When executed, the following output demonstrates the macro variable &COMPONENTS that could be parsed further (not shown) to determine dynamically whether all required components for specific software are available:

```
%get_components;

NOTE: PROCEDURE PRINTTO used (Total process time):
      real time         0.00 seconds
      cpu time          0.00 seconds

NOTE: The infile "c:\perm\components.txt" is:
      Filename=c:\perm\components.txt,
      RECFM=V,LRECL=32767,File Size (bytes)=1217,
      Last Modified=07May2016:23:13:25,
      Create Time=07May2016:23:10:46

NOTE: 32 records were read from the infile "c:\perm\components.txt".
      The minimum record length was 0.
      The maximum record length was 98.
NOTE: DATA statement used (Total process time):
      real time         0.01 seconds
      cpu time          0.00 seconds

%put SAS Components: &components;
SAS Components: Base SAS Software*SAS/Secure 168-bit*SAS/Secure
  Windows*SAS Enterprise Guide*SAS Workspace Server for Local Access
```

This methodology is extremely beneficial where code is intended to be ported to other and especially unknown environments, as is common when SAS code is published. For example, including the %GET_COMPONENTS macro in software that relies on the SAS/GRAPH license could enable the software to determine programmatically that SAS/GRAPH was (or was not) installed and dynamically alter program flow to prevent runtime errors.

SAS Environment Portability

Install the same SAS application and version on identical hardware in two different organizations, and SAS software may still run differently. SAS system options, configuration files, and Autoexec.sas files provide endless opportunities to customize not only the look and feel of SAS but moreover the way the SAS application functions. Another source of variability often stems from the hierarchy of logical drives, folders, and files that comprise the SAS infrastructure.

SAS System Options

Because SAS options can so profoundly affect software function and performance, while they are customizable, many cannot be modified within a SAS session and must be invoked through Autoexec.sas, configuration files, or at program invocation through command line statements. Some system options can also be restricted by SAS administrators to protect against accidental or malicious modification. Extensive information about system options and their respective limitations is included in *SAS® 9.4 System Options: Reference, Fourth Edition.*[13]

To further complicate the options landscape, the mode in which SAS executes—interactive or batch—can affect default options settings. To demonstrate these differences, the following code should be saved to the file C:\perm\options_batch.sas and executed:

```
* saved as c:\perm\options_batch.sas;
libname perm 'c:\perm';
%put SYSENV: &sysenv;
proc optsave out=perm.options_batch;
run;
```

The program uses the OPTSAVE procedure to write all option values to the PERM.Options_batch data set. When the following code is subsequently executed, it runs the Options_batch.sas program a second time in batch mode (with the SYSTASK statement) to demonstrate differences between system options:

```
libname perm 'c:\perm';
proc optsave out=perm.options_interactive;
run;

systask command """%sysget(SASROOT)\sas.exe"" -noterminal -sysin
   ""c:\perm\options_batch.sas"" -log ""c:\perm\options_batch.log"""
   wait status=rc_batch;

proc sort data=perm.options_batch;
   by optname;
run;
proc sort data=perm.options_interactive;
   by optname;
run;
data perm.options;
   merge perm.options_batch (in=a) perm.options_interactive
      (in=b rename=(optvalue=optvalue2));
   by optname;
   if a and not b then put optname 'not in batch';
   else if b and not a then put optname 'not in interactive';
   else if optvalue^=optvalue2 then do;
      put optname 'values differ';
      put '-- BATCH: ' optvalue;
      put '-- INTER: ' optvalue2;
      end;
run;
```

The good news is that identical system options are identified in both batch and interactive modes. However, the following abridged log does demonstrate subtle differences in option settings. Other installations and environments will of course demonstrate unique results:

```
DLDMGACTION values differ
-- BATCH: FAIL
-- INTER: REPAIR
FONT values differ
-- BATCH:
-- INTER: (Sasfont 8)
_LAST_ values differ
-- BATCH: _NULL_
-- INTER: PERM.OPTIONS
```

When SAS software requires substantial customization of SAS system options, including those defined both at software invocation and in the code itself, one best practice to ensure reliable execution is to implement a quality control early in the program that compares current system option settings to expected or required options. Thus, by establishing a baseline data set that contains the required system options settings for specific software, that software can validate current options against the baseline and prevent execution if options have been modified. Especially when running software in a new environment, this type of quality control can prevent disaster while elucidating environmental discrepancies.

SAS Folder Structure

Every SAS environment has an inherent structure that defines where programs, macros, saved formats, configuration files, control tables, data sets, and other files are located. When software is ported to and installed in a new environment, the new infrastructure may be vastly different and require software components that are missing. Even when necessary environmental structure and files do exist, portable software should flexibly reference those components so multiple copies of software don't need to be maintained. For example, server names or logical folder names will often differ from a development server to a production server, but through flexible software design, these attributes can be dynamically encoded to maximize maintainability and ensure a single software instance can suffice for all three environments.

Throughout this text, the C:\perm folder is repeatedly referenced and assigned to the PERM library. To facilitate readability, the library assignment is hardcoded:

```
libname perm 'c:\perm\';
```

However, this statement assumes that the folder C:\perm exists and, with the default option NODLCREATEDIR enabled, will produce a note to the SAS log if it does not exist:

```
libname perm 'c:\perm';
NOTE: Library PERM does not exist.
```

To overcome this exception, the DLCREATEDIR option can be specified, which will create the folder through the LIBNAME statement if C:\perm does not exist:

```
options dlcreatedir;
libname perm 'c:\perm';
```

```
NOTE: Library PERM was created.
NOTE: Libref PERM was successfully assigned as follows:
      Engine:        V9
      Physical Name: c:\perm
```

This code will still fail if the C:\ drive itself doesn't exist. Moreover, the code is biased toward a Windows environment. Another method to make library assignments more flexible is to assess and use the current library—that is, the folder in which a SAS program is saved. This method is demonstrated earlier in the "Environmental Variables" section, and can be used to create or access a subordinate folder structure. For example, if the %GETCURRENTPATH macro assesses that the current path is C:\perm and assigns this value to the global macro variable &PATH, then &PATH could be used subsequently to create or access subordinate folders like C:\path\data or C:\path\output. This functionality allows SAS practitioners to drop code into an environment and automatically populate its infrastructure, similar to dropping Sea-Monkey or Sea-People capsules into an aquarium.

Another way to eliminate hardcoded references is to enforce that SAS library assignments be made outside of the software—through the SAS Management Console, Autoexec.sas, SAS configuration files, the SAS Autocall Macro Facility, or other dynamic methods. These assignments establish a single, hardcoded reference to a logical folder and replace the necessity to hardcode common library assignments throughout various programs. This may not seem like a tremendous advantage until your information technology (IT) department unexpectedly decides to rename your SAS server, causing hundreds of programs to crash simultaneously as every library reference fails. I was once faced with this crisis and, while a team of developers scrambled to alter our entire software base to run in the "new environment," we also improved the portability of all software by removing all LIBNAME statements from production software in favor of dynamic library assignments.

SAS Servers

Some SAS environments maintain separate development, testing, and production servers to handle discrete phases of the SDLC. Portability becomes a common concern in these environments because of software transfer between systems that may maintain different versions of SAS, file structures, or even functional intent. For example, in development and test environments, SAS software is often run interactively so that SAS practitioners can modify it, which presumably will require repeated interrogation of the SAS log.

Production software, on the other hand, hopefully embraces some method to parse SAS logs automatically for warnings, runtime errors, and exceptions, but might not require maximum verbosity needed during earlier phases of the SDLC.

Because a unique SAS license is required for each installation, evaluation of the &SYSSITE automatic macro variable can be extremely useful in dynamically altering program flow based on execution environment. For example, the following %INIT macro assesses in which environment software is executing, appropriately modifies system options, dynamically assigns libraries, and even prints a "TEST" caveat on all output produced in the development environment:

```
%macro init();
%if &syssite=12341234 %then %do; * DEVELOPMENT SERVER;
   options fullstimer mlogic msglevel=i;
   libname perm 'c:\perm';
   title3 'TEST';
   %put "Maximum verbosity.";
   %end;
%else %if &syssite=56785678 %then %do; * PRODUCTION SERVER;
   options nofullstimer nomlogic msglevel=n;
   libname perm 'd:\perm';
   title3;
   %put "Brief.";
   %end;
%mend;
```

Software maintainability is substantially increased when a single version of software can be maintained that functions across all internal environments and servers. Without dynamic server recognition, SAS practitioners might be forced to implement error-prone hardcoding as software is transferred between environments. This unnecessary maintenance violates maintainability and stability principles discussed in the "Toward Software Stability" section in Chapter 11, "Security."

Data Portability

SAS practitioners who have operated within a single environment, or have switched environments but haven't stolen data from their previous employer, might be surprised to learn that data set encoding can also vary from one system to another. Encoding gives greater flexibility to how data are interpreted—including language and other considerations—but can be defaulted to different values across environments.

To demonstrate this phenomenon, the following code creates the TEST.Mydata data set in the SAS University Edition:

```
data test.university (drop=i);
   length num1 8;
   do i=1 to 1000;
      num1=round(rand('uniform')*10);
      output;
      end;
run;
```

However, when the data set is opened in the SAS Display Manager for Windows on the same computer, a note states that the data set encoding does not match the environment encoding:

```
data windows;
   set test.university;
NOTE: Data file TEST.UNIVERSITY.DATA is in a format that is native
to another host, or the file encoding does not match the session
encoding. Cross Environment Data Access will be used, which might
require additional CPU resources and might reduce performance.
313  run;


NOTE: There were 1000 observations read from the data set
TEST.UNIVERSITY.
NOTE: The data set WORK.WINDOWS has 1000 observations and 1 variables.
NOTE: DATA statement used (Total process time):
      real time           0.15 seconds
      cpu time            0.01 seconds
```

Even the COMPARE procedure finds no discrepancies between the SAS University Edition and SAS Display Manager data sets. However, through an investigation of the encoding for the two SAS interfaces with the OPTIONS procedure, the default encodings are shown to differ. The SAS University Edition defaults to UTF-8 while the SAS Display Manager for Windows defaults to WLATIN1.

```
proc options option=encoding; * run in SAS Display Manager;
run;


ENCODING=WLATIN1  Specifies the default character-set encoding for
the SAS session.
```

```
NOTE: PROCEDURE OPTIONS used (Total process time):
      real time           0.01 seconds
      cpu time            0.00 seconds

proc options option=encoding; * run in the SAS University Edition;
run;

ENCODING=UTF-8    Specifies the default character-set encoding for
the SAS session.
NOTE: PROCEDURE OPTIONS used (Total process time):
      real time           0.02 seconds
      cpu time            0.03 seconds
```

In this example, the data are portable between the two SAS environments thanks to a conversion that the SAS application automatically applies; no data were deleted or misrepresented. When encoding values differ more substantially, however, such as when porting data between SAS software that have been run on SAS applications with different default languages, discrepancies can emerge and corrupt data. To eliminate the risk of misrepresenting data or ingesting corrupt data, a quality control can be built (not demonstrated) that programmatically validates actual data set encoding against expected encoding values. This control can be especially beneficial to environments required to ingest third-party data over which they have no influence in data quality, construct, or encoding.

PORTABILITY IN THE SDLC

Within teams and organizations that maintain a single SAS server or infrastructure, portability will have little relevancy in software design. More universal aspects of portability include actions that SAS practitioners might undertake to implement external SAS code or ingest SAS data from third-party sources. However, for SAS practitioners who operate in diverse environments or intend to publish SAS code for general distribution, environmental aspects that influence portability should be considered and implemented commensurate to the diversity of the intended execution environments.

Requiring Portability

Portability requirements are typically straightforward and specify the environments in which software must operate. For a team operating in a single SAS infrastructure, a requirement could state that software should operate on the SAS Display Manager for Windows 9.4. Other beneficial statements could explicitly state that a single code base for software should be maintained to support batch and interactive modes, as well as throughout all phases of the SDLC. This approach demonstrates a commitment to software maintainability and stability and ensures that software will not diverge into separate development and production versions.

To the extent that teams operate on multiple SAS servers or infrastructures, their portability requirements should increase. Even if the environments are virtual mirror images of each other, SAS license numbers, server names, and other subtle variations may require that exception handling dynamically identify the execution environment to perform unique functions. Even in complex environments, however, successful portability requirements can often be expressed in a single statement.

In some cases, portability will need to be interwoven into other performance requirements. For example, given the substantial differences in execution speed between the SAS University Edition and the SAS Display Manager demonstrated in the "Contrasting the SAS University Edition" section, if an organization or team utilized both SAS interfaces, they might need to differentiate required execution speed based on the interface used. Moreover, given that the SAS University Edition is incapable of processing larger files that SAS Display Manager and SAS Enterprise Guide can easily handle, file size thresholds and other requirements might also need to be customized to each specific SAS interface.

Measuring Portability

The ultimate goal of portability is effectively to expand the range of environments in which SAS software successfully operates. In most cases, the outcome of software portability is dichotomous—software either executes in UNIX or it fails. To measure portability, however, software must be tested within the specific environments to which it will be deployed and against all variability that it is designed to handle. For example, a SAS practitioner developing software in the SAS University Edition cannot certify that the software will execute correctly in SAS Enterprise Guide until the software is tested within SAS Enterprise Guide. Thus, measuring portability essentially provides validation that software has been tested in a specific environment against established environmental diversity or variability.

While software functionality is fairly dichotomous, software performance such as execution speed can vary substantially based on variability in the execution environment. A SAS development environment might have a relatively tame infrastructure while its production environment includes grid architecture and other performance enhancements. These and other disparities between environments would necessitate that performance be measured in each environment to ensure portability specifications are met in both environments. In another example, development and testing environments might utilize test data sets with data volumes significantly lower than expected during operational conditions. Given this disparity, as described in Chapter 9, "Scalability," software would need to be tested to demonstrate it scales successfully to the volume and velocity of production data.

WHAT'S NEXT?

While the primary objective of software portability is to expand the functionality of software across diverse environments, a secondary goal is to ensure that software does not encounter an unknown (or untested) environment and subsequently act in an unpredictable manner. In other words, 'tis better to fail safe than to fail stupid. This second objective of portability focuses on software security, which is introduced in the next chapter. Secure software aims to ensure that it never hurts itself or its environment.

NOTES

1. ISO/IEC 25010:2011. *Systems and software engineering—Systems and software Quality Requirements and Evaluation (SQuaRE)—System and software quality models*. Geneva, Switzerland: International Organization for Standardization and Institute of Electrical and Electronics Engineers.
2. Id.
3. Id.
4. Id.
5. Id.
6. Id.
7. *SAS 9.4® Macro Language: Reference, Fourth Edition*. Retrieved from https://support.sas.com/documentation/cdl/en/mcrolref/67912/HTML/default/viewer.htm#titlepage.htm.
8. *Base SAS® 9.4 Procedures Guide, Fifth Edition*. Retrieved from https://support.sas.com/documentation/cdl/en/proc/68954/HTML/default/viewer.htm#titlepage.htm.
9. *SAS® 9.4 Functions and CALL Routines: Reference, Fourth Edition*. Retrieved from https://support.sas.com/documentation/cdl/en/lefunctionsref/67960/HTML/default/viewer.htm#titlepage.htm.
10. *SAS® 9.4 Companion for UNIX Environments, Fifth Edition*. Retrieved from http://support.sas.com/documentation/cdl/en/hostunx/67929/HTML/default/viewer.htm#titlepage.htm.
11. *SAS® 9.4 Companion for Windows, Fourth Edition*. Retrieved from http://support.sas.com/documentation/cdl/en/hostwin/67962/HTML/default/viewer.htm#titlepage.htm.
12. Troy Martin Hughes, 2013. *Winning the war on terror with waffles: Maximizing GINSIDE efficiency for Blue Force Tracking big data*. Retrieved from http://analytics.ncsu.edu/sesug/2013/PO-18.pdf.
13. "Using SAS System Options." *SAS® 9.4 System Options: Reference, Fourth Edition*. Retrieved from http://support.sas.com/documentation/cdl/en/lesysoptsref/68023/HTML/default/viewer.htm#n1ag2fud7ue3aln1xiqqtev7ergg.htm#n1243vjtfty6pan1ramcaeho4jte.

CHAPTER 11
Security

"¡Señor! ¡Señor! ¡Señor!"

Emerging from behind a pile of rubble, having just swung down from a semi-prostrate ramón nut tree (*Brosimum alicastrum*), I was mid-selfie when a bellowing baritone brashly overtook the still breeze.

The agitated words were undeniably Spanish, but I was too preoccupied with the grandeur of the moment to attempt a translation.

"*Que? Como?!*" I returned with a perfunctory cry, still focused solely on capturing the panoramas—if no longer the stillness—that the outcrop provided. After all, this was Tikal, Guatemala, famed pre-Columbian Maya metropolis and UNESCO World Heritage Site—and some squawker was not about to ruin my ruins!

Pocketing the camera and reeling around to more bellows, I at last saw the gentleman, a miniature security guard who had worked up quite a sweat to summit the pyramid, apparently in haste. Standing behind the security cordon—the chain-link fence to which he was wildly gesticulating—the guard's guttural screeches began to draw a congregation of onlookers.

As I clambered over, sacrificing vistas to confront the commotion, the situation grew somewhat clearer. I stood on rough, precarious terrain, while the guard and tourists stood on smooth, hewn stone. Yes, there were clear boundary markings at the top of Temple IV but, having scampered up the impassable backside, how was I to know I was out of bounds?

Reaching for what I assumed only could be a concealed machete, the guard pulled a radio from his hip, alerting the authorities that all was in order. At that moment, I hoped that my ascent up the back country at least had been mistaken for a troop of howler monkeys, as I half-leapt, half-crept up the tree-laden terrain, branches shaking and breaking in my wake—but the howlers unfortunately were absent from his radio report.

Now "safely" on the right side of the chain-link, I could see how from his vantage it might appear as though I'd illegally crossed the barrier (giving hope to wide-eyed tourists looking to do the same), but I wasn't about to attempt to explain in Spanish that Tikal needed better signage or security at the bottom of the pryamid.

After enjoying a few moments atop the temple taking in unparalleled views (and fondly reminiscing that my archaeology professor had once told me that the Maya built pyramids solely because mosquitos can't fly that high), I descended the contemporary wooden stairway, the guard mirroring my every step.

■ ■ ■

Countless security guards roam the expansive Tikal landscape—and thank God for their service. There have been periods in Tikal's distant past where *bandidos* (bandits) or even troops of howler monkeys have given tourists quite a scare, but Guatemala has invested significant resources in eliminating and mitigating threats to tourists to provide security.

Security aims to protect assets, and at Tikal, assets comprise not only tourists but also the environment, natural habitat, and untold archaeological treasures that may still lie unearthed. Thus, while the guard may have been bellowing because he thought I had intentionally crossed a security cordon, was he protecting me or the decrepit pile of 1,500-year-old rubble I had ascended? It was probably a little of both.

Software security is similarly poised not only to protect software but also to protect the environment in which it operates. While the guard didn't want me to fall and break myself, he was also acting instinctively to protect a site historically ravaged by treasure hunters.

The security cordon and accompanying signage—to which the guard was pointing, albeit the unlettered side that I could not view—acted as security controls, a barrier limiting the interaction between tourists and the environment to protect both. The chain-link and guard didn't eliminate but helped mitigate the risk of falling to one's death. Where threats to software can't be eliminated, software security controls act to reduce or eliminate vulnerabilities that threats might exploit, thus reducing risk.

Security and security controls also aren't intended to be static devices but should mature over time as threats or risks evolve. Each time I've returned to Tikal, there's been at least one additional temple that's been cordoned off and which I'm no longer able to scale due to safety concerns. Software security, too, should continue to incorporate new security measures as additional threats or vulnerabilities are identified, or as the level of acceptable risk changes over time.

Implementing security often comes with a hefty price tag. At Tikal, guard salaries, chain-link fences, signage, ancillary wooden staircases, and lighting all aim to promote security—none of which are free. Stakeholders in software development must also prioritize what combination of security controls will be emplaced to achieve an acceptable level of risk, thus essentially valuating security.

Designing and developing SAS software with security in mind helps stakeholders identify threats, calculate risks, and eliminate or mitigate known vulnerabilities. Similar to security cordons and security guards, SAS best practices can help facilitate secure software execution in a secure environment.

DEFINING SECURITY

Security is "the protection of system items from accidental or malicious access, use, modification, destruction, or disclosure."[1] Information security typically is described using the CIA triad that supports the confidentiality, integrity, and availability of software and resultant data products. Confidentiality ensures that only authorized users have access to software and data and, in SAS environments, is generally maintained through physical security, user authentication, and other access controls. Integrity ensures that software does what it says it will do, through development best practices, testing, and validation. Availability, discussed most extensively in Chapter 4, "Reliability," ensures that software and its components (including external macros, formats, data, and other files) are accessible whenever needed.

Software security is important because of the tremendous investment made to produce software within the software development life cycle (SDLC). After planning, design, development, testing, validation, acceptance, and a possible beta operational phase, stakeholders need to have confidence that software will function correctly when needed. Through integrity principles, stakeholders can be assured that software will not be altered after validation, thus preserving the benefits of quality assurance gained through formalized testing and validation. Moreover, integrity facilitates the development of software that does not damage itself or its environment. Through availability principles, SAS software can more flexibly adapt to locked data sets, missing data, invalid configuration files, and other exceptions that could otherwise cause software failure.

This chapter introduces and applies the CIA information security triad to SAS software development. Because SAS software is commonly executed within end-user development environments protected by access controls at the infrastructure and organizational level, the chapter focuses on integrity and availability rather than confidentiality. Best practices are introduced, such as the generation of checksums for stabilized SAS software, the encapsulation of SAS macros to prevent leakage, and fail-safe process flows. Because of the precisely choreographed interaction that must occur between SAS processes running in parallel, SAS techniques are demonstrated that facilitate secure parallel processing.

CONFIDENTIALITY

Confidentiality is "the degree to which the software product provides protection from unauthorized disclosure of data or information, whether accidental or deliberate."[2] While confidentiality typically refers to securing data, in software development environments that don't produce open-source code, software itself must be secure from penetration. Through compilation and encryption, code is rendered unreadable to users to prevent modification or reverse engineering of proprietary methods and techniques.

Base SAS, however, is much more likely to be utilized in end-user development environments in which software is developed and used by the same cadre of SAS practitioners. Because many of the users of SAS software are themselves—or work in conjunction with—developers, maintaining software confidentiality is typically not a priority. In fact, in data analytic development environments, code transparency is often more desirable than code confidentiality because stakeholders may need to review data modeling, transformation, and analytic techniques. In more extreme environments, such as is required in clinical trials and other medical research, software readability is paramount because code additionally must be audited.

Data confidentiality is more likely to be a concern or requirement in SAS environments than software confidentiality. SAS/SECURE provides both proprietary and industry-recognized encryption methods (e.g., RC2, RC4, DES, TripleDES, AES) which can be implemented manually as well as automatically through organizational policy. Passwords can also be encrypted using the PWENCODE procedure. Starting in SAS 9.4, the SAS/SECURE module is included in Base SAS rather than being licensed separately. For more information on data encryption, please reference *Encryption in SAS® 9.4, Fifth Edition.*[3]

INTEGRITY

Integrity is "the degree to which the accuracy and completeness of assets are safeguarded."[4] In a software development sense, the term *assets* typically refers to software itself, the data it utilizes, and the data or other products that it produces. Data integrity is not a focus of this text, but is discussed extensively in the comprehensive must-read *Data Quality for Analytics.*[5] Many of the practices that facilitate higher quality data, however, can be implemented as quality controls for metadata (e.g., macro variables) utilized by SAS software.

Another aspect of integrity is the extent to which the asset has been safeguarded. In other words, has it been protected from intentional or accidental alteration? *Data pedigree* demonstrates the source of data to the specificity of the data steward, data set, or even observation. For example, if an observation requires further interrogation, metadata in the pedigree can direct stakeholders toward the original, raw data source for comparison or validation. The SDLC similarly seeks to ensure software pedigree, in that stakeholders questioning some aspect of software can review requirements documentation, versioning documentation, a software test plan, test cases, and test results to understand and gain confidence in SAS software.

Pedigree operates akin to chain-of-custody documentation, demonstrating transfer of ownership—of data through data processes or software through the SDLC. Checksums provide an alternative method to demonstrate that assets—whether software or data—have not been altered. A checksum uses a cryptographic hash function to reduce a data set, SAS program, or other file to a hexadecimal string that uniquely identifies the file. Thus, while checksums cannot demonstrate the source of files or speak to their pedigree, they can validate software or data products by demonstrating that the checksum of the current version of the file matches the checksum of the original file. Pedigree and checksums together can instill confidence in stakeholders that software and its required components demonstrate integrity.

Data Integrity

Data are often said to have integrity if they are valid, accurate, and have not been altered. Data accuracy is often difficult to determine or demonstrate without comparison to the data source or real-world constructs, which supports the importance of data pedigree. For example, to determine whether a data transformation procedure is operating correctly, developers often use test cases to validate input and expected output. However, the original data still could be inaccurate (e.g., missing, incorrect) despite an accurate transformation process.

Data validity, on the other hand, describes data that are of an expected type, structure, quantity, quality, and completeness. Thus, data validity doesn't speak to whether data are accurate but rather whether data prima facie appear as they should. Thus, a test data set created to test software functionality should be valid but cannot be accurate because the observations don't reflect any real-world constructs. Data integrity is typically enforced through data integrity constraints and quality controls that flag, expunge, or delete invalid data. While not demonstrated with regard to SAS data sets, quality controls that validate macro parameters and macro variables are demonstrated later in the "Macro Validation" section.

To demonstrate that a data set has not been altered, both data and accompanying metadata must be analyzed. It is not enough solely to show that data are equivalent because this leaves open the possibility that data were altered for some purpose and subsequently changed back to their original structure, content, and values. While the COMPARE procedure compares the contents of a data set (i.e., structure, variables, and values), it does not examine SAS metadata and thus cannot sufficiently demonstrate data integrity. Rather, metadata contained within SAS data sets and available through the DICTIONARY.Tables data set must be interrogated.

The following code creates a PERM.Dict data set that contains metadata describing the PERM.Mydata data set:

```
libname perm 'c:\perm';
data perm.mydata;
   length quote $50;
   quote="Because that's how you get ants!";
run;

proc sql;
   create table perm.dict as
   select *
   from dictionary.tables
   where upcase(libname)="PERM" and upcase(memname)="MYDATA";
quit;
```

If the code is rerun, the *data* contained in Mydata will be identical to the data produced in the first run; however, the *metadata* from Mydata (contained in Dict) will differ because the creation date and modification date (CRDATE and MODATE, respectively) will have been updated. Because SAS metadata are contained within SAS data sets and cannot be separated from the data themselves, cryptographic checksums (discussed later in the "Checksums" section) unfortunately cannot be used to validate new copies of SAS data sets. For example, if you change a data set name, library, or create date—yet leave all actual data unchanged—the checksum will still be altered despite the identical data. This characteristic differs from SAS programs because the metadata in program files (e.g., file name, create date) are not included in the binary contents of the file but are maintained by the operating system (OS). The validation of SAS program files with checksums is demonstrated in the "Checksums" and "Implementing Checksums" sections.

Therefore, when comparing two data sets, validation of data and metadata must occur through SAS rather than through external methods. One limited use of checksums with SAS data sets is to validate that a static data set remains unchanged. For example, a team might want to demonstrate integrity of a SAS format STATES that it had created, which converts state acronyms to their full names. SAS data sets can be used to generate SAS formats with the CNTLIN option in the FORMAT procedure, as demonstrated in the "SAS Formatting" section in Chapter 17, "Stability." For example, the following code builds an abbreviated STATES format from the PERM.States data set:

```
data perm.states;
   input fmtname: $char8. type: $char1. start: $char4. label: $25.;
   datalines;
states c CA California
states c OH Ohio
states c PA Pennsylvania
states c MD Maryland
states c VA Virginia
;
run;

proc format cntlin=perm.states;
run;
```

Of note, the TYPE variable "c" is required for character data when using the CNTLIN option. In this scenario, once the format is created, the team wants to ensure that the format is not modified thereafter. One method to help provide this integrity would be to build the SAS format (using the FORMAT procedure) before using the format in software, but only after validating the checksum of the PERM.States data set. If the checksum matches the original checksum that was produced when the PERM.States data set was initially created, this demonstrates that the data set has not been modified, and further validates the integrity of the STATES format.

If the DATA step that produces PERM.States is ever rerun, however, the data set metadata will be modified, and a new checksum value will need to be generated and recorded. Thus, while this method demonstrates a theoretical use of checksums to facilitate data integrity, in practice, data integrity should be validated through SAS data and metadata. For example, a much simpler solution to ensure the integrity of the STATES format would be to run the COMPARE procedure on the current PERM.States data set and a baseline data set like BASELINE.States.

Macro Encapsulation

Encapsulation and loose coupling are principles of modular software design, discussed extensively in Chapter 14, "Modularity." Encapsulation strives to protect software from unintended results of individual code modules, as well as to isolate code modules from each other except where necessary. Therefore, encapsulation essentially builds black boxes around code modules (demonstrating why unit testing is referred to as black-box testing), while loose coupling ensures that tiny pinholes in those boxes prescribe limited and highly structured communication between various boxes.

To the extent possible, when a SAS parent process calls a child process, the child process should receive only the information and data required to function. The child process correspondingly should return only the information (and results, if applicable) that are required. Because the parent process in theory cannot see inside the black box (that surrounds the child process), SAS macros should contain return codes that demonstrate success, failure, and possibly gradations thereof that can be interpreted by the parent to drive program flow dynamically. While the use of return codes to support a quality assurance exception handling framework is discussed extensively in Chapter 3, "Communication," and Chapter 6, "Robustness," the following sections focus on ensuring that only necessary and valid information is passed back and forth between parent and child processes.

Leaky Macros

Anyone who has ever taken the SAS Advanced Programmer certification exam will be familiar with some variation of the following question type:

```
When the following SAS code is submitted, what is the final value of the
macro variable &MAC?

%macro test(mac=3);
%let mac=4;
%mend;

%global mac;
%let mac=1;
%test(mac=2);

A) 1
B) 2
C) 3
D) 4
```

In this variation of the question, the answer is A (1) because the value of the parameter MAC (2), the default parameter value (3), and the local macro variable (4) within %TEST do not influence the global macro variable &MAC that was assigned outside the %TEST macro. Thus, because the %TEST macro contains a parameter MAC, references inside the macro to &MAC reflect the local macro variable &MAC that exists only inside the macro.

However, when the parameter MAC is removed from the macro invocation, references to &MAC inside %TEST now reference the global macro variable &MAC. This causes the assignment of &MAC inside %TEST to change the global macro variable &MAC to 4:

```
%macro test();
%let mac=4; * intended to be a local variable but read as global;
%mend;

%global mac;
%let mac=1;
%test;
%put MAC: &mac;
MAC: 4
```

To remedy this interference between local and global macro variables of the same name, the %LOCAL macro statement must be used to initialize all local macro variables that are not identified as parameters in the macro definition. By defining &MAC as a local macro variable inside %TEST, the assignment of the value 4 to the local macro variable &MAC now no longer conflicts with the value of the global macro variable &MAC;

```
%macro test();
%local mac;
%let mac=4;
%mend;

%global mac;
%let mac=1;
%test;
%put MAC: &mac;
MAC: 1
```

This potential conflict illustrates the importance of initializing all local macro variables inside macros with the %LOCAL statement. Macro variables that are required to be used by the parent process or other parts of the program external to the macro must be explicitly initialized with the %GLOBAL statement to ensure they persist outside the child process. While it's also good form to uniquely name macro variables to avoid potential confusion and conflict, use of %LOCAL and %GLOBAL statements remains a best practice.

Variable Variable Names

No, it's not a misprint, although Microsoft Word does despise this section heading. A security risk posed by global macro variables occurs when they are defined dynamically—that is, the macro variable name itself is mutable. For example, the %OHSOFLEXIBLE macro allows its parent process to specify the name of the return code that will be generated as a global macro variable:

```
* child process;
%macro ohsoflexible(rc=);
%let syscc=0;
%global &rc; * dynamic assignment;
%let &&rc=FAILURE;
* do something;
%if &syscc>0 %then %let &&rc=you broke it!;
%else %let &&rc=it worked!;
%put &&rc: &&&rc;
%mend;
```

By first assigning and later testing the value of &SYSCC, the macro can determine whether warnings or runtime errors occurred, and if the macro completed correctly, this is communicated through the dynamically named return code that is specified in the parameter RC:

```
* parent process;
%ohsoflexible(rc=MyReTuRnCoDe);
MyReTuRnCoDe: it worked!
```

At first glance, this appears to be a viable solution, but its flexibility can be its downfall. If the parent process needs to reference the macro variable &MYRETURNCODE, it cannot because while the macro variable value (it worked!) is encoded in the macro variable &MYRETURNCODE, the macro variable itself is not a global macro variable. In fact, a separate global macro variable would need to be created inside %OHSOFLEXIBLE just so the return code macro variable could be globally referenced.

Aside from the unnecessary complexity that this creates, dynamically named macro variables pose very real risks to software integrity. The first risk is that the variable name will be invalid, which can occur if a global macro variable initialization is attempted inside a macro in which a local variable of the same name already exists. A second risk is that an existing macro variable—typically a global macro variable—will be overwritten. For example, the previous parent process is not necessarily aware of individual return codes used by other macros or processes within the software or SAS session, so it could easily overwrite other return codes or values in the global symbol table.

One example of dynamically named global macro variables is demonstrated in the "Passing Parameters with SYSPARM" section in Chapter 12, "Automation," which flexibly parses the SYSPARM parameter to create one or more global macro variables. SAS doesn't get any more flexible than this, but the risks must be understood:

```
* accepts a comma-delimited list of parameters in VAR1=parameter one,
  VAR2=parameter two format;
%macro getparm;
%local i;
%let i=1;
%if %length(&sysparm)>0 %then %do;
   %do %while(%length(%scan(%quote(&sysparm),&i,','))>1);
      %let var=%scan(%scan(%quote(&sysparm),&i,','),1,=);
      %let val=%scan(%scan(%quote(&sysparm),&i,','),2,=);
      %global &var;
      %let &var=&val;
      %let i=%eval(&i+1);
      %end;
   %end;
%mend;
```

When %GETPARM is saved to Getparm.sas and executed in batch mode from the command line, if the SYSPARM value "tit1=Title One, tit2=Title Two" is passed, then the program creates and assigns two global macro variables, &TIT1 and &TIT2, to Title One and Title Two, respectively. The risks are identical to those described earlier—an invalid macro variable name could be supplied, or the variable could already exist and be overwritten. However, if macro variables truly need to be created dynamically, then input controls should be implemented to minimize potential risk.

Input controls should demonstrate that the intended name of the macro variable being dynamically created meets SAS variable naming conventions.

This can be accomplished through a reusable macro (not demonstrated) that ensures the proposed macro variable name is 32 characters or less and contains no special characters. Additional quality controls could ensure that SAS automatic macro variable names were not included and that existing user-defined macros were not overwritten. Input validation is discussed more in the "Macro Validation" section, in which parameters can also be made more reliable through this quality control technique.

The %SYMEXIST macro function demonstrates whether a local or global macro variable exists and can be utilized to ensure that duplicate macro variables are not dynamically created. For example, in the %GETPARM macro, the %SYMEXIST function could be implemented immediately before the %GLOBAL statement to cause the macro to abort if the macro variable name passed through the parameter is already in use.

Macro Validation

In modular software design, SAS macros can exhibit childlike behavior, in that a parent process invokes a macro and temporarily transfers program control to that child process. Where communication is required from the parent, parameters can be passed through the macro invocation. Because most parent processes require some form of validation that the child completed correctly, return codes can be generated within the child and passed to the parent. Thus, the role of validation ensures that parameters have valid structure, format, and content, and that return codes accurately depict macro success, failure, or exceptions that may have been encountered.

Parameter Validation

End-user development is often immune to the extensive input validation required by more traditional software applications. For example, an entry form on a website must not only validate user responses, but also ensure that attackers are not attempting to gain access to the site through buffer overflow, SQL injection, or other hacking techniques. At the other end of the spectrum, software products with no front-facing components typically have no need for comprehensive user input controls because the threat of malice does not exist. Notwithstanding, as macro complexity grows and especially where modular software design is espoused, some level of parameter validation may be warranted in SAS software.

The following %DOSOMETHING macro alters output based on the parameter VAR and is intended to represent dynamic processing that might occur based on some parameterized value:

```
%macro dosomething(var=);
%if &var=YES %then %put Path1;
%else %if &var=NO %then %put Path2;
%mend;

%dosomething(var=YES);
```

Because the VAR parameter is hardcoded in the macro invocation, no attempt to validate the input is made. However, if the %DOSOMETHING macro is intended to have only two paths (initiated by the value of either YES or NO), then the lack of parameter validation introduces a vulnerability because the current conditional logic intimates a third path if neither YES nor NO is selected. In relatively static software in which parameters are rarely or never modified, the risk of failure is low and this vulnerability—the lack of validation—is often ignored. However, where parameter values are numerous, often modified, relatively complex, or generated dynamically from macro variables, the risk increases and parameter validation becomes more valuable.

Parameter validation becomes increasingly more important as modular software is developed, reused, and repurposed. As complexity grows, the likelihood increases that parameter values may represent macro variables passed from a parent to a grandchild through the child process. And, while this type of modular design can greatly increase software reuse, it also requires higher levels of validation to ensure that parameters being passed are valid. The following code demonstrates the typical relationship between parent, child, and grandchild processes in which the parameter VAR is passed from the parent through the child (where it is modified) and on to the grandchild:

```
* grandchild;
%macro print(printvar=);
%put VAR: &printvar;
%mend;

*child;
%macro define(var=);
%if &var=YES %then %print(printvar=&var2);
```

```
%else %if &var=NO %then %print(printvar=NO &var2);
%else %return;
%mend;

*parent;
%define(var=YES);
```

In this example, the child process (%DEFINE) is responsible for validating the parameters it passes to the %PRINT macro. This validation would be based on business rules that, for example, might require that unless the VAR parameter was YES or NO the macro would be aborted, as depicted. Other business rules might instead accept YES as valid and default all other parameter values to the NO path. Because parent processes have limited visibility into the inner workings of child processes, and typically no visibility into grandchild processes, validation is essential to achieving integrity in complex, modular design. In actual production software, in addition to aborting a process when an exception occurs, a return code should be generated that can be used by the parent process to drive program flow dynamically and possibly terminate the program or perform other actions.

Perseverating Macro Variables

Because the Base SAS language does not inherently provide a method to generate return codes from macros, return codes must be faked through global macro variables, as described and demonstrated in the "Faking It in SAS" section in Chapter 3, "Communication." In third-generation languages (3GLs), return codes are limited in scope and are passed directly from child process to parent. However, because SAS return codes must hitch a ride on global macro variables, the return codes effectively are broadcast to the entire program, thus violating both encapsulation and loose coupling principles of modularity, discussed throughout Chapter 14, "Modularity."

For example, the following code demonstrates two macros run in series in the same program. Each macro aims to provide integrity by validating software completion with a return code. However, contamination occurs because the same return code is used in both macros:

```
libname perm 'c:\perm';
data perm.mydata;
run;

%macro one();
%global rc;
```

```
%if %sysfunc(exist(perm.mydata)) %then %let rc=found it!;
%else %let rc=missing!;
%mend;

%macro two();
%global rc; * intended to be a separate return code;
%put RC inside two: &rc;
%mend;

%one;

%two;
```

When executed, despite the global macro variable &RC being defined within each macro, the results are confounded because of interference between the macros. The following output demonstrates this error, as "found it!" has no business appearing inside the %TWO macro:

```
%one;
%put RC: &rc;
RC: found it!
%two;
RC inside two: found it!
```

To remedy this type of leaky macro, global macro variables should always be initialized to an empty string or some other meaningful value immediately after the %GLOBAL macro statement. These two additional statements now prevent contamination from the %ONE macro into the %TWO macro:

```
libname perm 'c:\perm';
data perm.mydata;
run;

%macro one();
%global rc;
%let rc=;
%if %sysfunc(exist(perm.mydata)) %then %let rc=found it!;
%else %let rc=missing!;
%mend;

%macro two();
%global rc; * intended to be a separate return code;
%let rc=;
%put RC inside two: &rc;
%mend;
```

While the value of &RC in %ONE now can no longer interfere with the value of &RC in %TWO, the initialization of &RC in the %TWO macro does overwrite the value of &RC from %ONE. One vulnerability has been eliminated, but the threat still exists of macro variable collision. Thus, in addition to the best practice of always immediately initializing global macro variables after creation, return codes and other global macro variables should ideally have unique names to prevent possible confusion and collision. One method to eliminate this risk is to append _RC or RC after the macro name if a single return code is generated. In some cases, the return code can be imbedded in another macro variable by using the in-band signaling technique introduced in the "In-Band" section in Chapter 3, "Communication."

Even worse than global macro variables that perseverate between macros are global macro variables that perseverate between separate programs. When SAS programs are run in interactive mode, it's common to execute software, check the log, and then execute subsequent (possibly unrelated) programs without first terminating the SAS session. All global macro variables created by the first program will still exist and potentially contaminate the execution environment for all subsequent software run in that session. An example of this is demonstrated in the "Failure" section in Chapter 4, "Reliability." To eliminate this vulnerability, when production software is executed manually in interactive mode, a best practice is always to run software in a fresh SAS session in which no other code has been previously executed.

Default to Failure

A common pattern when operationalizing out-of-band return codes in SAS is to represent successful macro completion with an empty macro variable. This is preferred because the %LENGTH macro function can be used to test for process success—if the macro variable is empty, the length will be zero, and no exception, warning, or runtime error will have occurred. In the following code, a parent process calls a child process and must determine whether the child process succeeded or failed:

```
* child process;
%macro child();
%global childRC;
%let childRC=;
* do something;
%if 2=2 %then %do; * simulate failed process;
   %let childRC=you broke me!;
   %end;
%mend;
```

```
* parent process;
%macro parent();
%global parentRC;
%let parentRC=;
%child;
%if %length(&childRC)>0 %then %do;
   %let parentRC=blame my child!;
   %return;
   %end;
* later processes that depend on success of the child;
%mend;
```

Because 2 always equals 2, the child process will always fail (as currently hardcoded), which produces the following output when executed:

```
%parent;
%put child: &childRC;
child: you broke me!

%put parent: &parentRC;
parent: blame my child!
```

This technique is very effective and, as demonstrated in the "Exception Inheritance" section in Chapter 6, "Robustness," facilitates multiple levels of parent–child relationships so that exceptions, warnings, and runtime errors can be inherited appropriately through modular software. One caveat when using zero-length return codes to represent process success is the possibility of returning a false positive return code—that is, claiming victory when the process in fact failed. Thus, if a zero-length return code is intended to represent process success, it should only be assigned at process completion when it can be demonstrated that no exceptions or runtime errors were encountered.

For example, in the previous child process, if a runtime error or fault occurs in the "*do something;" line, the return code &CHILDRC already would have been assigned a value indicating macro success. If the macro abruptly exited and was unable to assign an appropriate error code to &CHILDRC, then &CHILDRC would be invalid, essentially lying to the parent and stating that no error had occurred when in fact one had. One best practice when using return codes to assess process success or failure is to assign a value like GENERAL FAILURE (rather than blank or SUCCESS) initially, as demonstrated in the revised child process:

```
* child process;
%macro child();
```

```
%global childRC;
%let childRC=GENERAL FAILURE;
* do something;
%if 2=2 %then %do; * simulate failed process;
   %let childRC=you broke me!;
   %end;
%else %let childRC=;
%mend;
```

Only in the final step of the macro (once success is confirmed) should the return code reflecting success be assigned. This often incorporates analysis of the &SYSCC macro variable to demonstrate that no warning or runtime error occurred within the macro or module. Thus, if &SYSCC is 0 and other prerequisites (or business rules) have been met, the return code is set to a value that represents successful completion.

Toward Software Stability

When software is tested, validated, and accepted by a customer, it effectively is certified to be accurate and meet all functional and performance requirements. A formalized test plan (discussed in Chapter 16, "Testability") can further enhance software integrity by demonstrating how the software was shown to meet stated testing requirements. Moreover, a risk register (discussed in Chapter 1, "Introduction") can demonstrate known vulnerabilities at the time of software release—essentially risk that has been accepted by stakeholders. These artifacts can collectively present a more comprehensive view of software reliability to stakeholders and represent a certification that software will perform as stated. Stakeholders in part place confidence in software commensurate with the level of quality with which it was developed and tested.

Once production software is modified, its integrity is depreciated until it can be retested, revalidated, and rereleased. If modified software is not subsequently tested and validated commensurate with the level of quality assurance with which it was originally developed, errors or latent defects may unknowingly introduce vulnerabilities and risk. Especially in development environments that espouse modular, reusable code, subtle modifications to one module could have much broader implications to other programs that rely on that code base. Even when modifications do not adversely affect software, until revised software can be retested to demonstrate this truth, software integrity is inherently diminished.

Software stability is a common objective and often a prerequisite for production software that is automated and scheduled. From a security standpoint, stakeholders need some guarantee that software has not been modified since its last testing and acceptance. In many end-user development environments, the guarantee is a SAS practitioner stating that the software has not been modified or that the modifications were insignificant. In more formalized development environments, stored processes, macros, and programs are saved in central repositories on servers rather than client machines to prevent ad hoc modification. When additional integrity is required to guarantee that software has not been modified in any way, checksums can be generated to validate bit-level comparison of SAS program files.

Stored Macros and Stored Processes

When software is developed through modular design, as individual modules are developed and tested, they can be hardened and placed into a central repository for general use. This practice greatly supports code reuse and repurposing because the code modules have integrity, having already been thoroughly tested and vetted. Code reuse libraries and reuse catalogs can help organize central code repositories and are described and demonstrated in Chapter 18, "Reusability." The SAS application facilitates stability of code modules through the SAS Autocall Macro Facility, the SAS Stored Compiled Macro Facility, and stored processes.

The SAS Autocall Macro Facility enables SAS macros to be stored centrally but compiled locally at execution time. Stored macros, on the other hand, are precompiled, thus saving some time at execution, but their open-source code must be maintained separately from the executable files, resulting in duplication. While use of these facilities is not demonstrated in this text, each supports software security, as production code can be segregated from the development environment to prevent accidental modification, as discussed in the *SAS® Macro Language Reference*.[6] SAS stored processes offer similar advantages in that they store SAS programs centrally on a server rather than on distributed client machines, thus providing greater stability and security.[7]

Checksums

A *checksum* is defined as a "fixed-length string of bits calculated from a message of arbitrary length, such that it is unlikely that a change of one or more bits in the message will produce the same string of bits, thereby aiding detection of accidental modification."[8] Checksums—sometimes referred to as hashes or hash algorithms—are strings of hexadecimal characters that uniquely represent files or text in a substantially reduced format. One-way cryptographic hash functions (e.g., MD5, SHA-256) evaluate a file or text and produce the checksum. Checksums provide a reliable method of demonstrating software integrity by validating when two files are identical and, conversely, by elucidating corrupt files when checksums do not match.

For example, when you download software from the Internet, a checksum is typically performed at download completion to determine whether the entire file was downloaded correctly. A bit-level comparison is made to ensure that the downloaded copy of the software is identical to the original. The checksum of the original file on the hosting server already would have been produced, thus the download process needs to generate a checksum only for the downloaded file. If the checksums are found to differ, this represents file corruption because the downloaded software does not match the original. However, when the checksums are identical, file integrity is quickly and confidently demonstrated.

SAS software can benefit from this same methodology, and by comparing a current SAS program to its verified original, stakeholders are assured that the certified operational instance has not been altered since software testing and customer acceptance. However, rather than inefficiently comparing each line of code between two programs, comparison of their two respective checksums provides equivalent integrity.

Base SAS hashing functions include MD5, SHA256, and SHA256HEX, each of which can create a checksum. For example, the following code computes the SHA-256 hash (in hexadecimal format) for the provided text:

```
%let sha=%sysfunc(sha256hex(It is pitch black. You are likely to be
   eaten by a grue.));
%put &sha;
5A3F8C31D8D529D52DE843FB5C5B9B5EA9EFFF3CC04B4F3C21880B6C9615B797
%put LEN: %length(&sha);
LEN: 64
```

To determine whether a new string matches the text string, the hash value of the new string can be created and compared to the previous hash value. Because hash values uniquely reference content, if the hash values are identical, then the strings they represent are also identical. This same hashing technique, when applied to files such as SAS programs, produces checksums that can be used to compare and validate software. While this example represents two sentences as a 64-character checksum (which is not tremendously efficient), a 10,000-line SAS program would also be represented by a 64-character checksum, demonstrating the tremendous efficiency of comparing two checksums to validate file integrity rather than having to compare 10,000 lines of code.

While Base SAS hash functions such as SHA256HEX are useful in producing checksums for text strings, they are not intended to produce checksums for binary files. Although it is theoretically possible to read a SAS program into a character variable and create a checksum, language and encoding issues could produce unexpected or invalid results. A much more straightforward method is to use any object-oriented programming (OOP) language to produce the file checksum natively.

To demonstrate this technique, the string "It is pitch black. You are likely to be eaten by a grue." is saved to the text file C:\perm\grue.txt and the following Python code is run:

```
# PYTHON CODE!!! NO SEMICOLONS HERE!!!
import hashlib
block = 65536
sha = hashlib.sha256()
with open(r'c:\perm\grue.txt', 'rb') as f:
    buff = f.read(block)
    while len(buff) > 0:
        sha.update(buff)
        buff = f.read(block)
print(sha.hexdigest())
```

The Python code produces the same checksum (SHA-256 hash value) as the previous SAS SHA256HEX function:

```
5A3F8C31D8D529D52DE843FB5C5B9B5EA9EFFF3CC04B4F3C21880B6C9615B797
```

Once software has been developed, tested, hardened, and accepted by a customer, immediate computation of the program file's checksum can be utilized as a baseline for later software validation. When subsequent checksums are calculated for the program file and are validated against the baseline checksum, this demonstrates that the original version of the software is still being executed and confirms that no modifications have been made (which would reduce software integrity).

Implementing Checksums

Unlike SAS data sets, SAS program files contain no metadata such as file name or creation date. Thus, when a hash function produces a checksum of the binary contents of a SAS program file, only the program itself is analyzed. This is critical because it ensures that malleable file metadata do not influence checksum values. For example, a SAS program can be renamed, moved to a different folder, and saved with a newer date, and so long as the contents of the file are unchanged, the checksum will remain static. The discrepancy between SAS data sets and program files, introduced briefly in the "Data Integrity" section, results because SAS data sets contain proprietary metadata that cannot be removed (but which are highly variable), whereas SAS programs are text files whose metadata are managed by the OS, not the SAS application.

To implement checksum integrity checking within a production environment, a checksum should be calculated on SAS software (including SAS programs, associated external macro modules, or static configuration files) after it has been tested, validated, and accepted by the customer. Only once the code has been hardened should the baseline checksums be generated, because even a single additional semicolon in a program will change and thus invalidate its checksum. As discussed in the "Reuse Catalog" section in Chapter 18, "Reusability," checksums are popularly included in reuse catalogs because they provide additional validation to SAS practitioners that code preserved in reuse libraries has not been modified since it was tested and validated.

Thus, one method to validate the integrity of production software is to programmatically calculate the checksum of the program file immediately before execution. When the calculated checksum is found to match the checksum located in the reuse catalog or other artifact, stakeholders are assured that the software running has not been modified since software testing and acceptance. The quality control process (not demonstrated) adds only a couple seconds to software execution but tremendous integrity to the operational environment. And, if a checksum does not validate, business rules can specify that the software should not execute or that a report be generated stating that software integrity could be compromised and that the software should be retested.

The technical implementation of checksums is straightforward and, as demonstrated, requires very little code to operationalize. Thus, the only actual labor involved in enforcing integrity constraints is the maintenance of the roster of checksums. Every time software is tested and accepted by the customer, the software should be hardened and the checksum must be regenerated and recorded in the reuse catalog or other operational artifact. After the checksum is generated, no modifications can be made to the software. A library of current checksums can be maintained in a SAS data set and, where a reuse library and reuse catalog are in use, linked to software modules listed in the reuse catalog. Thus, before reusing an existing module of code within new software, developers can be immediately assured they are utilizing the most recent (and securely tested) version of software by validating its checksum against the reuse catalog.

Checksums for Enterprise Guide Projects

Don't go there! This section is included only as a caveat. Like SAS data sets, SAS Enterprise Guide project files (.egp) are riddled with SAS metadata. To illustrate, open SAS Enterprise Guide and create a new program that contains the following text, so the checksum value can be compared to those in the previous "Checksums" section:

```
It is pitch black. You are likely to be eaten by a grue.
```

Don't worry that this text doesn't represent SAS syntax. Save the SAS Enterprise Guide project as C:\perm\test_checksum_EG.egp and exit SAS Enterprise Guide. Note the date-time stamp of the project file in your OS, because it will be changing. To generate the project file's checksum, execute the following Python code, which can be saved as C:\perm\test_checksum_EG.py:

```
# PYTHON CODE!!! NO SEMICOLONS HERE!!!
import hashlib
block = 65536
sha = hashlib.sha256()
with open(r'c:\perm\test_checksum_EG.egp', 'rb') as f:
    buff = f.read(block)
    while len(buff) > 0:
        sha.update(buff)
        buff = f.read(block)
print(sha.hexdigest())
```

The checksum output follows, but yours will differ:

```
8746d07fc11753a1037c66b671961310585ff0d531796c5c9fd5dddd3ea61f54
```

Now run the SAS Enterprise Guide project (which will generate runtime errors since SAS can't figure out how to survive a grue attack). Save the project file, exit SAS Enterprise Guide, and rerun the Python code to generate the new checksum:

```
edc769ecd734991213051671b2b243612485b2d422a1882c40c0af606c741285
```

Again, your checksum will differ from this one but, more importantly, the two checksums differ from each other—demonstrating that the project file was modified simply by executing and saving the file. As one final experiment, reopen SAS Enterprise Guide and the C:\perm\test_checksum_EG project file. Run the project (which again produces runtime errors) and then attempt to exit SAS Enterprise Guide. You'll be asked if you want to save changes—which is interesting because you didn't make any changes. Do not save changes, exit SAS Enterprise Guide, and rerun the Python code to generate a familiar checksum value:

```
edc769ecd734991213051671b2b243612485b2d422a1882c40c0af606c741285
```

The "changes" that SAS Enterprise Guide was referencing were actually changes to project metadata that are inherently saved as part of all SAS Enterprise Guide project files. Because the changes were not saved, SAS Enterprise Guide did not update those metadata, so the most recent checksum matches the previous value. To view these metadata, peel back the covers on the project file by renaming test_checksum_EG.egp to test_checksum_EG.zip so that the OS will recognize it instead as a compressed file. Open the zip file and go spelunking—the project.xml file is a text file that includes information about the create date, last modification date, and who was doing the modifying. In more complex project files, references in the XML file point to subordinate compressed directories that contain program files.

Due to the complexity and malleability of project files, checksums are useless as an integrity control to demonstrate project file stability. However, by *linking* to external SAS programs from inside projects (rather than creating programs inside projects), the type of integrity control demonstrated previously in the "Implementing Checksums" section can be implemented on individual SAS program files that a SAS Enterprise Guide project file utilizes.

AVAILABILITY

Availability is "the degree to which a software component is operational and available when required for use."[9] Availability requires that software as well as all required components (including data sets, configuration files, control tables, and other files) exist where they should, are valid, and are accessible to necessary stakeholders. As a central principle of software reliability, availability is discussed throughout Chapter 4, "Reliability." However, from a risk management perspective, best practices can facilitate more secure software by eliminating common threats to data set and file availability.

Inaccessible data sets are one of the leading causes of SAS software failure, even in production software that should be designed to be reliable and robust to known threats. For example, if the SORT procedure is executing and sorting data into the PERM.Mydata data set, the procedure holds an exclusive lock that restricts all other users and processes from accessing the data set. However, this locked data set can be seen from two perspectives—the internal perspective of the software executing the SORT procedure and the external perspective of users or processes waiting for (or attempting) access to the data set.

Chapter 6, "Robustness," discusses data set availability from the external perspective of processes trying to gain access to locked data sets, demonstrating techniques that can avoid runtime errors and software failure when data sets are locked. However, from the internal perspective of the software performing the SORT that has successfully gained access to the data set, security rather than robustness is the priority. For example, this security aims to maximize the availability of PERM.Mydata to other potential users or processes, dictating that the SORT procedure should act swiftly and responsibly. Swiftness encompasses execution efficiency principles and development best practices while responsibility moreover requires that the lock of PERM.Mydata should be released as quickly as possible, thus enabling other competing processes to access the data subsequently. The following sections demonstrate techniques that can increase the availability of SAS data sets from the security rather than the robustness perspective.

Fail-Safe Path

The fail-safe path is the secure conclusion to a failed macro, process, or program. In failing safe, software must first detect the fatal exception or runtime error and, through exception handling, terminate the affected code. Failing safe is described throughout Chapter 6, "Robustness," and demonstrated within an exception handling framework. Failing safe is also discussed in Chapter 5, "Recoverability," as the harmless principle of the TEACH mnemonic, which requires that software should not damage its input, output, data products, or other aspects of its environment during or as a result of failure.

Software can cause damage through a number of modalities when it fails. For example, if exceptions are not handled or are handled improperly, critical data sets can be overwritten with invalid data. This can cause loss of business value as customers and other stakeholders are forced to wait while data or data products are corrected or recreated. While failure may be inevitable in some circumstances, graceful failure paths can reduce or eliminate unnecessary damage that can otherwise occur.

While the causes of failure and types of resultant damage are legion and beyond the scope of this text, a few specific threats to security are discussed in the following sections. When file streams are opened or data sets are explicitly locked with the LOCK statement, these accesses must be terminated within the fail-safe path to ensure that the files or data sets are not rendered inaccessible by other users and processes. Other idiosyncrasies can occur when SAS logs are being directed to an external file (via the PRINTTO procedure) and software does not fail safe. The following sections demonstrate the importance of unlocking data sets, closing file streams, and terminating other unresolved accesses as part of graceful software termination.

Closing Data Streams

The OPEN and FOPEN input/output (I/O) functions can be used within a DATA step to open data streams to SAS data sets and other files, respectively. For example, to determine the number of variables in a data set programmatically, SAS practitioners can interrogate the SAS dictionary tables or they can use the OPEN function to retrieve these metadata. The following code opens the data set Mydata and assigns the value 2 to the macro variable &NVARS, representing the number of variables in the data set:

```
data mydata;
   length char1 $10 num1 8;
   char1='tacos';
   num1=5;
run;

data _null_;
   dsid=open('work.mydata','i');
   nvars=attrn(dsid,'nvars');
   call symput('vars',strip(nvars));
run;
```

When OPEN is utilized within a DATA step, the data stream lasts only as long as the DATA step. Even if the DATA step terminates abruptly with a runtime error, the data stream is always closed automatically and the shared file lock is released. For this reason, I/O functionality within the DATA step is secure and the corresponding CLOSE function is unnecessary.

However, OPEN, FOPEN, and other I/O functions are commonly invoked using the %SYSFUNC macro function to facilitate I/O functionality within DATA steps, SAS procedures, or anywhere else. For example, the following

%GETVARS macro calculates the number of variables in a data set and is much more flexible and reusable than the previous code because it does not require use of the _NULL_ data set to create &NVARS:

```
%macro getvars(dsn=);
%let syscc=0;
%global nvars;
%let nvars=;
%global getvarsRC;
%let getvarsRC=FAILURE;
%local dsid;
%let dsid=%sysfunc(open(&dsn,i));
%if &dsid>0 %then %do;
   %let nvars=%sysfunc(attrn(&dsid,nvars));
   %let close=%sysfunc(close(&dsn)); *required;
   %end;
%else %do;
   %let getvarsRC=data set could not be opened;
   %return;
   %end;
%if &syscc=0 %then %let getvarsRC=;
%mend;

%getvars(dsn=work.mydata);
%put VARS: &nvars;
%put RC: &getvarsRC;
```

For this reason, a similar methodology in which I/O functions are invoked with %SYSFUNC is typically preferred. The one disadvantage to this macro-driven technique is the lack of automatic file stream termination, as occurs when OPEN or FOPEN are invoked without %SYSFUNC from within a DATA step. The CLOSE function must be executed any time that the OPEN function is successfully invoked with %SYSFUNC or the file stream will remain open for the duration of the SAS session.

For example, if %SYSFUNC(OPEN) fails to open a data stream and returns a value of 0, then no stream is opened, so no closure is necessary. However, if %SYSFUNC(OPEN) conversely returns a value greater than 0, then %SYSFUNC(CLOSE) must explicitly be used to close the data stream. If the CLOSE function is not utilized, the data stream will remain open indefinitely, maintaining a shared file lock on the data set for the duration of the SAS session. Other users and processes would subsequently be able to open or view the data set, but would be unable to modify or delete the data set while the shared lock unnecessarily persisted.

To ensure that data streams are closed, the fail-safe path must ensure that all data streams opened with %SYSFUNC(OPEN) or %SYSFUNC(FOPEN) are closed with %SYSFUNC(CLOSE) or %SYSFUNC(FCLOSE), respectively. In more complex processes in which multiple I/O functions may be strung together through nested logic branches, multiple CLOSE or FCLOSE functions may be required to ensure that if a failure occurs, all opened data streams are closed before process or program termination.

Unlocking Data Sets

A file lock is created every time SAS accesses a data set, whether to view, modify, or delete the data set. The vast majority of SAS file locks are implicit—that is, they occur automatically, and SAS releases the lock as soon as the DATA step or procedure has terminated. In some cases, such as when a file lock is implicitly created by the OPEN or FOPEN I/O functions (when invoked through the %SYSFUNC macro function), an explicit CLOSE or FCLOSE statement is required to terminate the file lock. This idiosyncrasy is demonstrated earlier in the "Closing Data Streams" section. The LOCK statement is the only method through which SAS creates explicit file locks; thus, it requires an explicit LOCK CLEAR statement to release the lock, as demonstrated in the following code:

```
libname perm 'c:\perm';
data perm.mydata;
   length char1 $10;
   char1='tacos';
run;

%macro sortstuff;
lock perm.mydata;
%if &syslckrc=0 %then %do;
   proc sort data=perm.mydata;
      by char1;
   run;
   lock perm.mydata clear;
   %end;
%else %put somebody in my data!;
%mend;

%sortstuff;
```

The LOCK statement was historically used to prevent separate SAS sessions, processes, or users from accessing a data set that was in use, but has been largely deprecated due to errors in LOCK functionality. Within legacy code, however, whenever LOCK is implemented, a corresponding LOCK CLEAR statement must follow. A more reliable method to lock shared data sets exclusively is demonstrated in the "Mutexes and Semaphores" section in Chapter 3, "Communications."

Missing: One SAS Log

The PRINTTO procedure can be used to redirect the SAS log to a text file. Log files can be invaluable in ferreting out sources of failure, and some environments require that production software logs be maintained for posterity to validate the success of processes. Another use of PRINTTO is to redirect the SAS log temporarily to a file so that notes in the log can be parsed to alter program flow dynamically. For example, because the SETINIT procedure doesn't create a usable return code or output data set that can be programmatically assessed, the log results from SETINIT must be saved to a text file and immediately parsed to determine if the SAS environment contains specific licensed SAS modules. This technique is demonstrated in the "SAS Component Portability" section in Chapter 10, "Portability." A similar use of PRINTTO is demonstrated in the "FULLSTIMER Automated" section in Chapter 8, "Efficiency," in which FULLSTIMER metrics from the SAS log are saved to a text file and subsequently parsed and analyzed via the %READFULLSTIMER macro.

This use of PRINTTO is reprised and captures the FULLSTIMER metrics that depict system resource utilization during the DATA step:

```
libname perm 'c:\perm';
%let path=%sysfunc(pathname(perm));
options fullstimer;
proc printto log="&path/out.txt" new;
run;
data perm.subset;
   set perm.sortme (where=(num1<.1));
run;
proc printto;
%readfullstimer(textfile=&path/out.txt);
```

If the program is abruptly terminated during the DATA step, however, the log destination will still be directed toward the text file, not the SAS log window. The second PRINTTO procedure is required to redirect the log back to the default SAS window. In this scenario, at least two errors can occur when software terminates abruptly while the log is being sent to a file. First, if additional code is executed in the same SAS session, the log will continue to be sent to the log file, which may perplex SAS practitioners who can't find results of their code, while also adulterating the log file with irrelevant or unrelated results. Second, as long as the stream to the log file remains open, SAS maintains a file lock on the log file that prevents its deletion. Therefore, a later process intended to delete the log after parsing it (not shown) can also fail because SAS maintains the file lock until the second issuance of the PRINTTO procedure.

Thus, whenever PRINTTO is implemented to redirect the log destination, exception handling routines should ensure that the second PRINTTO procedure is implemented as part of the fail-safe path so that logs generated subsequent to the termination do not disappear into thin air or cause cascading failures.

Euthanizing Programs

In most instances, the fail-safe path is invoked because an exception, warning, or runtime error is encountered during execution which requires process or program termination. Thus, a functional failure precipitates the need to fail gracefully. However, occasionally a performance failure—notably, the failure to meet execution time requirements—necessitates program termination. For example, if a SAS process is expected to complete in less than ten minutes but is still running an hour later, some business rules might recommend or require that the process be terminated. However, production SAS software too often is terminated for this reason by manually killing the job rather than by building this logic into automated quality controls within a quality assurance exception handling framework. As demonstrated, Base SAS does provide all the tools to force processes to terminate when they exceed execution time thresholds.

Consider the following code that creates and sorts a data set and which simulates a process within production software:

```
libname perm 'c:\perm';

data perm.mydata (drop=i);
   length num1 8;
   do i=1 to 100000000;
      num1=round(rand('uniform')*10);
      output;
      end;
run;
```

```
proc sort data=perm.mydata;
   by num1;
run;
```

The code executes in approximately 90 seconds, which will vary by environment and can be modified by changing the number of observations created. But imagine if the same code were expected (and required) to execute instead in 10 seconds or less. In taking 90 seconds (nine times longer than expected), this performance failure would undoubtedly leave SAS practitioners with the difficult decision of whether to kill the job manually or allow it to continue unfettered on its slow and likely errant path. Moreover, allowing a rogue process to continue processing after it has already demonstrated substantial deviation from performance norms poses a security risk, especially where the cause of the delay remains unknown while the code continues to execute.

To improve both the performance and security of production software, one best practice is to modularize processes that can be invoked with the SYSTASK statement. The SYSTASK statement, when followed by the WAITFOR statement with the TIMEOUT option, kills the specified SAS program (i.e., batch job) if an execution time threshold is reached. This method ensures that the SAS session (in which the batch job is running) is immediately terminated, providing a fail-safe path when performance failures are automatically detected.

To demonstrate this method, the previous SAS code should be saved in the file C:\perm\makeandsort.sas. This program will be called as a batch job, enabling the WAITFOR statement to terminate if necessary. The parent process follows and should be saved as C:\perm\engine.sas.

```
libname perm 'c:\perm';

%put %sysfunc(putn(%sysfunc(datetime()),datetime17.));

systask command """%sysget(SASROOT)\sas.exe"" -noterminal -nosplash
   -nostatuswin -noicon -sysin ""c:\perm\makeandsort.sas"" -log
   ""c:\perm\makeandsort.log"""
   status=rc_makeandsort taskname=task_makeandsort;
waitfor task_makeandsort timeout=20;
%put SYSRC: &sysrc;

%put %sysfunc(putn(%sysfunc(datetime()),datetime17.));

systask kill task_makeandsort;
```

The SYSTASK statement is explained in detail in Chapter 12, "Automation," but the TASKNAME option is required in this instance because the WAITFOR statement uses it to reference this specific batch job. Because the SAS process (now saved as Makeandsort.sas) is required to complete in 10 seconds or less, the TIMEOUT option has been set at 20 seconds, arbitrarily specifying that the job should be killed if it has not completed in twice its expected execution time. Because the process in this scenario is taking more than 20 seconds to execute (approximately 90 seconds), it will be terminated automatically after 20 seconds.

The following output is produced by the Engine.sas program, which demonstrates that the batch job did time out. The passage of time is demonstrated in the log with the DATETIME function and the &SYSRC automatic macro variable is –1, demonstrating that the SYSTASK function did not complete correctly (due to the timeout):

```
libname perm 'c:\perm';
NOTE: Libref PERM was successfully assigned as follows:
      Engine:        V9
      Physical Name: c:\perm

%put %sysfunc(putn(%sysfunc(datetime()),datetime17.));
11MAR16:22:19:27

systask command """%sysget(SASROOT)\sas.exe"" -noterminal -nosplash
-nostatuswin -noicon
! -sysin ""c:\perm\makeandsort.sas"" -log ""c:\perm\makeandsort.log"""
status=rc_makeandsort
! taskname=task_makeandsort;
waitfor task_makeandsort timeout=20;
NOTE: WAITFOR timed out.

%put SYSRC: &sysrc;
SYSRC: -1

%put %sysfunc(putn(%sysfunc(datetime()),datetime17.));
11MAR16:22:19:47

systask kill task_makeandsort;
```

Because the SYSTASK statement specifies that a log file (for the batch job only) should be created as C:\perm\makeandsort.log, this output is demonstrated:

```
libname perm 'c:\perm';
NOTE: Libref PERM was successfully assigned as follows:
      Engine:        V9
      Physical Name: c:\perm
```

```
%put %sysfunc(putn(%sysfunc(datetime()),datetime17.));
11MAR16:22:19:27

data perm.mydata (drop=i);
   length num1 8;
        do i=1 to 100000000;
      num1=round(rand('uniform')*10);
      output;
      end;
run;

NOTE: The data set PERM.MYDATA has 100000000 observations and
1 variables.
NOTE: DATA statement used (Total process time):
      real time           11.55 seconds
      cpu time            11.03 seconds
```

Note that the batch job had time to complete the DATA step, which is demonstrated in the log, but did not complete the SORT procedure. Because the WAITFOR statement causes an abrupt termination of the batch job (and the SAS session in which it is running), whatever process is currently executing (when the KILL statement is received) is not written to the log. However, previous processes that have executed prior to the termination will be recorded and can be used to investigate the cause of the delay.

Because the &SYSRC return code is –1 when a timeout occurs (and 0 when execution occurs within the specified time limit), the &SYSRC value can be assessed programmatically to drive further exception handling routines. For example, upon discovering that a batch job has timed out, an email could notify stakeholders of this exception or any number of other actions could be performed, thus increasing the quality and responsiveness of software. And, because this technique can kill rogue processes that are likely wasting system resources and may be causing other harm, software security and efficiency can be substantially increased through euthanasia.

Complexities of Concurrency

Parallel processing is demonstrated throughout Chapter 7, "Execution Efficiency," Chapter 9, "Scalability," and Chapter 12, "Automation," as a best practice that can substantially increase software performance by decreasing execution time. The parallel processing model described most often involves spawning multiple SAS sessions concurrently—through manual execution, batch jobs, or the SYSTASK statement—and enabling them to communicate with each other to divide and conquer tasks or data sets. However, with this added communication and complexity comes the responsibility that parallelized software be secure to ensure parallel processes do not damage or interfere with each other.

A common parallel processing paradigm entails a SAS program engine (AKA controller or driver) that spawns batch jobs (child processes) or otherwise coordinates software executing in other SAS sessions. In this parallel processing model, software security must encompass not only the program engine but also its children or other concurrent software with which it is interacting. For example, if multiple batch jobs are spawned through SYSTASK, exception handling within the engine should monitor the success and failure of the respective batch jobs and ensure that their respective return codes and other information do not conflict within the engine. In addition, outside of the engine, additional measures must ensure that log files, data products, or other output from batch jobs or other concurrently executing programs do not conflict with each other.

Salt is "a variable incorporated as secondary input to a one-way or encryption function that derives password verification data."[10] Salting is used most commonly in cryptography, in which some unique value is internally added to a construct (such as a password) to strengthen it. Because the salted portion of the construct is typically unknown to users and only used for internal processing and validation, it facilitates passwords that are less vulnerable to attack. The use of salts in cryptography is beyond the scope of this text, but salts can be conceptually utilized in parallel processing to identify constructs uniquely to prevent collisions.

By appending a known, unique token—which can be as straightforward as an incremental number—to macro variable names, SAS data set names, log file names, data product names, and other relevant constructs, collisions are avoided because identically named entities are not created. Salts are used inside engines to ensure that batch job metadata and return codes do not conflict, and are used in child processes to ensure that their respective data products, log files, and other output do not conflict.

Salting Your Children

To eliminate the threat of collisions between child processes running in batch mode or other software running concurrently in different SAS sessions, salt can be applied to data set names, log file names, and other output. For example, the following program uses I/O functions to determine all variables in a data set and iteratively sorts the data set by each of the variables. That is, the data set Mydata is sorted first by the variable CHAR1 and next by the variable CHAR2. Because the output data sets need to be saved uniquely, a salt—the incremental macro variable &I—is appended to the output data set name to avoid collisions.

```
data mydata;
   length char1 $10 char2 $15;
   char1='one';
   char2='two';
run;

%macro serialsort(dsn=);
%local dsid;
%local vars;
%local i;
%let dsid=%sysfunc(open(&dsn));
%let vars=%sysfunc(attrn(&dsid,nvars));
%do i=1 %to &vars;
   proc sort data=&dsn out=&dsn._&i;
      by %sysfunc(varname(&dsid,&i));
   run;
   %end;
%mend;

%serialsort(dsn=mydata);
```

The code produces two data sets, Mydata_1 and Mydata_2, thus ensuring that file names are unique and do not conflict. With subtle modification of the code, these data sets instead could have been named after their respective sorted variable, or with any other dynamically generated salt values that would guarantee uniqueness.

Relevant to parallel processing, salts must be used in SYSTASK invocations wherever multiple copies of the same software are being asynchronously spawned. For example, the following two lines of code are demonstrated in the "Decomposing SYSTASK" section in Chapter 12, "Automation":

```
systask command """%sysget(SASROOT)\sas.exe"" -noterminal -sysin
   ""&perm\means.sas"" -log ""&perm\means1.log"" -print
   ""&perm\means1.lst""" status=rc_means1 taskname=task_means1;

systask command """%sysget(SASROOT)\sas.exe"" -noterminal -sysin
   ""&perm\means.sas"" -log ""&perm\means2.log"" -print
   ""&perm\means2.lst""" status=rc_means2 taskname=task_means2;
```

Note that the log files and output files must be uniquely named to ensure that they don't collide in the environment during execution. In more dynamic software, rather than generating log file names like Means1 and Means2 manually, the salt value would be dynamically generated and incremented, as is demonstrated in the following section.

Salting Your Parents

When batch jobs are spawned with the SYSTASK statement, the WAITFOR statement can be implemented to force the engine to pause while one or more batch jobs complete. Without WAITFOR implemented, SYSTASK continues processing and can asynchronously spawn other batch jobs that will execute concurrently. In addition to ensuring that the batch jobs themselves do not collide, batch job metadata within the engine may also need to be salted to ensure they are uniquely named. SYSTASK is discussed extensively throughout Chapter 12, "Automation."

To demonstrate this threat of collision, the %SERIALSORT macro from the "Salting Your Children" section is retrofitted so that it can be run in parallel. Because macro variables cannot be passed as parameters to external SAS sessions, the values for the data set name, variable name, and variable position—DSN, VAR, and I, respectively—must instead be supplied through the SYSPARM parameter. The engine that choreographs the respective SAS children is demonstrated and saved as C:\perm\parallelsort.sas:

```
libname perm 'c:\perm';

data perm.mydata;
   length char1 $10 char2 $15;
   char1='one';
   char2='two';
run;

%macro parallelsort(dsn=);
%local dsid;
%local vars;
%local var;
%local i;
%let dsid=%sysfunc(open(&dsn));
%let vars=%sysfunc(attrn(&dsid,nvars));
%do i=1 %to &vars;
   %let var=%sysfunc(varname(&dsid,&i));
```

```
   systask command """%sysget(SASROOT)\sas.exe"" -noterminal -nosplash
      -nostatuswin -noicon -sysparm ""&dsn &var &i"" -sysin
      ""c:\perm\serialsort.sas"" -log ""c:\perm\serialsort_&i..log"""
      status=rc_sort&i taskname=task_sort&i;
   %end;
waitfor _all_;
%put Sorting Done!;
%mend;

%parallelsort(dsn=perm.mydata);
```

Because the SYSTASK statement is executed iteratively—as many times as there are variables in the data set—its metadata (i.e., &STATUS and &TASKNAME macro variables) and output (i.e., log and print files) must be uniquely named utilizing the salt value &I. Without this convention, the batch job task names and return code variables could not be distinguished and would overwrite each other. For example, were Parallelsort.sas required to be more robust, exception handling would need to validate the individual return codes &TASK_SORT1 and &TASK_SORT2 to demonstrate that their respective batch jobs completed without error. While this exception handling is not demonstrated, the uniquely named task return codes facilitate this endeavor through salting.

The revised engine now invokes multiple simultaneous instances of the following batch job, saved as C:\perm\serialsort.sas.

```
libname perm 'c:\perm';

%let dsn=%scan(&sysparm,1,,S);
%let var=%scan(&sysparm,2,,S);
%let i=%scan(&sysparm,3,,S);

proc sort data=&dsn out=&dsn._&i;
   by &var;
run;
```

Because macro parameters cannot be passed from the parent process (Parallelsort.sas) to the child process (Serialsort.sas), the SYSTASK statement passes parameters via the SYSPARM parameter, which is read and parsed as the &SYSPARM automatic macro variable inside Serialsort.sas. And, because multiple copies of the child process execute concurrently, the salt &I is used again to ensure that output data sets and log files do not collide and overwrite each other. By parallelizing the serial process demonstrated in the "Salting Your Children" section, all SORT procedures run simultaneously and performance is greatly improved.

SECURITY IN THE SDLC

Direct references to security are less common in data analytic development environments and projects because malicious threats pose an unlikely source of risk. Thus, aspects of security more often are couched in terms of integrity and availability objectives during software planning and design discussions. For example, how will software integrity be demonstrated so that stakeholders are assured that production software has not been modified since testing and acceptance? Or, what methods will be implemented to ensure that data sets and other required software components are available when needed?

Notwithstanding the shift in focus that security may experience in data analytic development environments, many security objectives espouse best practices that should be implemented within all software regardless of other performance requirements. For example, leaky macros should never be tolerated because encapsulation principles can easily eliminate this vulnerability. In this sense, secure coding often equates to supporting software development best practices rather than requirements codified in project documentation.

Other security solutions may be more complicated to effect and are typically implemented at the organizational or team level rather than for one software product. For example, integrity constraints such as checksums are powerful tools that can demonstrate the integrity and stability of software, but which are implemented as a comprehensive solution that can support a corpus of software rather than one program. Thus, conversations about software security often occur at the team or organization level as opposed to during planning and design of a particular software product.

Requiring Security

Security requirements are typically organized around the CIA triad of confidentiality, integrity, and availability. Confidentiality is not a focus of this text because it is more commonly managed in the SAS application through the SAS administrator, metadata content, and OS controls and user privileges. SAS data sets can be encrypted programmatically, and this may be a requirement in some organizations, especially where data are transmitted beyond the organization through email, web services, or other electronic modality.

Due to the overwhelming focus in SAS literature on data quality, integrity requirements for SAS software products too often are focused narrowly on data integrity to the exclusion of code integrity. For example, requirements might state that a critical third-party data set ingested by software must demonstrate validity through a series of quality controls that validate the data or metadata. While data integrity is paramount in data analytic software, software integrity should be equally valued.

Where requirements seek to improve the quality of software products by introducing data quality control techniques, software integrity checks such as checksums that can instantaneously validate software and its components should also be included in software requirements. Integrity requirements often also speak to the level of stability required of software, for example, by stating that production software should be saved as stored processes or as separate macro modules.

This language can help demonstrate to developers and other stakeholders that software is intended to remain stable for some intended duration once placed into production. In a Waterfall environment, this could be several months, when a planned (or unplanned) software update might be released to eliminate defects or provide additional functionality or improved performance. In an Agile environment, on the other hand, that software stability might only last the length of one iteration, when upgrades to the SAS software are integrated as the next batch of working code is released. Thus, regardless of the type of development environment that exists—whether phase-gate or rapid development—an expectation of software stability must be demonstrated if software integrity is to be achieved.

Availability, the third facet of the CIA triad, is more commonly associated with software reliability and robustness requirements. Thus, where software availability is a stated requirement, technical specifications typically state the percent of the time that software must be operational, a maximum number of failures that software is permitted to have per month, or some other reliability-centric metric. While closing data streams, unlocking data sets, and salting dynamic constructs so they don't collide represent technical best practices that should always be followed, they typically are not included in requirements documentation. Availability as it relates to software requirements is further discussed in the "Reliability in the SDLC" sections in Chapter 4, "Reliability."

Measuring Security

Because security aims to eliminate threats, successful security policies are typically measured through reliability metrics. Thus, if a software product or its data products lack integrity, this will be demonstrated through functional failures that either terminate with runtime errors or produce invalid, inaccurate results. For example, if SAS practitioners fail to predict that data sets stored in persistent, shared SAS libraries may be locked, missing, or otherwise unavailable, then failures will result due to the lack of availability.

Security does differ from reliability in that security is more outward-focused. In measuring reliability, stakeholders may want to establish the percent of time that software was functional or the number of failures that occurred within some timeframe. Assessing software security, on the other hand, would additionally involve assessing how many times and to what extent software adversely affected the reliability or security of *other* software or the production environment as a whole. At the process level, security aims to measure not only if a code module is secure but moreover that the module does nothing to violate the functionality of *other* external software modules or programs.

This is an important distinction because a SAS module could appear to be very secure if it contained a macro definition that accepts a limited number of predefined parameters. Thus, the macro might appear to be encapsulated from and only loosely coupled to other components of the software. However, if the same macro also creates a plethora of global macro variables that overwrite existing macro variables, or if it overwrites SAS data sets or data products used by other components, then despite the reliability of the module, it doesn't demonstrate security because it harms its environment. Security, therefore, must always be considered from the perspective of not only the software being assessed but also the environment in which it interacts, which includes external software and data products.

WHAT'S NEXT?

A common objective of production software is to remove the human element so it can run unfettered, allowing SAS practitioners to perform more important and interesting tasks, such as analyzing data products or making data-driven decisions. Once software has achieved a necessary level of reliability, has been stabilized, and has been tested sufficiently to predict future performance, software can be automated. The next chapter demonstrates successful software automation—a well-deserved reward for SAS practitioners who have labored over a quality product.

NOTES

1. ISO/IEC 25010:2011. *Systems and software engineering — Systems and software Quality Requirements and Evaluation (SQuaRE)—System and software quality models*. Geneva, Switzerland: International Organization for Standardization and Institute of Electrical and Electronics Engineers.

2. Id.
3. *Encryption in SAS® 9.4, Fifth Edition*. Retrieved from http://support.sas.com/documentation/cdl/en/secref/68007/HTML/default/viewer.htm#secrefwhatsnew94.htm.
4. ISO/IEC 25010:2011.
5. Gerhard Svolba. 2012. *Data quality for analytics using SAS®*.
6. *SAS® 9.4 Macro Language: Reference, Fourth Edition*. Retrieved from http://support.sas.com/documentation/cdl/en/mcrolref/67912/HTML/default/viewer.htm#n0sjezyl65z1cpn1b6mqfo8115h2.htm.
7. *SAS® 9.4 Stored Processes: Developer's Guide, Third Edition*. Retrieved from http://support.sas.com/documentation/cdl/en/stpug/68399/HTML/default/viewer.htm#n024evv86ifv0wn1his51rrq9azu.htm.
8. ISO 16609:2012. *Financial services—Requirements for message authentication using symmetric techniques*. Geneva, Switzerland: International Organization for Standardization.
9. ISO/IEC 25010:2011.
10. IEEE Std 1363.2-2008. *IEEE standard specifications for password-based public-key cryptographic techniques*. Geneva, Switzerland: Institute of Electrical and Electronics Engineers.

CHAPTER 12

Automation

Cajeros automáticos—automatic teller machines, or ATMs.

When you're backpacking from Guatemala to Patagonia, there's nothing like a *cajero automático* to save the day.

I recall backpacking "back in the day" before the ubiquity of ATMs and before global acceptance of credit cards. Running out of money was a real threat and replenishing the stash always a hassle.

The production began with staking out a bank that had reasonable lines. On Monday mornings, for example, many Central and South American bank lines wrap around the block, easily surpassing a two-hour wait. So you definitely didn't want to run out of green stuff over the weekend.

Once inside, past the posted sentries with shotguns, I always seemed to end up in the wrong line. I'd reach the counter, utter my trite phrase "*Necessito cambiar algun dinero, por favor*" (I need to change some money), and often be shuffled off to another line or some obscure corner desk.

Bank officers are some of the most patient people I've ever met, dealing with broken Spanish all day as tourists attempt to exchange currency. Their first request is always for the passport, so out it comes, instinctually, like your driver's license when you look in the rearview mirror and watch the fuzz walking up.

After your credentials have been validated, you have to dig out the stack of U.S. bills or traveler's checks or some combination thereof from whatever coffer or compartment you've hidden them, often half undressing in the process while trying to avoid eye contract.

The banks are extremely finicky about foreign currency—subtle marks, maculae, tears, or even wrinkles will cause a bill to be rejected. And then you're left carrying useless bills for the duration of your trek or until some hapless individual or bank finally accepts them. Oh, but try rejecting some of *their* bills sometime and wait for the contemptuous looks!

After examining each bill—front and back, literally holding it to the light—the bank officer counts, double counts, and triple counts the stash. After your rejected bills have been returned, the total of the thrice-counted bills goes into an adding machine.

In a whir of calculations, after a minute you're finally shown a number. With a nod of approval, you think you're about to get paid, but no—this is roughly the halfway point.

Next, forms are printed—sometimes from a computer, but often on a typewriter—that describe the transaction, showing the *cambio* (exchange rate), as well as associated bank fees that were charged, and finally the amount in the local currency.

The forms are always produced in triplicate or quadruplicate. When the bank officer breaks out a large desk stamp and moistens it with ink, every page is stamped multiple times with so much force that the desk shakes.

Finally the cash drawer is unlocked, which sometimes requires an additional trip to the back office by the officer to retrieve extra currency or a supervisor. The currency is counted—again in triplicate, and again with the precision and handling of Las Vegas poker dealers.

But the bills are never counted at once. Each denomination is counted individually after which the figure is entered into the adding machine, and this back-and-forth continues until the last pesos or centavos have been added. And the count occurs—you guessed it—at least three times.

At long last, after the hullaballoo has concluded, you're handed the stack of currency and a copy of the receipt. No effort ever is made to count the money to the customer—the fanfare is only to demonstrate to the bank officer that the correct amount has been received and dispersed.

At this point, having spent an hour exchanging money for the week, you're finally on your way. With this significant level of effort, it's easy to understand the appeal that *cajeros automáticos* have for tourists backpacking through countless countries, who can in seconds securely withdraw local currency from their foreign accounts.

■ ■ ■

The inefficiency of and delays in Central and South American banks are notorious, but these are made more frustrating because a simpler, more efficient solution exists for cash withdrawal—ATMs. While the end result—some pesos, lempiras, or quetzales—is identical, the process of changing money in person can be painful.

Just as ATMs are taken for granted in the United States—and increasingly so in Central and South America—automation is taken for granted in software applications development. User-focused applications are compiled, encrypted, and run with the press of a button or the click of a mouse.

SAS software, however, is often run manually by opening the SAS application and executing a program. While necessary during development, testing, and empirical research, manual execution should be eliminated in most production programs that are executed regularly.

Banking in person can be slow and frustrating, but typically that's where the criticisms end. The bank officers are polite and professional, the transactions accurate, and the shotgun-wielding guards imbue patrons with a definite sense of security. But shirking automation in SAS not only can increase software execution time but also can introduce defects, unintended (or unapproved) modification, and ultimately variability that can cripple reliability.

Banking in person requires numerous checks and balances to ensure and validate that the correct amount is dispersed, all of which require time and effort. ATMs conversely automate these validation processes, and while I still count the bills I receive, I've never once received an incorrect amount. SAS software executed manually similarly must rely on manual inspection of the SAS log to validate process success or demonstrate failure, whereas a hallmark of automated software is its ability to identify and communicate automatically when exceptions or errors occur.

On occasion, there are banking activities that do require visiting a physical bank branch—although I can't think of any right now, I'm sure they exist. But for routine banking activities such as weekly withdrawals to ensure you can buy daily pupusa rations or bribe a Belizean zoo guard to look the other way while you scale a fence into an exhibit, ATMs are a necessity. All SAS software similarly is not destined for automation, but where practicable, it can be a tremendous advantage to software performance.

DEFINING AUTOMATION

The International Organization for Standardization (ISO) defines *automatic* as "pertaining to a process or equipment that, under specified conditions, functions without human intervention."[1] Similar to the automatic transmission in a car, which allows a driver to accelerate and decelerate without manual shifting, automated software performs necessary functions without developer intervention. This typically includes not only execution, but also detection of exceptions and errors, communication of failures to stakeholders, and ultimately validation of process success. In data analytic development, a common further objective of software automation is the facilitation of software that can be scheduled to run recurrently.

Automation is taken for granted in traditional software development because third-party software users typically obtain software whose code is stable, compiled, encrypted, and impenetrable. Thus, although I'm interacting with Microsoft Word now as I type this paragraph, this interaction occurs only within a graphical user interface (GUI), and I'm prevented from interacting with or modifying the underlying code. End-user development environments differ substantially because they encourage interaction with underlying code, which decreases code stability and increases the unfortunate reliance on SAS log parsing to validate software success. However, where stable SAS software must exhibit high reliability and availability, automated, scheduled batch jobs can facilitate these objectives.

This chapter introduces and differentiates SAS interactive and batch modes, the latter of which can be used to execute and schedule software directly from the operating system (OS) environment. Because batch jobs can also be initiated from within SAS software, techniques are demonstrated that enable software to spawn simultaneous batch jobs in separate SAS sessions through parallel processing. Through the implementation of automation principles, SAS practitioners not only gain independence from software during its execution but also can more creatively and efficiently prescribe the program flow of modularized software.

AUTOMATION IN SAS SOFTWARE

From a software applications perspective, software that is being developed represents the final product and ultimate business value to a user. For example, end users purchase Microsoft Office understanding that they'll neither have access to nor interact with its source code. However, they also trust that the high reliability and customizability of Microsoft Word—coupled with the complexity of its source code—ensure that no additional benefit would be provided through access to source code. Consumers thus expect to receive an impenetrable product that is reliable, secure, stable, and high quality.

As discussed in Chapter 1, "Introduction," data analytic software development differs substantially from applications development in that the final product bearing ultimate business value is not software itself but rather a data product, data analysis, information, knowledge, or data-driven decision. Because data models must often be repetitively refined and, even once stabilized, often need to be accessible to SAS practitioners, the paragon of quality SAS software is transparent and frequently more malleable than software applications intended for third-party users.

The downside of the accessibility and malleability prevalent in data analytic software is the lack of stabilization this encourages. Software is less likely to be hardened or finalized because it can be modified in seconds and rerun, contributing to a preference in many environments to execute SAS software through interactive sessions rather than stable, automated batch jobs. While continually modified software has its place in empirical research, data modeling, and other applications of Base SAS, central and critical data processes should be built through software development best practices within the SDLC that necessitate stabilizing and automating software. With this automation can come increased reliability and efficiency as developers are freed from painful, repetitive software execution and manual process validation.

SAS PROCESSING MODES

The *Step-by-Step Programming with Base SAS® Software* distinguishes two processing modes for SAS software:[2]

- *Foreground processing (AKA the SAS Windowing Environment)*—This includes the interactive mode and non-interactive mode, the latter of which is not discussed. Manually running programs in the SAS Display Manager, SAS Enterprise Guide, or SAS Studio (including the SAS University Edition) demonstrates the interactive mode.
- *Background processing*—This includes batch jobs that can be initiated from within software, or initiated or scheduled directly from the OS environment without manually starting the SAS application. SAS Display Manager, SAS Enterprise Guide, and SAS Studio (among others) all have batch processing available, but only SAS Display Manager batch processing is demonstrated in this chapter.

Because only interactive foreground processing is discussed, all foreground processing is referenced as the *interactive mode*, or *running SAS interactively*. Interactive mode is arguably the most common environment in which SAS is learned and in which development and testing occur while batch mode best facilitates software automation. Interactive mode includes the familiar SAS GUI interfaces—SAS Display Manager, SAS Enterprise Guide, and SAS Studio—from which code can be typed and executed, with results viewable in the SAS log, output window, or data set viewer. Batch mode instead spawns a new SAS session and executes a SAS program before terminating, at which point results can be viewed externally. This chapter demonstrates methods to run SAS software in batch mode within the SAS Display Manager, emphasizing increased performance gained through automated parallel processing.

Introduction to Interactive Mode

To initiate SAS interactive mode, the SAS executable file must be executed through the Windows start menu, a desktop icon, the command prompt, or

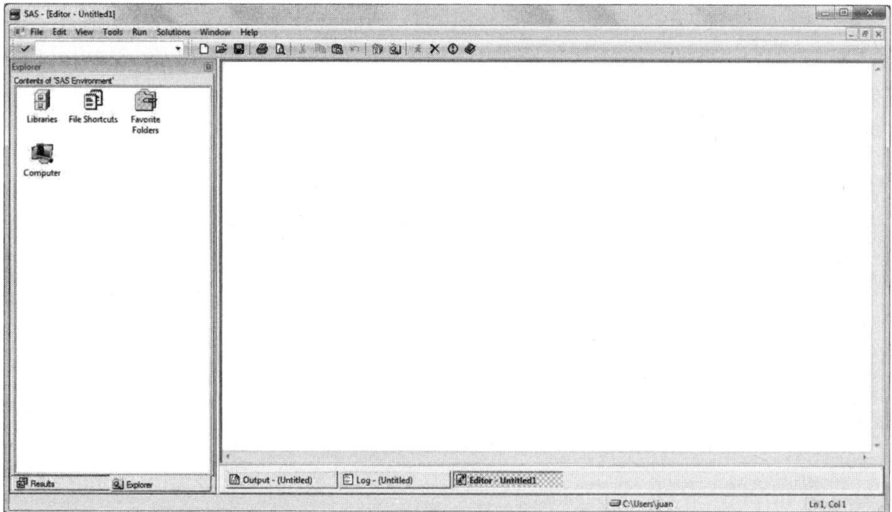

Figure 12.1 SAS Display Manager at Startup

other method. For example, in a Windows environment, the SAS Display Manager is launched with SAS.exe and Enterprise Guide is launched with SEGuide.exe. Figure 12.1 demonstrates the SAS Display Manager startup window, from which SAS practitioners can open, write, edit, or run code, or perform a multitude of menu-driven functions that require no software development whatsoever.[3]

Other common interactive modes include SAS Enterprise Guide and SAS Studio, the latter of which provides the backbone for the SAS University Edition. While the look and feel differ for these SAS interfaces, the functionality of Base SAS software is largely equivalent across all SAS interfaces. The "SAS Interface Portability" section in Chapter 10, "Portability," identifies very few differences when interactively running SAS. However, because batch processing inherently involves interacting directly with the OS environment to spawn SAS sessions, significant variations in functionality exist, such as the inability to spawn SAS sessions with the SYSTASK statement in SAS Studio. For this reason, use of batch is demonstrated solely within the SAS Display Manager.

Introduction to Batch

Batch mode automates the spawning, execution, and termination of not only the SAS application (e.g., SAS.exe) but also SAS programs that are run within it. Because SAS programs are always compiled at the time of execution (except when using the SAS Stored Compiled Macro Facility, which is briefly introduced in Chapter 11, "Security"), the SAS application must be opened before SAS software can be executed. Thus, batch jobs effectively combine the actions of spawning a new SAS session and executing SAS software into one seamless activity.

SAS Display Manager batch jobs can be spawned from interactive mode, (with the SYSTASK statement that can execute OS command line syntax), or in batch mode directly from the OS (from the command prompt or Windows Task Scheduler). With scheduling enabled, batch jobs can be fully automated for regular, reliable, recurrent execution without human interaction, freeing developers to pursue far more interesting endeavors. Both the OS and SAS programs enable batch jobs to be run in series or parallel. Both methods of spawning batch jobs can also implement business logic that enables program flow to pause until one or more batch jobs have completed.

Unlike running SAS software interactively, in which the SAS log can be viewed in real-time within the SAS interface, batch mode executes in the background and thus saves—rather than displays—all logs, data sets, and other output. To facilitate quality assurance and exception handling, batch jobs launched from either an interactive SAS session or the OS environment provide return codes that can demonstrate exceptions, warnings, and runtime errors. Given that production software aims to be self-sufficient and should not require independent, manual review of logs to determine program success, exception handling and program validation should be integral components of all batch processing.

Synchronicity and Asynchronicity

Data analytic programs typically execute in a serialized manner in which processes have clear prerequisites, dependencies, and ordered program flow. Serial program flow is sometimes described as *synchronous* because processes occur one after another. For example, the following code demonstrates a data transformation module (simulated with the DATA step) that must be run before an analytic module (simulated with the MEANS procedure). The MEANS procedure will not execute until after the DATA step has completed.

```
data final;
   length char $10 num 8;
   do num=1 to 100;
      char=put(num,10.);
```

```
      output;
      end;
run;

proc means data=final;
   var num;
run;
```

While some SAS software will require serialized steps, in other cases processes or programs can be run in parallel to reduce execution time and dramatically improve performance. In this example, if a separate analytic module (represented by the FREQ procedure) is appended to the program flow, the new module can be run in parallel with the MEANS procedure because each requires read-only access to the Final data set. While the following code demonstrates a serialization of the three modules, the two analytic modules in theory could be run in parallel:

```
data final;
   length char $10 num 8;
   do num=1 to 100;
      char=put(num,10.);
      output;
      end;
run;

proc means data=final;
   var num;
run;

proc freq data=final;
   tables char;
run;
```

In synchronous execution, the FREQ procedure will not execute until the MEANS procedure terminates, a fact we take for granted in the Base SAS language. However, because the MEANS and FREQ procedures in theory could be executed at the same time (but are not), their serialization represents a false dependency, described in the "False Dependencies" and "Parallel Processing" sections in Chapter 8, "Efficiency." An *asynchronous* solution, on the other hand, would execute the MEANS procedure and immediately thereafter execute the FREQ procedure, enabling them to run in parallel.

The following code implements an asynchronous solution that launches and executes two modules (Means.sas and Freq.sas, demonstrated in the later "Modularity" section) in parallel via the SYSTASK statement:

```
%let perm=c:\perm;
libname perm "&perm";

data perm.final;
   length char $10 num 8;
   do num=1 to 100;
      char=put(num,10.);
      output;
      end;
run;

systask command """%sysget(SASROOT)\sas.exe"" -sysin
   ""&perm\means.sas"" -log ""&perm\means.log"" -print
   ""&perm\means.lst""" status=rc_means;

systask command """%sysget(SASROOT)\sas.exe"" -sysin
   ""&perm\freq.sas"" -log ""&perm\freq.log"" -print ""&perm\freq.lst"""
   status=rc_freq;

waitfor _all_;
```

In this revised example, the DATA step executes synchronously because it must wait for the LIBNAME statement to complete before starting. However, the two SYSTASK statements execute asynchronously because they don't require subsequent processes to wait for them to finish before starting. Thus, the return code &RC_MEANS hasn't been generated yet when the second SYSTASK statement executes, because the Means program is still executing. The asynchronicity of SYSTASK is described further in the "Decomposing SYSTASK" section, including exception handling that should accompany batch jobs both to detect failure and validate success.

This example demonstrates the tremendous flexibility in combining SAS interactive and batch modes. The unnamed program is run in interactive mode through an open SAS session. However, the interactive session is able to spawn two batch jobs with SYSTASK, each of which completes in batch mode while the unnamed program also continues to execute. If the unnamed program were instead saved to a SAS program C:\perm\ETL.sas and run from the operating system as a batch job, the code would not change but its processing would represent a program (ETL.sas) running in batch mode, spawning two additional batch jobs, Means.sas and Freq.sas. This versatility is demonstrated further in the following sections.

STARTING IN INTERACTIVE MODE

As demonstrated in the "Synchronicity and Asynchronicity" section, software executing in interactive mode can spawn batch jobs (with the SYSTASK statement) that execute in batch mode. When the interactive mode launches a batch job, the interactive session and software remain open while the batch job independently opens, executes, and terminates a separate SAS session in which the batch job is executed. This independence is beneficial because it ensures that the new SAS session will be fresh, thus not contaminated by modified system options, global macro variables, open data streams, files in the SAS WORK library, or other artifacts that could cause unwanted variability during execution.

While this independence is required to facilitate consistency and reliability of production jobs, it also limits communication between parent and child processes. For example, a parent process running interactively could spawn a child process via SYSTASK that will execute in batch mode. However, macro variables cannot be passed to the child, and return codes cannot be passed back to the parent. SAS practitioners accustomed to reviewing SAS logs after software execution within the log window will also quickly discover that batch process logs are saved instead to log files. Notwithstanding these complexities of communication, the benefits of spawning batch jobs from the interactive mode are profound.

A Serialized Program Flow Example

The SYSTASK COMMAND statement executes OS environment commands from SAS software.[4] However, SYSTASK examples in this text only demonstrate execution of the SAS executable file (SAS.exe) to spawn a SAS session and batch job. The SYSTASK statement by definition spawns only asynchronous tasks. Thus, ten consecutive SYSTASK statements would attempt to run ten separate programs in ten parallel SAS sessions. In high-resource environments, this might be an effective solution that provides substantially increased performance, but low-resource environments might choke and demonstrate slowed processing, inefficiency, or outright failure.

To implement more advanced program control, SYSTASK includes a WAIT option that pauses the parent process until one or more child processes have completed, thus effectively prescribing synchronicity. With the WAITFOR and SYSTASK KILL statements, additional dynamic processing can be implemented. Both WAITFOR and KILL are demonstrated in the "Euthanizing Programs" section in Chapter 11, "Security."

The following serialized code simulates ETL software in which one program includes separate Trans, Freq, and Means modules:

```
%let perm=c:\perm;
libname perm "&perm";

* transformation;
data perm.final;
   length char $10 num 8;
   do num=1 to 100;
      char=put(num,10.);
      output;
      end;
run;

* means;
proc means data=perm.final;
   var num;
run;

* freq;
proc freq data=perm.final;
   tables char;
run;
```

As demonstrated in the "Synchronicity and Asynchronicity" section, this code includes a false dependency because the FREQ procedure is required to wait for the MEANS procedure to complete, but in theory these could complete in parallel. To execute these two modules in parallel, the PERM.Final data set must be accessible by both child processes. Thus, referenced data sets should always be in shared libraries and never in the WORK library, which will be inaccessible to other SAS sessions. Because the PERM library is utilized, each child process must utilize the LIBNAME statement to initialize the PERM library (unless it is globally defined), as demonstrated in the following section on "Modularity." Modular software design is often the first step toward software automation that can support parallel processing models, so these advantages and techniques are demonstrated in the following section.

Modularity

A technical requirement should first exist before embarking on a crusade to needlessly infuse existing software with batch processing. That requirement should demonstrate what function or performance will be improved through batch processing, balancing this additional value with the inherent complexity that batch brings. In the previous example, two analytic modules—Means and Freq—could in theory be run in parallel because each requires the prior DATA step to have completed and a shared file lock on the PERM.Final data set. To increase performance, the original monolithic program could be enhanced with modular software design to include a parent process that asynchronously spawns two SAS batch jobs (children): Means and Freq. With this vision of how parallel processing will drive higher software performance, technical requirements can be modified and development can commence.

The first step toward parallel processing through automation must modularize code into discrete chunks, after which prerequisites, dependencies, inputs, and outputs can be identified. Modularity is discussed throughout Chapter 14, "Modularity," while its central role in functional decomposition of software (to facilitate critical path analysis and eventual parallel processing) is described throughout Chapter 7, "Execution Efficiency." One common paradigm is to maintain one parent program that operates as the engine (AKA controller or driver) and dynamically spawns and validates all subsequent batch children. This program flow facilitates flexible exception handling that can detect and dynamically respond to exceptions and runtime errors not only in the parent but also its child processes.

Child processes essentially suffer from amnesia, having no awareness of their parent environment, macro variables, library assignments, or other alterations to the environment. If the PERM library is not globally or permanently defined, it must be assigned in each child process. SAS system options in the parent session may also need to be established in batch sessions. The now independent Means module is saved as C:\perm\means.sas:

```
* saved as c:\perm\means.sas;
%let perm=c:\perm;
libname perm "&perm";
%put SYSENV: &sysenv;

proc means data=perm.final;
   var num;
run;
```

The global macro variable &SYSENV, which demonstrates the SAS application processing mode, has also been added to the code to distinguish batch from interactive modes. In interactive mode (or whenever the default SAS system option TERMINAL is active), &SYSENV will be FORE, representing that SAS is running in the foreground. When a SAS program runs in batch mode with the NOTERMINAL option, &SYSENV will be BACK, representing that SAS is running in the background. Without the NOTERMINAL option explicitly specified, the &SYSENV value will indicate FORE whether software is executing interactively or in batch mode.

The Freq module is similarly saved as C:\perm\freq.sas. It, too, requires definition of the PERM library, demonstrating the benefits of global library assignments.

```
* saved as c:\perm\freq.sas;
%let perm=c:\perm;
libname perm "&perm";
%put SYSENV: &sysenv;

proc freq data=perm.final;
   tables char;
run;
```

The final component enabling asynchronous program flow is the parent process, saved to C:\perm\etl_engine.sas.

```
* saved as c:\perm\etl_engine.sas;
%let perm=c:\perm;
libname perm "&perm";

data perm.final;
   length char $10 num 8;
   do num=1 to 100;
      char=put(num,10.);
      output;
      end;
run;
%put SYSENV: &sysenv;

systask command """%sysget(SASROOT)\sas.exe"" -noterminal -sysin
   ""&perm\means.sas"" -log ""&perm\means.log"" -print
   ""&perm\means.lst""" status=rc_means taskname=task_means;

systask command """%sysget(SASROOT)\sas.exe"" -sysin
   ""&perm\freq.sas"" -log ""&perm\freq.log"" -print ""&perm\freq.lst"""
   status=rc_freq taskname=task_freq;
```

```
waitfor _all_ task_means task_freq;
%put RC_MEANS: &rc_means;
%put RC_FREQ: &rc_freq;
```

When the ETL_engine.sas is executed, it first creates the PERM.Final data set, after which the Means and Freq modules now execute in parallel. While the twin return codes—&RC_MEANS and &RC_FREQ—demonstrate the highest warning or error code generated in the respective child process, at this point, no actual exception handling has been implemented in either the parent or the child processes. In the next section, the functionality of the SYSTASK and WAITFOR statements is explored.

Decomposing SYSTASK

While the examples of SYSTASK in this chapter demonstrate only calling SAS.exe from the OS, SYSTASK is far more robust and can be utilized to pass other instructions to the OS. For example, in a Windows environment, SYSTASK can implement the DIR command to pipe the contents of a directory to a text file. The following statement compiles a list of all SAS programs in the C:\perm folder and pipes this list to the text file Allmysas.txt:

```
systask command "dir c:\perm\*.sas /b > c:\perm\allmysas.txt";
```

In this example, the SYSTASK statement includes instruction only for the OS, but more complex implementation can additionally include instruction for the parent and child process. For example, the first SYSTASK statement from the "Modularity" section example shells to the OS, opens the SAS application (SAS.exe), spawns the Means module (C:\perm\means.sas), and terminates SAS.

```
systask command """%sysget(SASROOT)\sas.exe"" -noterminal -sysin
   ""&perm\means.sas"" -log ""&perm\means.log"" -print
   ""&perm\means.lst""" status=rc_means taskname=task_means;
```

The triple quotation marks signify the beginning and end of instruction to the OS (on how to invoke SAS.exe) while the STATUS and TASKNAME parameters are instructions to the parent process to create macro variables representing the batch job return code and task name, respectively. If the SYSPARM parameter is present (as demonstrated in the "Passing Parameters" section), it represents instruction for the child process. The previous SYSTASK statement is equivalent to typing the following statement at the command prompt:

```
"C:\Program Files\SASHome\SASFoundation\9.4\sas.exe" -noterminal -sysin
   "c:\perm\means.sas" -log "c:\perm\means.log" -print
   "c:\perm\means.lst"
```

The %SYSGET macro function extracts the SASROOT operating environment variable, the location of the SAS.exe executable file. Two double quotations are used in the SYSTASK invocation to represent one double quotation within the OS. Quotations are required only when spaces exist in folder or file names, thus the following quotation-free invocation from the command prompt is equivalent because "Progra~1" is the 8.3 legacy filename format for the "Program Files" folder:

```
C:\progra~1\SASHome\SASFoundation\9.4\sas.exe -noterminal -sysin
   c:\perm\means.sas -log c:\perm\means.log -print c:\perm\means.lst
```

SAS system options such as NOTERMINAL are preceded with a dash, as are the batch parameters SYSIN, LOG, and PRINT, regardless of whether implemented from within a SAS program or from the command prompt. The SYSTASK COMMAND statement itself, as well as SYSTASK options such as STATUS, TASKNAME, or WAIT, are never used at the command prompt. Rather, they are used by the parent process to initiate, control, and track the status of the batch job.

The SYSIN parameter represents the SAS program that will be executed, the LOG parameter its log, and the PRINT parameter its output. If either the LOG or PRINT parameters are omitted, the respective log or output files are created with the same name and in the same directory as the SAS program being executed. Thus, both the LOG and PRINT parameters could be omitted from the previous SYSTASK statement to produce identical functionality.

Omitting the LOG and PRINT parameters is typically acceptable unless a SAS program is intended to be run concurrently—that is, multiple instances of the same program run in different SAS sessions in parallel. This complexity is also demonstrated in the "Complexities of Concurrency" section in Chapter 11, "Security." With the identical program launched through two or more SYSTASK statements, each batch session will default to identically named log and output files, thus causing data access collisions because both sessions cannot write to the same file. For example, if two SYSTASK statements attempt to launch two concurrent instances of the Means.sas program, the LOG and PRINT parameters are required and need to specify uniquely named files between the two SYSTASK statements. The STATUS and TASKNAME parameters would also need to be unique between statements:

```
systask command """%sysget(SASROOT)\sas.exe"" -noterminal -sysin
   ""&perm\means.sas"" -log ""&perm\means1.log"" -print
   ""&perm\means1.lst""" status=rc_means1 taskname=task_means1;

systask command """%sysget(SASROOT)\sas.exe"" -noterminal -sysin
   ""&perm\means.sas"" -log ""&perm\means2.log"" -print
   ""&perm\means2.lst""" status=rc_means2 taskname=task_means2;
```

The STATUS statement is not passed to the OS and specifies a macro variable to represent the return code of the spawned child process. SYSTASK STATUS return codes—in this example represented by &RC_MEANS and &RC_FREQ—behave differently than the more common automatic macro variables such as &SYSERR and &SYSCC. Empty status codes represent that the return has not been generated while 0 indicates no warning or error was encountered. A value of 1 demonstrates a warning and 2 a runtime error. As further demonstrated in the later "Batch Exception Handling" section, STATUS return codes are created as local macro variables if the SYSTASK statement is inside a macro definition and as global macro variables if not.

Because SYSTASK always executes asynchronous batch jobs that run in the background, the WAITFOR statement—which causes the parent process to halt until child processes have completed—must be invoked before the value of return codes can be assessed. Without the WAITFOR statement, both the Freq and Means modules will still be executing when &RC_MEANS and &RC_TASK are assessed, so the values of each will be empty. For example, the following code demonstrates when the STATUS macro variable is created and when it is assigned:

```
%macro test;
%if %symexist(new_status)=0 %then %put NEW_STATUS does not exist before;
systask command """%sysget(SASROOT)\sas.exe"" -noterminal -sysin
   ""&perm\means.sas"" -log ""&perm\means.log"" -print
   ""&perm\means.lst""" status=new_status taskname=new_task;
%if %symexist(new_status)=0 %then %put NEW_STATUS does not exist after;
%put STATUS Before: &new_status;
waitfor _all_ new_task;
%put STATUS After: &new_status;
%mend;

%test;
```

The output demonstrates that the STATUS return code &NEW_STATUS (a local macro variable) does not exist before the SYSTASK statement, exists but is not assigned after the SYSTASK statement, and is assigned only after the WAITFOR statement has determined that the child process has terminated:

```
%test;
NEW_STATUS does not exist before
STATUS Before:
STATUS After: 0
```

The TASKNAME statement names the asynchronous task, also known as the batch job or child process. TASKNAME can be finicky because if not closed with either a WAITFOR or KILL statement, the task will remain open in the SAS parent session, even if the child process has long since completed. For example, invoking the following duplicate SYSTASK statements in series produces an error that the "Specified taskname is already in use":

```
systask command """%sysget(SASROOT)\sas.exe"" -noterminal -sysin
   ""&perm\means.sas"" -log ""&perm\means.log"" -print
   ""&perm\means.lst""" status=new_status taskname=new_task;

systask command """%sysget(SASROOT)\sas.exe"" -noterminal -sysin
   ""&perm\means.sas"" -log ""&perm\means.log"" -print
   ""&perm\means.lst""" status=new_status taskname=new_task;
```

The WAITFOR statement references task names of processes for which the parent must wait while the SYSTASK KILL statement terminates referenced processes. The KILL statement is demonstrated in the "Batch Exception Handling" section later in the chapter.

The NOTERMINAL SAS system option specifies that dialog boxes are not displayed during the SAS session.[5] For example, when running automated processes overnight, you don't want to discover that some window has popped up, halting software execution. NOTERMINAL prevents popups during batch jobs, and is also required so that the &SYSENV automatic macro variable will accurately represent BACK for background processing. With the default option TERMINAL enabled, &SYSENV will always reflect FORE, even when batch jobs are running.

Other SAS system options can be included in SYSTASK invocations as necessary. System options that are modified programmatically in the parent process will not percolate to a child process unless explicitly invoked in its SYSTASK statement. Thus, as with global SAS library assignments that are preferred to redundant, manual assignments in child processes, the extent to which SAS options can be standardized will benefit automated software while reducing error-prone, manual invocations of SAS system options via SYSTASK.

Passing Parameters with SYSPARM

The Means program (demonstrated in the "Modularity" section) simulates an analytical module that can be optimized and automated through parallel processing. Modules are often constructed as SAS macros, so they regularly include parameters that are passed during macro invocation to facilitate dynamic processing. Because a new SAS session is spawned when SYSTASK invokes a batch job, neither global nor local macro variables in the parent process can be passed directly to the child. Rather, a single command line parameter—SYSPARM—can pass a text string. Through creative engineering, tokenization, and parsing, multiple macro variables and other information can be passed via SYSPARM to a child process.

Consider a process that runs the MEANS procedure and requires two levels of titles, Title1 and Title2. This is a straightforward task in static code not performed in batch mode:

```
title1 'Title One';
title2 'Title Two';
proc means data=perm.final;
   var num;
run;
```

To create these titles dynamically, macro parameters for Title1 and Title2 can be passed through the %MEANS invocation. The output (not shown) is identical to that produced by the previous code but now can be varied by modifying the macro parameters at invocation:

```
%macro means(tit1=, tit2=);
title1 "&tit1";
title2 "&tit2";
proc means data=perm.final;
   var num;
run;
%mend;

%means(tit1=Title One, tit2=Title Two);
```

The dynamic macro works well when run interactively, but if it is saved as a separate program and invoked in batch mode, dynamism is lost because the necessary parameters can no longer be passed. To overcome this limitation, the SYSPARM command option is implemented within SYSTASK and used to pass the parameters.

```
systask command """%sysget(SASROOT)\sas.exe"" -noterminal -sysin
   ""&perm\means.sas"" -log ""&perm\means.log"" -print
   ""&perm\means.lst"" -sysparm ""tit1=Title One, tit2=Title Two"""
   status=rc_means taskname=task_means;
```

To execute the SYSTASK statement, the updated Means module is saved as C:\perm\means.sas:

```
%let perm=c:\perm;
libname perm "&perm";
```

```
%macro means(tit1=, tit2=);
title1 "&tit1";
title2 "&tit2";
proc means data=perm.final;
   var num;
run;
%mend;

%means(&sysparm);
```

When the Means module is spawned via SYSTASK, it interprets the &SYSPARM value, parses this into separate macro variables &TIT1 and &TIT2, and dynamically produces the output. While this solution is functional, it lacks flexibility, which diminishes the solution's reusability and extensibility. The %GETPARM macro represents a more reusable solution that accepts a comma-delimited list of parameters and assigns these parameters to dynamically named global macro variables. If the &SYSPARM macro variable is empty (e.g., if the SYSPARM parameter was not passed), exception handling directs the macro to terminate via the %RETURN statement.

The %GETPARM macro does pose some security risks, discussed in the "Variable Variable Names" section in Chapter 11, "Security." Despite these risks, its reusable design is beneficial, and the following code is saved to the SAS program C:\perm\getparm.sas:

```
* accepts a comma-delimited list of parameters;
* in VAR1=var 1, VAR2=var 2 format;
%macro getparm;
%local i;
%let i=1;
%if %length(&sysparm)=0 %then %return;
%do %while(%length(%scan(%quote(&sysparm),&i,',','))>1);
   %let var=%scan(%scan(%quote(&sysparm),&i,','),1,=);
   %let val=%scan(%scan(%quote(&sysparm),&i,','),2,=);
   %global &var;
   %let &var=&val;
   %let i=%eval(&i+1);
   %end;
%mend;
```

The Means module (saved as C:\perm\means.sas) can now be updated to include a reference to the %GETPARM macro. This dynamism enables the

SYSTASK statement to pass a virtually unlimited number of parameters to the Means module or other software:

```
%let perm=c:\perm;
libname perm "&perm";
%put SYSENV: &sysenv;

%let tit1=;
%let tit2=;
%include 'c:\perm\getparm.sas';
%getparm;
title1 "&tit1";
title2 "&tit2";
proc means data=perm.final;
   var num;
run;
```

However, the Means module is now dependent on the &SYSPARM value; as a result, the code is no longer backward compatible with the interactive mode. In other words, if SAS practitioners need to execute the Means module manually during further development, debugging, or testing, the values for &TIT1 and &TIT2 will be missing when run interactively. This issue is discussed and a solution demonstrated in the "Batch Backward Compatibility" section later in the chapter.

Recovering Return Codes

In production software, return codes are necessary to demonstrate successful completion of macros and programs and to identify exceptions, warnings, or runtime errors that were encountered. Because Base SAS does not provide inherent return code functionality within macros, as discussed in the "Faking It in SAS" section in Chapter 1, "Introduction," global macro variables are commonly utilized to pass return codes and other information from child processes back to the parent process that invoked them. However, when the child is invoked as a batch job in a separate SAS session, macro variables cannot be passed back to the parent, creating a huge obstacle in exception handling and exception inheritance within batch processing.

Very rudimentary exception handling can be achieved by altering the value of the global macro variable &SYSCC within a child process because this value is communicated back to the optional STATUS parameter invoked in the SYSTASK statement. To demonstrate this exception inheritance from child to parent, the following single-line SAS program is saved as C:\perm\child_exception.sas:

```
%let syscc=0;
```

The child process is called by the parent, saved as C:\perm\parent_exception.sas:

```
libname perm 'c:\perm';

systask command """%sysget(SASROOT)\sas.exe"" -noterminal -sysin
   c:\perm\child_exception.sas -log c:\perm\child_exception.log"
   status=RC taskname=task;

waitfor _all_;
systask kill task;
%put RC: &rc;
```

When &SYSCC is 0 (representing no warnings or runtime errors) in the child process, this communicates successful child completion to the parent, causing the STATUS option to set the &RC return code to 0, as demonstrated in the following log:

```
systask command """%sysget(SASROOT)\sas.exe"" -noterminal -sysin
   c:\perm\child_exception.sas -log c:\perm\child_exception.log"
   status=RC taskname=task;
NOTE: There are no active tasks/transactions.
waitfor _all_;
systask kill task;
%put RC: &rc;
RC: 0
NOTE: Task "task" produced no LOG/Output.
```

However, if &SYSCC instead is assigned a value of 1 through 4 in the child process, STATUS will be 1 in the parent process, representing that a warning occurred in the child. Assigning &SYSCC a value greater than 4 will change the STATUS value to 2, representing a runtime error in the child process. Thus, three nondescript return codes—representing successful completion, warning, or runtime error—can be inherited from child to parent without much difficulty.

However, many SAS modules may need to provide much more descriptive information through exception handling return codes. For example, a module testing the existence of a data set might create one return code value if the data set is missing, another if it is locked, and a third if the data set is present and unlocked but has an invalid format. For this level of specificity, Base SAS is unfortunately ill-equipped to transmit information from child processes back to their parents.

In addition to return codes, actual data may need to be passed from a child process back to its parent. SAS macros are often utilized as functions, effectively performing some operation to return some value or values (or dynamically generated SAS code) as one or more global macro variables. This works well when both parent and child exist in the same SAS session; however, when a parent process calls a child batch process, the parent cannot retrieve information or data directly from that child, just as it cannot retrieve descriptive return codes.

The workaround for returning information from a child process to its parent is to write those data either to a SAS data set or to a text file. The data set or text file acts as a control table, storing the information from the child and enabling the parent process to query the control table to receive the information. While messy, this methodology greatly extends the functionality of SAS batch jobs, ensuring bidirectional communication between parent and child processes and apt identification and handling of exceptions and runtime errors. This functionality is discussed in the "Control Tables" section in Chapter 3, "Communication."

Batch Backward Compatibility

While software *testing* is depicted as a single phase of the SDLC, software *testability* is a quality characteristic that should be considered throughout the entire SDLC. As code is hardened and operationalized through automated batch jobs, batch-specific techniques (such as passing parameters via SYSPARM) may have been incorporated. These techniques can sometimes reduce the testability of SAS programs, especially when they prevent them from running interactively. Because development and testing occur in the interactive—not batch—mode, all automated processes that incorporate batch processing should ensure backward compatibility to the interactive mode to facilitate inevitable maintenance and testing activities.

The Means.sas program depicted in the "Passing Parameters with SYSPARM" section might at some point need to be modified, requiring development and testing in the SAS interactive mode. But when the Means module is opened and executed interactively, it fails because &SYSPARM is empty, as the SYSPARM parameter has not been passed.

One way to overcome this obstacle is to first test the length of &SYSPARM and, if this is zero, to assign a default value to &SYSPARM that will allow the Means module to execute in test mode. The updated Means module now first tests the content of &SYSPARM and, if it is empty, assigns test values to both &TIT1 and &TIT2:

```
* saved as c:\perm\means.sas;
%let perm=c:\perm;
libname perm "&perm";

%let tit1=;
%let tit2=;

%macro parm;
%if %length(&sysparm)>0 %then %do;
   %include 'c:\perm\getparm.sas';
   %getparm;
   %end;
%else %do;
   %let tit1=Test 1;
   %let tit2=Test 2;
   %end;
%mend;

%parm;

title1 "&tit1";
title2 "&tit2";
proc means data=perm.final;
   var num;
run;
```

When the updated code is saved to C:\perm\means.sas, it can now run in both batch and interactive modes with values supplied for the title statements either through the actual SYSPARM parameter or through the test data. It is critical, however, that test values can be distinguished from actual values. The risk, of course, is that with these modifications, the program could now execute even if the SYSPARM were accidentally omitted from a batch invocation, which should cause program failure. However, this risk could easily be overcome through exception handling that further limits functionality in test mode, for example, by causing the program to terminate by throwing a specific exception.

Another method to mitigate the risk of failure caused by accidental invocation without the SYSPARM option would be to assess the &SYSENV automatic macro variable, discussed earlier in the "Modularity" section. For example, if &SYSENV is FORE, this indicates that the program is executing interactively rather than in batch mode, which could signal the code is being developed or tested. And if the value of &SYSENV is BACK, this indicates the program is running in a production environment as a batch job. Thus, programmatic assessment of &SYSENV could also be used to assign test values to macro variables that otherwise would have been assigned via the SYSPARM parameter.

To be clear, however, testing thus far has described only *unit testing* in which the batch job is tested in isolation, irrespective of the parent process calling it. Because the Means.sas program is ultimately intended to be run in production as a batch job (as opposed to interactively), *integration testing* should also be performed that assesses the performance of the batch job in respect to the parent process. Unit testing and integration testing are described and demonstrated in Chapter 16, "Testability."

Batch Exception Handling

The prior-referenced invocations of SYSTASK allude to, yet don't demonstrate, exception handling to validate software success. Because errors can occur inside both the parent and child processes, exception handling must separately address these risks. As with all exception handling routines, business rules first must specify what courses of action to follow when exceptions or runtime errors occur. For example, the following sample business rules prescribe program flow under exceptional circumstances:

- If an exception occurs in the *invocation* of the first SYSTASK statement (calling Means.sas), the second SYSTASK statement should not be executed and the parent process should terminate.
- If an exception occurs in the *invocation* of the second SYSTASK statement (calling Freq.sas), the first SYSTASK statement should be stopped and the parent process should terminate.
- If an exception occurs inside the Means module (but its SYSTASK invocation succeeds), the second SYSTASK statement should execute and the parent process should continue.
- If an exception occurs inside the Freq module (but its SYSTASK invocation succeeds), the first SYSTASK statement should continue to execute and the parent process should continue.

These rules demonstrate a common program flow paradigm in which the parent process acts as an engine and spawns separate child processes (batch jobs) that execute in parallel. Inherent in this program flow complexity, however, is a commensurately complex exception handling framework that must dynamically detect and handle exceptions not only in the parent process but also in each of its children. The sample business rules demonstrate this complexity and also attempt to maximize business value in the face of potential failure. For example, if something goes awry inside the Means.sas program, the Freq.sas program may be unaffected, so it continues to execute. The best practice of maximizing business value despite exceptions or failures is discussed throughout Chapter 6, "Robustness." However, when a more serious exception occurs in the parent process (while invoking the SYSTASK statement itself), the parent process is terminated.

The following parent process implements the sample business rules and represents a more reliable and robust solution:

```
%let perm=c:\perm;
libname perm "&perm";
%global rc_means;
%let rc_means=;
%global rc_freq;
%let rc_freq=;

data perm.final;
   length char $10 num 8;
   do num=1 to 10000000;
      char=put(num,10.);
      output;
      end;
run;

%macro spawn_analytic_modules;
systask command """%sysget(SASROOT)\sas.exe"" -noterminal -sysin
   ""&perm\means.sas"" -log ""&perm\means.log"" -print
   ""&perm\means.lst""" status=rc_means taskname=task_means blah;
%let rc_systask_means=&sysrc;
%if &sysrc^=0 %then %put failed1;
systask command """%sysget(SASROOT)\sas.exe"" -noterminal -sysin
   ""&perm\freq.sas"" -log ""&perm\freq.log"" -print ""&perm\freq.lst"""
   status=rc_freq taskname=task_freq;
%put SYSRC: &sysrc;
%if &sysrc^=0 %then %put failed2;
```

```
data _null_;
   put 'before sleep';
   call sleep(1,1);
   put 'after sleep';
run;

%put RC_MEANS INSIDE: &rc_means;
%put RC_FREQ INSIDE: &rc_freq;
%if %eval(&rc_means>0) or %eval(&rc_freq>0) %then %do;
   systask list;
   systask kill task_means task_freq;
   %end;

waitfor _all_ task_means task_freq;
%mend;

%spawn_analytic_modules;

%put RC_MEANS OUTSIDE: &rc_means;
%put RC_FREQ OUTSIDE: &rc_freq;

%macro test_analytic_modules;
%if %length(&rc_means)=0 %then %put The MEANS module did not complete;
%else %if &rc_means=1 %then %put A warning occured in the MEANS module;
%else %if &rc_means>1 %then %put An error occured in the MEANS module;
%if %length(&rc_freq)=0 %then %put The FREQ module did not complete;
%else %if &rc_freq=1 %then %put A warning occured in the FREQ module;
%else %if &rc_freq>1 %then %put An error occured in the FREQ module;

%if &rc_means=0 and &rc_freq=0 %then %do;
   %put Yay! Everything worked!;
   %end;
%mend;

%test_analytic_modules;
```

The DATA _NULL_ step is required to wait a second so that the return code from the two STATUS options will have had time to be generated. Without this convention, STATUS return codes will always be blank unless they are assessed after a WAITFOR statement, because the STATUS values will not have been generated.

The %TEST_ANALYTIC_MODULES macro prints text descriptions of various warnings and errors that could have occurred. In actual production software, this logic would dynamically alter program flow rather than printing less-than-useful messages (i.e., exception reports) to the SAS log. The benefits of exception handling over exception reporting are discussed in the "Exception Handling, Not Exception Reporting!" section in Chapter 6, "Robustness."

Although not demonstrated, the WAITFOR statement TIMEOUT option is useful in exception handling. TIMEOUT forces a SYSTASK job—including its associated SAS session and SAS program—to terminate if execution time exceeds a parameterized threshold. The TIMEOUT option can be used to initiate a fail-safe path because it can kill rogue or runaway batch jobs that have exceeded execution time thresholds, thus preserving the integrity (and control) of the parent process. TIMEOUT is demonstrated in the "Euthanizing Programs" section in Chapter 11, "Security."

Starting Batch from Enterprise Guide

Batch jobs can be spawned by Enterprise Guide projects just as easily as from SAS programs. To demonstrate this, open the Enterprise Guide environment, drag the SAS program C:\perm\etl_engine.sas into the Process Flow window, and run the process. In many environments, this will execute identically to running the program in the SAS application. In other environments, based on local policy set by SAS administrators, the SYSTASK statement may be prohibited because it shells directly to the OS. SAS practitioners should consult their SAS administrator if failures occur during attempted SYSTASK invocation within the SAS Display Manager or SAS Enterprise Guide.

STARTING IN BATCH MODE

Throughout this text, *batch jobs* represent the SAS programs (child processes) that execute in *batch mode*. As demonstrated in "Starting in Interactive Mode," batch jobs can be spawned through the SYSTASK statement within software executing interactively. While only one batch mode exists, various paths lead to it. Several methods to spawn batch jobs are discussed in the *SAS® 9.4 Companion for Windows, Fourth Edition*, including:[6]

- From SAS software executing interactively, with the SYSTASK or X statement
- From SAS software executing in batch mode, with the SYSTASK or X statements

- From the command prompt, by invoking SAS.exe and referencing the specific SAS program to be run with the SYSIN parameter
- From a saved batch file, by double-clicking the icon, typing the file name at the command prompt, or by otherwise executing it
- From the OS, by right-clicking a SAS program file (or icon) and selecting "Batch Submit with SAS 9.4"

Thus far, only the first method of spawning batch jobs has been demonstrated. This list also does not include additional methods that can run batch jobs in SAS Enterprise Guide or SAS Studio. The tremendous advantage of the first two methods—initiating batch jobs with SYSTASK—is the ability to incorporate information and macro variables from the parent process and pass them via the SYSPARM parameter to child processes. Another advantage is the ability of SYSTASK statements to be dynamically generated and invoked. For example, as demonstrated in the "Salting Your Engine" section in Chapter 11, "Security," a parameterized number of SYSTASK statements can be dynamically invoked through a macro loop that iteratively calls SYSTASK. While the SYSTASK statement can spawn batch jobs from both Windows and UNIX environments, some installations of SAS Enterprise Guide locally restrict use of SYSTASK and X statements for security reasons, eliminating the ability to run batch jobs by invoking SYSTASK.

The remaining three methods of spawning batch jobs each initiate a SAS session from scratch—that is, directly from the OS rather than from within an open SAS session. Thus, the primary limitation of these methods is the inability to pass macro variables and other information dynamically from the parent process to the child. A secondary limitation is the inability to execute SYSTASK statements dynamically, for example, by invoking a parameterized number of SYSTASK statements in a macro loop. However, parameters can still be passed to batch jobs via the SYSPARM option, which is demonstrated in the "Passing Parameters with SYSPARM" section. The principal advantage of initiating SAS batch jobs directly from the OS is the ability to schedule execution through the OS, eliminating the need for SAS practitioners to be present when production software is executed or to babysit it while it runs.

While syntax differences between batch job spawning methods exist, these are minor, and the capabilities of each are similar. Each method provides a way to invoke batch jobs in series or parallel, can halt a spawned job until other processes have completed, and can generate return codes that can be utilized to facilitate quality assurance through an exception handling framework. Notwithstanding these largely equivalent capabilities, the extent to which program flow business logic can be included within SAS software (as opposed to command syntax found in batch files) will add versatility to program flow and exception handling paths.

Starting Batch from Command Prompt

Batch mode is always invoked when SAS is executed from the command prompt by calling SAS.exe. In the interactive mode, the %SYSGET(SASROOT) macro function supplies the location of the SAS executable file, but when launching from the OS, it must be supplied manually. In SAS 9.4 for Windows, the default location for the SAS executable follows:

```
C:\Program Files\SASHome\SASFoundation\9.4\sas.exe
```

Because "Program Files" contains a space, quotes must surround the command when invoked. If no other options or parameters follow the command, the SAS application opens in interactive mode. Specifying the NOTERMINAL option enables batch mode, causing the SAS Display Manager not to open and modifying the &SYSENV value from FORE (foreground) to BACK (background). Additional useful options include NOSPLASH, NOSTATUSWIN, and NOICON, which can prevent windows from opening unnecessarily while in batch mode.

Other command parameters are identical to those discussed throughout the previous "Starting in Interactive Mode" sections. SYSIN references the SAS program being executed, LOG its log file, and PRINT its output file. In interactive mode, each of these parameters could be generated through macro code, such as dynamically referencing the folder or program name. When launching batch jobs directly from the OS, however, these attributes each must be hardcoded, as demonstrated in the following command statement, which launches the Means.sas program from the command prompt:

```
"C:\Program Files\SASHome\SASFoundation\9.4\sas.exe" -noterminal -sysin
   c:\perm\means.sas -log c:\perm\means.log -print c:\perm\means.lst
```

This batch job invocation from the OS does not differ from launching the batch job from within the SAS application (via SYSTASK), as demonstrated throughout the "Starting in Interactive Mode" sections. The main differences are the lack of STATUS and TASKNAME options used by the SYSTASK statement itself to control program flow and enable exception handling. However, because Means.sas is now spawned directly from the OS, when it completes, program control returns to the OS. Thus, if warnings or runtime errors are encountered in the batch job, because no SAS engine is driving program control, no exception handling framework exists to validate program success or detect its failure. For this reason, to the extent possible, program flow should be initiated by the OS but controlled by SAS software.

Had the Means.sas program failed with a warning or runtime error, this would have been communicated to the log file Means.log, although no analysis of this file would have occurred. Because one of the primary responsibilities of automated software is automatic detection and handling of exceptions and runtime errors, while the previous command line invocation does spawn a batch process, it cannot truly be said to be automated. A more reliable method of invoking a batch job from the command line entails executing a parent process that can act as an engine by spawning, controlling, and validating subsequent child batch processes. The following command line statement runs the parent process ETL_engine.sas (which in turn invokes the Means.sas batch process via SYSTASK) from the OS:

```
"C:\Program Files\SASHome\SASFoundation\9.4\sas.exe" -noterminal -sysin
   c:\perm\etl_engine.sas -log c:\perm\etl_engine.log -print
   c:\perm\etl_engine.lst
```

The parent process (ETL_engine.sas) is a much better candidate to be invoked as a batch job directly from the OS because it contains exception handling routines that monitor not only its own exceptions but also exceptions generated by its child processes. This method also enables children to be terminated if necessary via the WAITFOR statement TIMEOUT option—an especially useful feature if they exceed execution time thresholds.

Starting Batch from a Batch File

Within this text, *batch jobs* represent SAS programs that are run in batch mode, regardless of whether they are invoked interactively through SAS software, through SAS software running in batch mode, or directly from the OS. *Batch files,* on the other hand, represent text files that contain command line syntax interpreted by the OS environment, not by the SAS application. Thus, while batch jobs are always SAS programs, batch files are never SAS programs but can invoke the SAS application via command line statements to run SAS programs. Batch files typically have the .bat file extension.

Starting a batch job from a batch file can overcome some of the limitations that occur when batch jobs are initiated directly from the command prompt. For example, because multiple child processes can be spawned from a single batch file, a batch file can effect rudimentary control over its children, such as by halting batch file execution until child processes have completed. Batch files can also prescribe whether child processes execute in series (synchronously) or in parallel (asynchronously).

The downside of launching batch jobs from a batch file, like launching batch jobs from the command prompt, is the lack of dynamic program control and robust exception handling that are available within SAS software. The following batch file spawns both the Means and Freq modules and should be saved as C:\perm\etl_engine.bat:

```
start "job1" "c:\program files\sashome\sasfoundation\9.4\sas.exe"
   -noterminal -sysin c:\perm\means.sas -log c:\perm\means_batch.log
   -print c:\perm\means_batch.lst

start "job2" "c:\program files\sashome\sasfoundation\9.4\sas.exe"
   -noterminal -sysin c:\perm\freq.sas -log c:\perm\freq_batch.log
   -print c:\perm\freq_batch.lst
```

This batch program can be executed in several ways. The file name (including .bat extension) can be typed at the command prompt, or, from a Windows environment, the batch file can be double-clicked or right-clicked with "Run as administrator" selected. The START command signals that the batch jobs should be run asynchronously—that is, the Freq module does not wait for the Means module to complete. When START is used, the specific job must be named, thus the first batch job is named Job1 and second Job2.

To instead run these processes in series (i.e., synchronously), the following code should be saved to the batch file C:\perm\etl_engine.bat and executed:

```
"c:\program files\sashome\sasfoundation\9.4\sas.exe" -noterminal -sysin
   c:\perm\means.sas -log c:\perm\means_batch.log -print
   c:\perm\means_batch.lst

"c:\program files\sashome\sasfoundation\9.4\sas.exe" -noterminal -sysin
   c:\perm\freq.sas -log c:\perm\freq_batch.log -print
   c:\perm\freq_batch.lst
```

In this revised example, the Freq.sas program now executes only after Means.sas completes. Additional program control can be implemented with the WAIT option that can be specified after START, but the level of sophistication still does not begin to approximate what can be accomplished when the parent process is a SAS program rather than a batch file.

To further illustrate the inherent limitations in batch files, consider the business rules prescribed in the earlier "Batch Exception Handling" section. These rules require handling of exceptions generated not only in the parent process but also in child processes. Yet none of these business rules are operationalized in the previous batch file invocation, again because the parent process is the OS itself—now a batch file and not the command prompt—rather than SAS software. To eliminate these limitations and deliver a solution that maximizes the benefits of both the OS environment and SAS software, the following batch file demonstrates the best practice of invoking a SAS parent process from a batch file and should be saved as C:\perm\etl_engine.bat:

```
"c:\program files\sashome\sasfoundation\9.4\sas.exe" -noterminal -sysin
   c:\perm\etl_engine.sas -log c:\perm\etl_engine.log -print
   c:\perm\etl_engine.lst
```

In this final solution, the parent process ETL_engine.sas is invoked directly from the OS via the batch file. This enables the batch file to be scheduled through the OS so it can be executed as necessary without further human interaction. However, the dynamic program flow inside ETL_engine.sas is preserved, ensuring that prescribed business rules are followed. Thus, as the respective child processes execute, they can do so dynamically from within the parent process, ensuring that success is validated and runtime errors are identified and handled. This parent–child paradigm is arguably the most commonly implemented method to achieve robust, reliable automation in SAS software.

AUTOMATION IN THE SDLC

If the decision to automate is made during software planning, this will tremendously influence software design and development. SAS practitioners will be able to understand whether a central engine or parent process is necessary to drive program flow and what parameters, options, or other information will need to be passed to child processes via SYSPARM. Because substantial performance gains can sometimes be achieved through executing child processes as parallel batch jobs, this performance may spur the decision to modularize software into discrete chunks that can be spawned as concurrent batch jobs.

In other cases, the decision to automate software is made after the software is already in production. This decision is sometimes made due to unexpected success of software that has become increasingly critical or valuable over time and thus requires commensurate reliability and availability. To achieve this increased performance, including automated exception handling and validation of process success, business rules should prescribe program flow under normal and exceptional conditions and can be implemented via a robust exception handling framework. While it is always easier to design and develop software initially with automation in mind, the transition is commonly made, and in some cases will require only a few lines of extra code to implement.

Requiring Automation

Regardless of when the decision is made to automate SAS software, automation requirements should clearly state the intent and degree of automation. The intent is essentially the business case for what value should be gained through automation, which should surpass the cost of additional code complexity inherent in automation. For example, one intent might state that SYSTASK is being implemented so that parallel processes can be asynchronously spawned to facilitate faster parallel processing. This differs substantially from another common intent that aims to create batch jobs and batch files that can be scheduled to run recurrently on the SAS server. The first intent inherently requires modularization of software and critical path analysis, while the second intent could be achieved with a one-line batch file that invokes a SAS program from the OS.

The intent in the first example aims to increase software performance by reducing execution time, a metric that can be measured against pre-automation baselines (if those exist) and contrasted against the effort involved in automating the process. In the second example, the intent aims instead to decrease personnel workload during O&M activities, so these decreased operational hours can be measured against the development hours expended to automate and test the software. As the frequency with which software is intended to be run increases, the relative benefit gained from automation increases because the cumulative effects of greater software performance (or decreased personnel workload to support software operation) are multiplied with every execution. In other words, the more often stable software is executed, the more valuable automation of that software becomes.

The degree of automation should also accompany automation intent. For example, if child processes are being automated with SYSTASK so that they can be run in parallel, this doesn't necessitate that the parent process must also be automated. Thus, the SAS application might still have to be started manually, the parent process opened and executed, and the various parent and child logs reviewed for runtime errors. While somewhat inefficient, the software would still benefit from the limited automation of its child processes, despite not being a fully automated solution. Another common partial automation paradigm involves full automation of software execution, but no (or limited) automation of exception handling. Thus, software might be scheduled through the OS and run automatically each morning at 6 AM before SAS practitioners have arrived at work yet require someone to review SAS logs manually later in the day.

In other cases, automation truly denotes the total lack of human interaction during recurrent software operation. The OS environment executes a batch file at scheduled intervals, which in turn runs a SAS program. That program may in turn execute subsequent batch children either in series or parallel before terminating. Exception handling detects exceptions, warnings, and runtime errors during execution—in both parent and child processes—and a separate process may validate the various SAS logs after execution to provide additional assurance that execution was successful. Because stakeholders can pick and choose which aspects of software to automate and which to perform manually, and because they can increase or decrease automation levels throughout a software's lifespan, automation will differ vastly between organizations and specific software products.

Measuring Automation

Assessing automation performance will depend expressly on the degree to which software was intended to be automated. One of the most telling metrics of software automation is the degree to which exception handling and failure communication are automated. A first grader could schedule SAS software to run recurrently through a batch file that executes SAS software. However, if automation only automates execution and does nothing to automate quality controls and process validation, automation can actually make software *less* reliable when SAS file logs are overwritten or ignored, or when scheduled tasks execute despite failures of prerequisite tasks. Therefore, while automation does not necessarily imply that software should be *more* reliable or robust than manually executed software, automation should not make software *less* reliable or robust by masking exceptions, runtime errors, and possible invalid data products.

When scheduled jobs are executed in short succession, a performance failure might occur if one instance of the software takes too long to run, causing functional failures when the next scheduled run—quick on the heels of the first—executes before the first has completed. In fully automated software, robust processes should predict and appropriately handle these common exceptions through business rules and dynamic exception handling. In many cases, the performance of fully automated software can be assessed by evaluating software reliability and availability metrics, given that software failure often incurs human interaction to recover and restart software. Reliability and availability are discussed throughout Chapter 4, "Reliability."

Recoverability metrics, described in Chapter 5, "Recoverability," can also be utilized to measure the degree and success of automation. The TEACH mnemonic describes that recovery should be autonomous, in that recovery should not only be automated to the extent possible but also facilitate intelligent recovery to an efficient checkpoint that maximizes performance. Thus, while some software will be expected to run without human interaction, SAS practitioners may be expected to play a large role in software recovery from failure. In other cases, even software failure detection and recovery should be automated (to the extent possible), and automation additionally will be measured by the success and speed with which software recovers from failure without human intervention.

WHAT'S NEXT?

Software automation is typically the last step in hardening production software, so this chapter concludes Part I, which describes dynamic performance requirements that can be observed and measured during software execution. What occurs beneath the software covers, however, can be just as relevant to software quality as dynamic characteristics. While not directly observable during execution, static performance requirements aim to ensure that code can be easily maintained and, through modular software design, that code is organized sufficiently to benefit dynamic performance techniques such as parallel processing. The next six chapters demonstrate static performance requirements that should be incorporated to improve the maintainability of production software.

NOTES

1. ISO/IEC/IEEE 24765:2010. *Systems and software engineering—Vocabulary*. Geneva, Switzerland: International Organization for Standardization, International Electrotechnical Commission, and Institute of Electrical and Electronics Engineers.
2. *Step-by-Step Programming with Base SAS(R) Software*. Retrieved from http://support.sas.com/documentation/cdl/en/basess/58133/HTML/default/viewer.htm#a001103607.htm.
3. "Introduction to the SAS Windowing Environment." *SAS® 9.4 Language Reference: Concepts, Fifth Edition*. Retrieved from http://support.sas.com/documentation/cdl/en/lrcon/68089/HTML/default/viewer.htm#p1kcewwv8r36lun1nvr2ryx9ks9h.htm.
4. "SYSTASK Statement: Windows." *SAS® 9.4 Companion for Windows, Fourth Edition*. SYSTASK Statement: Windows. Retrieved from https://support.sas.com/documentation/cdl/en/hostwin/67962/HTML/default/viewer.htm#p09xs5cudl2lfin17t5aqvpcxkuz.htm.
5. "TERMINAL System Option." *SAS® 9.4 System Options: Reference, Fourth Edition*.
6. "Starting SAS." *SAS® 9.4 Companion for Windows, Fourth Edition*. Retrieved from http://support.sas.com/documentation/cdl/en/hostwin/67962/HTML/default/viewer.htm#p16esisc4nrd5sn1ps5l6u8f79k6.htm.

PART II

Dynamic Performance

CHAPTER 13
Maintainability

Having sporadically slept overnight on the harrowing bus ride from Cusco, Peru, I arrived in Puno on the western coast of Lake Titicaca at daybreak. Yet another stolid bus station, weary children asleep on their parents, the wide-eyed vigilance of some guarding their luggage, the apathy of others content to sleep blissfully on the cold, cement floor.

Still half asleep, massaging the seemingly permanent creases in my arms from having bear-hugged my backpack all night long, I stumbled through the bus terminal, physically exhausted from sleep. My legs ached, but I couldn't tell if that was due to my cramped sleeping quarters or the jaunt up Huayna Picchu two days earlier, consummating a four-day trek across the Inca Trail. I'd had porters then—what I wouldn't have given to have had one of those porters carrying my rucksack in Puno.

I was in search of relaxation, but somehow the throng of overnighters strewn about just made me wearier. In 14 hours I had a date with a different bus for yet another overnight adventure down the perilous mountainside to the Spanish colonial city of Arequipa. In other words, given 14 hours to live, how would you want to spend it?

The Inca Lake tour company stood nearby, and a subtle glance toward the orchidaceous display was enough to summon shouts from the attendant. Within minutes, I was in a cab headed to the dock to board a water taxi for a day of exploring Lake Titicaca.

Outstretched on the hull and having been lulled to sleep within seconds of departing, I awoke to shadows and shouts as tourists clambered over me snapping pictures. *Ah, there must be something to look at,* I thought, as I craned to see we'd arrived at the island of Uros.

Stepping out of the boat, I immediately sank several inches, as though wading through biscuit dough. Everything in sight was constructed entirely of the totora reeds. Their small sailboats: reeds, bound with reeds, having reed masts. Their thatched houses: reed floors, walls, and roofs. Even the communal island kitchen: reeds surrounding a stone fireplace where the standard fare of tortilla, beans, and rice was cooking. And thank God, because I was starving!

Renowned for constructing floating reed islands, the Uru people have inhabited the lake for centuries, once using the islands as defensive bastions. The Uru haven't lived on the islands for decades—despite what Wikipedia may tell you—but now reside in Puno and neighboring pueblos. Notwithstanding, their ability to maintain their culture for centuries is astounding.

The Uros islands are impeccably and constantly maintained. They are several feet thick, and like icebergs, the majority of the island and reeds lie unseen beneath the water, slowly disintegrating and continually being absorbed into the lake. The Uru combat this slow death by adding layers of new reeds to the top of the island every few weeks, ensuring their villages remain afloat.

Centuries of honing island maintenance has instilled precision in the Uru, and every aspect of island maintenance is planned and choreographed. For example, rather than hastily adding water-logged reeds to fortify islands, the reeds are dried for weeks, allowing them to harden and become moisture-resistant.

After they are dried, small bushels of reeds weighing 10 to 15 pounds are tightly bound together before being added to the top layer of the island. This modular practice builds smaller bundles that are more impenetrable to moisture, can bear more weight, and represent strong, cohesive units.

As I finished my Titicaca tacos—somehow, I was the only one to have cajoled lunch from the Uru—the guide concluded his scintillating rendition of island life and preservation. Taking a fresh bundle, he carefully wove it into the top layer of the island, and jumping to their feet suddenly and with clog-like movements, several Uru danced the new patch into place, demonstrating its seamless integration into the island fabric.

■ ■ ■

The Uru's adherence to maintainability principles has driven their continued success and ability to preserve their culture and islands. But the first step toward maintainability is the realization that maintenance is inevitable and the subsequent embracing of and planning for that inevitability.

Within the software development life cycle (SDLC), maintenance can be perceived as an onus but is required to ensure software is updated, receives patches, and continues to meet customer needs and requirements. Maintenance always entails effort and, like grooming the Uros reed islands, is often both necessary and continual. If you don't maintain your software, it will eventually sink into oblivion.

Like everything else in life, maintenance should be about working smarter, not harder. By planning for maintenance activities and by being proactive rather than reactive, the Uru forecast when they will need additional reed bundles and harvest and dry them accordingly. Software maintenance should similarly espouse maintainability principles during planning and design—long before software actually requires maintenance.

But maintenance also can be emergent—just ask the Uru after a storm has swept through the region. In software development, emergency and other corrective maintenance will also be required, but it, too, can be facilitated by modular software design, readability, and other maintainability principles.

Modular design tremendously benefits maintenance by limiting and focusing the scope of work when modifications must be made. The Uru not only bind reed bundles into manageable modules, but also ensure these integrate completely with surrounding reeds to produce a more robust veneer against weight from above and moisture from below. The final step in software maintenance is also integration and should incorporate testing to validate maintenance has met and will continue to meet its objectives.

DEFINING MAINTAINABILITY

Maintainability is "the degree to which the software product can be modified."[1] The International Organization for Standardization (ISO) goes on to specify that maintenance includes changes that correct errors as well as those that facilitate additional or improved performance or function. Moreover, maintainability refers to maintenance not only of code but also of the technical specifications in software requirements. Thus, the ISO definition of maintainability represents a far greater breadth than some definitions that speak only to the ease with which software can be repaired, or *repairability*.

Like other static performance attributes, maintainability doesn't directly affect software function or performance, but rather enables developers to do so through efficient understanding and modification of software and requirements. For example, maintainability can indirectly improve the mean time to recovery (MTTR) of failed software, by enabling developers to more quickly detect deviations and deploy a solution, thus increasing both recoverability and reliability. Especially in software intended to have an enduring lifespan, maintainability principles should be espoused to quietly instill quality within the code, benefiting developers who maintain the software in the future. Thus, maintainability can be viewed as an investment in the enduring use of software.

Because maintainability facilitates maintenance, this chapter first describes and distinguishes software maintenance types, patterns, and best practices. Failure to maintain software proactively can lead to unhealthy operational environments in which SAS practitioners are constantly "putting out fires" rather than proactively designing stable, enduring software products. The second half of the chapter demonstrates the shift from maintenance-focused operation to maintainability-focused design and development that facilitate more effective and efficient maintenance.

MAINTENANCE

Since maintainability represents the ability to maintain and modify software and software requirements, an introduction to software maintenance is beneficial. Without maintenance, maintainability is unnecessary and valueless. And, with poor maintenance or a poor maintenance plan that is defensive and reactionary, maintainability principles will never be prioritized into software design because developers will be focused on fixing broken software rather than building better software.

Software maintenance is "the totality of activities required to provide cost-effective support to a software system ... including software modification, training, and operating a help desk."[2] ISO goes on to define several categories of software maintenance, including:

- *Adaptive maintenance*—"The modification of a software product, performed after delivery, to keep a software product usable in a changed or changing environment."
- *Corrective maintenance*—"The reactive modification of a software product performed after delivery to correct discovered problems."
- *Emergency maintenance*—A subset of corrective maintenance, "an unscheduled modification performed to temporarily keep a system operational pending corrective maintenance."
- *Perfective maintenance*—"The modification of a software product after delivery to detect and correct latent faults in the software product before they are manifested as failures."
- *Preventative maintenance*—"The modification of a software product after delivery to detect and correct latent faults in the software product before they become operational faults."

While software repair represents a component of maintenance, software can be modified for several other reasons. Modifications can be made to correct existing problems, prevent new failures, or improve software through additional functionality or performance. Thus, *repairability*, a software quality attribute sometimes erroneously used interchangeably with *maintainability*, represents the ability of software to be repaired for corrective, preventative, or perfective maintenance only, but speaks nothing of adaptability to a changing environment or shifting software requirements.

Corrective Maintenance

Corrective maintenance, including emergency maintenance, addresses software failures in function or performance. Functional failures can be as insignificant as SAS reports that are printed in the wrong font or as deleterious as reports that deliver spurious data due to failed data models. Performance failures can similarly span from processes that complete just seconds too slowly to processes that exceed time thresholds by hours or catastrophically monopolize system resources. That is to say that some failures can be overlooked for a period of time, perhaps indefinitely, while others require immediate resolution.

Where software failure occurs due to previously unidentified vulnerabilities or errors, the failure also denotes discovery, at which point developers must assess how, when, and whether to correct, mitigate, or accept those vulnerabilities and risk of future failure. When the failure comes as a total surprise to developers, corrective maintenance is the only option. Thus, it is always better to have invested the time, effort, and creativity into imagining, investigating, documenting, and mitigating potential sources of failure than to be caught off guard by unimagined failures and be forced to assume defensive posturing.

Since software failure—unlike the end-of-life failures of a tangled Yo-Yo or twisted Slinky—doesn't represent a destroyed product, software can often be restarted and immediately resume functioning. This occurs in SAS when a program uses too much memory, stalls, and needs to be restarted. In many cases the software will execute correctly the second time, perhaps in this example if other SAS programs are first terminated to free more memory. This quick fix may be sufficient to restore functionality, but doesn't eliminate the underlying vulnerability or code defects that enabled the failure to occur.

All software inherently has risk, and stakeholders must determine an acceptable level of risk. As defects or errors are discovered, risk inherently increases. That risk can be accepted or software can be modified to reduce or eliminate the risk. Preventative maintenance is always favored to corrective maintenance because it proactively reduces or eliminates risk before business value has been lost.

Preventative Maintenance

Preventative maintenance aims to control risk by facilitating software reliability and performance through the elimination of known defects or errors before business value is adversely affected. However, if failure has already occurred, although the maintenance activities may be identical, they are *corrective*, not *preventative*. Under ideal circumstances, theorized failures are predicted, their sources investigated and documented, and underlying defects corrected before they can cause operational failure.

Preventative maintenance that predicts and averts failure is always preferred to corrective action that responds to a failed event. It allows developers to fully utilize the SDLC to plan, design, develop, test, and implement solutions that are more likely to be stable, enduring, and require minimal future maintenance. This contrasts starkly with emergency maintenance, which often must be implemented in waves, such as a quick fix that patches busted software but must be implemented in several phases as longer-term solutions are delivered. The ability to plan and act while not under the duress of failed software and looming deadlines is always a benefit, as is the ability to prioritize and schedule when preventative maintenance should occur.

The unfortunate reality of preventative maintenance is that stakeholders tend to deprioritize and delay maintenance that would correct theorized, unrealized failures that have never occurred. Effort that should be directed toward preventative maintenance is either reinvested in the SDLC to add more software functionality or, especially in reactionary environments, is channeled toward emergency maintenance activities to address broken software. However, if development teams don't proactively interrogate and eliminate vulnerabilities from software, they can create and succumb to an environment in which they are indefinitely defensively poised under insurmountable technical debt. "Technical Debt," described in the corresponding section later in the chapter, can be eliminated only through consistent corrective and preventative maintenance.

Adaptive Maintenance

Inherent in data analytic development is variability that must be counterbalanced with software flexibility. Variability can include changes in the data source, type, format, quantity, quality, and velocity that developers may have been unable to predict or prevent. Variability can also include shifting needs of the customer and resultant modifications to technical requirements. Variability is the primary reason why SAS data analytic software can require continuous maintenance even once in production and despite adherence to the SDLC and software development methodology best practices. In some environments in which data injects, needs, or requirements are constantly in flux out of necessity, the line between adaptive maintenance and future development activities may blur. While an acceptable paradigm in end-user development, this represents a significant departure from development of software products that are expected to be more robust, stable, and support a third-party user base with minimal modification.

For example, if a SAS extract-transform-load (ETL) infrastructure relies on the daily ingestion of a transactional third-party data set, and the format of the data set changes unexpectedly overnight, developers may need to scramble the next morning to rewrite code. Although espousing maintainability principles such as modularity and flexibility can reduce this workload, it doesn't eliminate the adaptive maintenance that must occur to accommodate the new format. In a best-case scenario, developers would have predicted this source of data variability and developed quality controls to detect the exception and prevent unhandled failure, possibly by halting dependent processes and notifying stakeholders. However, because the data environment has changed, developers must immediately restore functionality through adaptive and emergency maintenance.

Adaptive maintenance is a real necessity, especially in data analytic development, but because SAS software must often be continually modified to flex to data and other environmental changes, SAS practitioners can learn to espouse the unhealthy habit of never finalizing production software. The importance of maintaining static code that flexibly responds to its environment is the focus of Chapter 17, "Stability," while Chapter 12, "Automation," demonstrates the importance of automating production software, which can only be accomplished through some degree of code stability. In combining these principles, the intent is to deliver mature software that reliably runs and does not require constant supervision or modification.

Requirements Maintenance

While the heart of software maintenance is code maintenance, requirements maintenance is a critical yet sometimes overlooked component. As the needs of a software product change over its lifetime, its technical requirements may be modified by customers or other stakeholders. Thus, the aggregate concept of *maintenance* includes both code and requirements maintenance and acts to balance actual software performance with required performance through modification of both software and its requirements. Because the majority of maintenance-related activities will entail modifying software to meet established technical specifications, only maintainability principles that support code modification are discussed in this text.

In many cases, requirements—whether stated or implied—are modified in lieu of software modifications. For example, as the volume of data from a data source grows over time, performance failures may begin to occur as the software execution time begins to surpass performance thresholds specified in technical requirements. Either SAS practitioners can implement programmatic or nonprogrammatic solutions that increase execution speed and realign actual performance with expected performance, or stakeholders can accept the risk of increased execution time and do nothing about the slowing software. The latter choice is often an acceptable solution, but it effectively modifies requirements because stakeholders are acknowledging they are willing to accept a lower level of software performance given the increased data volume.

MAINTENANCE IN THE SDLC

Within the SDLC, software is planned, designed, developed, tested, validated, and launched into production. When development has completed, the software product should meet functional and performance requirements for its intended lifetime or until a scheduled update. In Agile development environments, while the expected lifespan of software may be several years, software is likely to be updated at the conclusion of each iteration, which may last only a couple of weeks. In Waterfall environments, on the other hand, software is updated infrequently at scheduled software releases or when an emergency patch is required. Regardless of the software development methodology espoused, maintenance will always be an expected onus of the development process and will typically continue throughout software operation.

At the core of maintenance activities lie software modifications that correct defects, improve functionality or performance, and flex to the environment. Maintenance activities can be prioritized by production schedule milestones, internal detection of defects, external detection of failures, or feedback by users and other stakeholders. Thus, while the maintenance phase of the SDLC is often depicted as a discrete phase, maintenance can compete fiercely for resources with other phases, most notably development. For example, SAS practitioners may need to choose between delivering additional functionality or correcting existing software defects—competing objectives that each require development resources.

The SDLC adapts to a broad range of software development methodologies and practices, from phase-gate Waterfall to Agile rapid development. It moreover adapts to traditional applications development environments in which customers, developers, testers, and users maintain discrete roles, as well as to end-user development environments in which developers are also the principal users. The following sections highlight the importance of software maintenance within the SDLC and across these diverse environments. The "Software Development Methodologies" section in Chapter 1, "Introduction," introduces and distinguishes Agile and Waterfall development methodologies while end-user development is described in the "End-User Development Maintenance" section.

Maintenance in Waterfall Environments

Waterfall software development environments typically cycle through the SDLC only once. Where software patches, updates, or upgrades are required, these are released in the maintenance phase while users operate the software in the operational phase, with both phases occurring simultaneously. However, even software patches—which represent typically rapid responses to overcome software defects, errors, vulnerabilities, or threats—require planning, design, development, and testing activity to ensure the patches adequately resolve the issues. A mini-SDLC is sometimes conceptualized as occurring within the maintenance phase alone, as developers strive to maintain the quality and relevance of software but may have to return to the drawing board to design, develop, and test a solution.

In a Waterfall development environment, software maintenance commences after software has been released into production. Because Waterfall-developed software is by definition delivered as a single product (rather than incrementally as in Agile), all required functionality and performance should be included in the original release. At times, however, it becomes necessary to release software despite it lacking functional elements or performance that can be delivered through subsequent upgrades. In some cases, latent defects are known to exist in software but are included in the initial software release with the intention of correcting them in future software patches or upgrades. But in all cases of Waterfall development, the vast majority of development activities occur in the development phase, with only minor development occurring during maintenance.

Therefore, in Waterfall environments, maintenance activities are less likely to be in direct competition with development activities for personnel resources. Because development, testing, and deployment of software have already concluded, developers don't have to make the difficult choice between developing more functionality in a software product or maintaining that product. Often in Waterfall environments, the developers who maintain software—commonly referred to as the operations and maintenance (O&M) team—are not necessarily the authors of the software, which can further prevent conflict and competition of personnel resources. This contrasts sharply with Agile environments, discussed in the next section, in which stakeholders often must prioritize maintenance activities against development activities. Notwithstanding this complexity in Agile development, a wealth of other benefits drives the overwhelming appeal of Agile to software development teams.

Maintenance in Agile Environments

Agile methodologies differ from Waterfall in that they are iterative and incremental, releasing updated chunks of software on a scheduled, periodic basis rather than as a single product. For example, users might receive upgraded software releases every two weeks or every month, depending on the length of the development cycle, a time-boxed period often termed an *iteration* or *sprint*. Because the entire SDLC occurs within each iteration, despite being released with a higher frequency, software releases should evince commensurate quality delivered through design, development, testing, and validation activities.

Within Agile development environments, a single release of software at the close of an iteration doesn't signal project or product completion, because a subsequent iteration is quick on the heels of the first. Thus, as developers are delivering the first release of software, they may already be preparing for the next release in another two weeks. As unexpected vulnerabilities or defects that require maintenance are discovered, developers must balance this added maintenance with their primary development responsibilities. Users will want their software to function correctly, customers will want software delivered on time, and developers can get caught in the middle if all stakeholders don't adopt a uniform valuation and prioritization of software maintenance with respect to software development activities.

Because maintenance and development activities compete directly for system resources in Agile development environments, developers are sometimes encouraged to minimize maintenance activities. Maintenance reduction in some environments occurs because maintenance activities do not get prioritized, thus technical debt builds and software grows increasingly faulty over time until it is worthless. In more proactive environments, however, maintenance reduction occurs because higher-quality software is developed that requires less maintenance. Espousing software maintainability principles during design and development facilitates software that requires less extensive maintenance and, therefore, software that more quickly frees developers to perform their primary responsibility—software development.

Applications Development Maintenance

In a traditional software development environment, software developers design and develop software. They often bear additional software testing responsibilities and may work in concert with a dedicated quality assurance team of software testers. Once software is developed, tested, and validated, it's deployed to third-party users, who operate the software while an O&M team may separately maintain the software. In some cases, the customer is equivalent to the user base and has sponsored and paid for the software to be developed. In other development models, the customer directs software development for a separate group of users. In nearly all cases, however, the developers of the software product are not its primary intended users.

Whether operating in a Waterfall or Agile environment, maintenance can be complicated in applications development because developers do not actually use the software. This is sometimes referred to as the *domain knowledge divide*, in which software is designed to be used within a certain domain, such as medicine or pharmaceuticals, that requires specific knowledge in that field. Formalized education, training, licensure, or certification additionally may be required to certify practitioners as domain experts. For example, medical doctors must pursue years of schooling, including residency and board certifications, to gain domain expertise, similar to many other professions.

However, software developers authoring medical software are unlikely to be physicians themselves. A very few may be and, while most will have gained some medical domain knowledge while working on related software development projects, a significant domain knowledge divide will exist between these software developers and software users (physicians or medical practitioners). With the exception of a few ubiquitous applications like Gmail, which is no doubt also utilized by the Google developers who develop it, the majority of software developers never actually use the software they write.

This context is critical to understanding and implementing software maintenance in traditional applications development environments. *User stories* are often written (by developers or customers, but from the perspective of users) to identify deficits in function or performance, in an attempt to place developers in the shoes of software users. In these environments, business analysts often act as the critical go-between, ensuring that business needs are appropriately transferred to developers and that software products meet the needs of their intended user base. Business analysts or a similar cadre are often essential translators because they possess sufficient technical expertise and domain knowledge to communicate in both of these worlds.

Although SAS data analytic development rarely describes software intended to be sold or distributed to external, third-party users, an applications development mentality (or team organization) can exist in which SAS practitioners write software to be used internally by separate analysts or consumers. Especially in situations where analysts lack the technical expertise or user permissions to modify the SAS software that they utilize, formalized maintenance planning and prioritization become paramount and should be in place to ensure software users don't adopt a learned helplessness mentality as they languish waiting for software updates.

End-User Development Maintenance

End-user development is a common paradigm in data analytic development and arguably the development environment in which the vast majority of all SAS software is written. This occurs because SAS software is developed not to be sold or distributed but rather to process data, answer analytic questions, and impart knowledge to internal customers. The resultant data products are often distributed outside the original team, environment, and organization, but the software itself is typically not distributed. I have written thousands of SAS programs that have been executed once and subsequently discarded, not because they failed, but because they succeeded in conveying information necessary for my team to produce some data product that had ultimate business value for my customer.

End-user development effectively eliminates the domain knowledge divide discussed in the "Applications Development Maintenance" section. When a technically savvy doctor is developing SAS code to support medical research or other medically related functions, no interpretation between developers and users is required because they are one and the same. The doctor has domain knowledge (in medicine), understands the overarching need for the software, can design and develop the software solution in SAS, and ultimately understands when all product requirements have been met. This isn't to suggest that end-user development occurs in isolated silos of individuals working alone, but rather that professionals are comfortable with technology and qualified to function as software developers, testers, and users.

Maintenance in end-user development environments in theory is straightforward. As users, SAS practitioners understand which functionality and performance of the software is most valuable, and thus can prioritize maintenance that corrects or perfects those areas. No user stories, business analysts, or other intermediaries are required to facilitate maintenance, because a user can often in seconds open the code and begin correcting deficiencies or adding functionality or performance. Straightforward in theory, yes—but often abysmal in practice.

When an end-user developer completes software, his priority typically turns immediately to using that software for some intended purpose. In SAS development, this can entail data analysis, creation of data products, and often subsequent code development empirically driven by results from the original code. When defects are discovered in software, their maintenance competes against analytic and other endeavors, typically future functionality that is envisioned or subsequent data products to be created. Defects are often deprioritized because savvy SAS practitioners can learn to overcome—while not correcting—these vulnerabilities.

For example, a coworker once approached me and tactfully mentioned that a macro I had developed would fail if a parameter was omitted. My flippant response was something akin to "Do you want me to write you a sticky note so you don't forget the parameter?" In many end-user development environments, because the cohort using our software is intimate, we expect them to overcome software threats and vulnerabilities through intelligence and avoidance rather than through robust software design and apt documentation. This attitude can be effective in many environments and reduce unnecessary maintenance but does increase software operational risk.

End-user developers also often have no third-party development team to which they can report software failures because they themselves are responsible for maintenance activities. For example, if fixing a glaring software vulnerability—especially one that I think I'm intelligent enough to avoid when executing the software—means that my data playtime will be curtailed, I may choose to forgo documentation and maintenance in lieu of more interesting tasks. Especially in data analytic development environments, because business value is conveyed through data products, end-user developers (and their customers) are often more concerned with resultant data products than perfecting software through maintenance.

Thus, in end-user development environments, an unfortunate paradigm often persists in which only the most egregious defects are ever resolved. Because users intimately understand the underlying code, they are often able to avoid accidental misuse that would result in failure were others to attempt to use the software. As a result, vulnerabilities are often identified but not mitigated or eliminated in any way. Espouse this practice within an organization and allow it to flourish over the lifespan of software products, and you start to understand why end-user development often gets a bad rap in software development literature.

In many cases, maintenance that would have improved software quality is instead traded for additional functionality. For example, SAS practitioners may be able to produce an extra data product or analytic report rather than performing corrective maintenance to eliminate vulnerabilities. If greater business value is conferred through this functionality, then this is often a wise choice. However, over time, if additional functionality, earlier software release, or lower software costs are repeatedly prioritized above maintenance activities, then the quality of software can diminish, failing to meet not only performance but also functional intent. To remedy this imbalance, quality must constitute both functional and performance aspects and maintenance activities should be valuated to facilitate inclusion in the SDLC.

Because end-user development environments are both more permissive and forgiving, SAS practitioners can often get away with development shortcuts and diminished maintenance without increased risk of software failure or reduced business value. However, to navigate this funambulism successfully, two principles are required. First, the decision not to maintain some aspect of software should be intentional and made in consideration of the risks and benefits, just as the original decision not to include software performance should be an intentional one. Second, consensus among stakeholders should drive the valuation of maintenance as well as its opportunity cost—that is, other activities (or profits or happy hours) that could be achieved if maintenance is not performed. With this assessment and prioritization, end-user developers will be well-placed to understand how and when to best maintain software.

A basic tenet of the SDLC requires that production software is used but never modified in situ. Thus, maintenance occurs on copies of software in a development environment wholly separate and distinct from the production software operated by users or run as automated batch routines. In reality, end-user development environments often combine these discrete ideals into a single development–test–production environment. While the physical commingling of SDLC environments does not portend doom and is often a necessity in many organizations, an unfortunate consequence is the commingling of SDLC phases that results when so-called production software is continually modified. Especially when software is modified without subsequent testing and validation, and when that software is immediately made available to users, this cowboy coding approach diminishes the integrity of the software and tremendously reduces the effectiveness of any software testing or validation that predated later maintenance. Thus, regardless of the structure of the development environment, segregation of development, test, and production *phases* can greatly improve the quality of end-user development and maintenance.

FAILURE TO MAINTAIN

Developing software with maintainability in mind does not describe immediately fixing every identified defect or error. Some defects should remain in place because they represent minor vulnerabilities and the effort to fix them could be better invested elsewhere. As demonstrated previously, end-user developers may also be able to produce software with more latent defects because they can intelligently navigate around them, as they themselves operate the software. Notwithstanding the development environment and methodology espoused, technical debt can insurmountably grow if SAS practitioners fail to perform maintenance activities causing an imbalance between expected and actual software performance.

Technical Debt

Technical debt describes the accumulation of unresolved software defects over time. Some defects represent minor errors in code and relatively insignificant vulnerabilities to software performance while others can pose a much greater risk. Regardless of the size or risk of individual defects, their accumulation and compounding effects can cripple software over time. Technical debt, like financial debt, also incurs interest. As the volume of defects grows, software function and performance typically slip farther away from established needs and requirements, leading stakeholders to lose faith in the ability of the software ever to return to an acceptable level of quality.

Technical debt can be used as a metric to demonstrate the current quality of software as well as to anticipate the trajectory of quality. Only defects found in production software should be included in technical debt; thus, in Agile environments in which some modules of software have been released while others are still in development, debt should be assessed as defects only in released software. A risk register can capture technical debt and essentially inform developers of the totality of work required to bring their software from its current quality to the level required through technical specifications.

While a static representation of technical debt is important, more telling can be a longitudinal analysis demonstrating debt velocity. By assessing the volume of defects on a periodic basis, stakeholders can better understand whether they are gaining or losing ground toward the desired level of quality. This type of analysis can demonstrate unhealthy software trajectories in which technical debt is outpacing the ability of developers to maintain software and can enable teams to more highly prioritize maintenance activities. Technical debt conversely can demonstrate to stakeholders which software products consume too many resources to maintain and would be better abandoned in lieu of newly designed or newly purchased software. But to make these decisions, debt must be recorded and analyzed.

More than crippling software, technical debt can psychologically poison software development and operational environments. Where defects cause decreased performance or real failure, users may lose confidence in the software's ability to meet their needs. Developers will be frustrated by technical debt when it grows to become an obstacle from which recovery of quality seems unimaginable or improbable. And customers and other stakeholders can lose confidence in the development team for failing to remedy software defects and to prioritize maintenance activities into the SDLC.

While the elimination of technical debt is most commonly conceptualized as occurring through software maintenance, debt can also be eliminated by lowering expectations—that is, accepting that lofty or unrealistic software requirements should be abandoned or tempered in favor of obtainable goals. In the previous "End-User Development Maintenance" section, I told my coworker to put a sticky note on his monitor if he was baffled by undocumented macro parameters that could lead to software failure. He was hoping I would reduce technical debt with a programmatic solution to make the macro more robust and readable, while I was content to reduce debt by accepting riskier software; ultimately our customer made the decision. But this demonstrates that not all vulnerabilities, threats, or defects should be considered technical debt, especially those that have been accepted by the customer and that are never intended to be remedied. Yes, software deficiencies such as those my coworker exposed should be included in a risk register, but that register should also indicate the customer's acceptance of identified risks.

Maintainability plays an important role in reducing technical debt because it enables developers to perform maintenance activities more efficiently. Greater efficiency means more defects can be resolved in the same amount of time, facilitating software that can be operated at or above—rather than below—optimal performance levels. Moreover, when maintainability principles are espoused, developers tend to loathe maintenance less because their maintenance efforts are efficient and organized, rather than haphazard and defensive.

Measuring Maintenance

As with other static performance requirements, it's difficult to measure maintainability directly because it's assessed through code inspection rather than software performance. Thus, maintenance is often used as a proxy for maintainability, including both maintenance that has been performed and maintenance that has yet to be performed, often referred to as technical debt. Maintenance levels can generally be assessed relative only to a specific team or software product but, when tracked longitudinally, can yield tremendous information about the quality of software and even the quality of the software development environment.

A team that spends an inordinate amount of time maintaining software (as compared to developing software) could be producing poor-quality software that requires a tremendous amount of corrective maintenance. However, it could also be the case that standards or requirements are shifting rapidly, causing significant adaptive maintenance to ensure that software continues to meet the needs of its customers. Another possible cause of high maintenance levels is gold-plating, in which developers continue fiddling with and trying to perfect software, even to the point of unnecessarily exceeding technical specifications. Thus, while the volatility of technical needs and requirements will shape the maintenance landscape of a software product, the extent to which maintenance can be minimized through software development best practices will enable developers to invest a greater proportion of their time into development—be that programmatic development of software or subsequent nonprogrammatic development of data products.

Another way to assess maintenance levels is by inspection of the failure log, introduced in the "Failure Log" section in Chapter 4, "Reliability." All logged failures that were previously described in a risk register (described in Chapter 1, "Introduction") represent software vulnerabilities that have been identified, were not mitigated or eliminated, and that were later exploited to cause software failure. Software that is continually failing due to known defects or errors reflects a lack of commitment to maintenance and possibly a difficulty in maintaining software due to a lack of maintainability principles. On the other hand, if software is not required to be robust and reliable, it may often fail due to identified vulnerabilities, and these issues should not be corrected if they have not been prioritized.

Measuring Technical Debt

Technical debt, a measure of prospective maintenance, assesses maintenance irrespective of past software failure or past maintenance levels. Instead, technical debt represents the sum of maintenance tasks that have been identified, prioritized for completion, yet not completed. Technical debt is distinguished from outstanding development tasks that have not been completed or possibly even started. It essentially represents work that was "completed" yet found to contain defects or errors, or "completed" yet not really because it lacked expected function or performance. Technical debt amounts to writing an IOU note in code, and until the IOU is paid, the debt remains.

For example, the following code is reprised from the "Specific Threats" section in Chapter 6, "Robustness," in which numerous threats to the code are enumerated. One threat is the possibility that the Original data set does not exist, which exposes the vulnerability that the existence of Original is not validated before its use. If software requirements state that the software should be reliable and robust to failure, then whether by design or negligence, this code incurs technical debt:

```
data final;
   set perm.original;
run;
```

Maybe the SAS practitioner completed the functional components of the code and intended to incorporate robustness at a later time. Or perhaps the developer did not think to test the existence of Original, thus not perceiving the threat. Regardless of intent or motive, technical debt was incurred but can be eliminated in part by first validating that Original does exist:

```
%if %sysfunc(exist(perm.original)) %then %do;
   data final;
      set perm.original;
   run;
   %end;
```

Other vulnerabilities do still exist, but this has made the software slightly more robust, so long as business rules and additional exception handling (not shown) either enter a busy-waiting cycle or dynamically perform some other action based on the exception to drive program flow. While technical debt is typically eliminated through software maintenance (which mitigates or eliminates risk), it can also be eliminated through risk acceptance and modification of requirements. For example, a developer could assess that the original code could fail for a dozen reasons, but choose to accept these risks so he can press on to deliver additional functionality or other business value. This decision essentially lowers the expected reliability and robustness of the software but, in many environments and for many purposes, this would be advised because the corresponding risk is so low that it doesn't reduce software value.

It's also important to conceptualize technical debt in terms of the entire SDLC, not just development tasks. If a latent defect is discovered and the decision is made to correct it (at some point), the defect adds technical debt until the corrective maintenance is completed. However, corrective maintenance should include not only fixing the code, but also performing other related SDLC activities, such as updating necessary documentation, unit testing, integration testing, regression testing, and software acceptance. In many cases, technical debt is accrued not through errors made in code but rather because developers are more narrowly focused on delivering functionality than performance. In other cases, the technical debt may have been fully eliminated, but until software integrity is restored through testing and validation, the complete resolution of that debt is not yet realized.

The risk register remains the best artifact from which to calculate technical debt, but must itself be maintained. Vulnerabilities should not be removed until they have been eliminated or accepted, and elimination always should denote that software modifications were appropriately tested and, if necessary, sufficiently documented. The risk register is introduced and demonstrated in Chapter 1, "Introduction." In actuality, entries in the risk register are not deleted when they are resolved, but rather amended somehow to demonstrate completion and inactivity. Historical entries in a risk register—including both resolved and unresolved entries—can paint a picture of the development environment and its ability to manage, prioritize, and perform maintenance effectively.

MAINTAINABILITY

That software must be maintained is a rather salient reality of all software products. That software can be intentionally designed to facilitate more efficient and effective maintenance is somewhat more of an enigma, unfortunately even among some developers and their customers. The effects of maintenance are sometimes immediately perceived, especially where corrective or emergency maintenance were required. Building maintainable software, on the other hand, doesn't fix anything, but rather makes software more readily fixed in the future. Thus, maintainability, like other static performance requirements, always represents an investment in software and has intrinsic benefits that can't be immediately demonstrated.

From Maintenance to Maintainability

The mind-set shift from maintenance-focused to maintainability-enabling requires a shift from a reactive "putting out fires" mentality to one of proactive fire prevention. Part of the reticence to implement maintainability and other static performance requirements occurs because while the efforts and effects of dynamic performance requirements are readily identified through software performance, those of static performance requirements are not. Making software more maintainable doesn't add function, make it run faster, or improve other measurable characteristics that can be assessed when software executes. The inherent advantages of maintainability are observable only in the future and only once software has failed or otherwise needs to be modified. Notwithstanding, as teams perceive that they are investing too great a proportion of development effort into software maintenance, a sure way to reduce future maintenance is to increase software maintainability.

Principles of Maintainability

If maintainable software is the goal that will lessen maintenance woes, how do you get there, and how do you know when you've arrived? As a composite attribute, maintainability isn't easy to pin down because it represents an aggregate of other static performance characteristics discussed throughout later chapters. These include several other aspects of software quality:

- *Modularity*—Modular software can be modified more easily because it is inherently composed of discrete chunks that are more manageable and understandable. Because modules should be loosely coupled and interact with a limited segment of software, maintenance requires less modification and testing to effect change.
- *Stability*—Stable software resists the need for frequent modification because it flexibly adapts to its environment. Because stable software

requires less maintenance, when maintenance is performed, it can be done so in a proactive, enduring fashion.

- *Reusability*—When software can be reused, including the reuse and repurposing of discrete modules of code in other software, maintainability is improved because software maintenance can be accomplished by changing code in one location to affect all software that uses those shared modules.
- *Readability*—Software that is straightforward and easily understood can be modified more readily because less time is required to investigate and learn intricacies of code. This can be accomplished through comments and documentation, clear syntax, standardized formatting, and smaller single-function modules.
- *Testability*—Testable software can be more readily validated to demonstrate its success as well as to show the results of exceptions or errors. Because software that is modified during maintenance activities must be subsequently tested and validated before being rereleased for production, the ease with which it can be tested expedites maintenance.

Complexities of Maintainability

Static performance attributes such as maintainability have no immediate benefit to software although they bear substantial long-term benefit. Stakeholders are more likely to disagree about the intrinsic value of software maintainability because it is difficult to define and quantify. Developers who best understand the technical intricacies of underlying code might want to instill maintainability principles to allay future maintenance when modifications are necessary. Yet they might face opposition from customers who instead value greater functionality that provides immediate, observable benefit. Or, conversely, customers who best understand the vision and long-term objectives of software might want to instill maintainability principles, but developers might be more focused on short-term technical requirements. Thus, a path to maintainable code requires clear communication between all stakeholders to identify its true value in software.

A tremendous psychological factor may also discourage stakeholders from discussing maintainability during software planning and design, when requirements are formulated. At the outset of a software project, discussion is forward-leaning and positively focused on creating excellent software.

Discussing developer errors, software defects, software failure, and its eventual demise can be an uncomfortable conversation before even one line of SAS code has been written. Despite this discomfort, maintainability principles should at least be considered in requirements discussions to identify whether they can be implemented to facilitate later successful and efficient maintenance. If no other conversation is had, the discussion and decision about intended software lifespan will at least provide SAS practitioners and other stakeholders with some understanding of the role that maintainability should have within the software.

A further complexity of maintainability is its location at the intersection of software development and operational service. Technical requirements documentation typically specifies the required function and performance that software must achieve. In many environments, however, software operations are managed separately from development by an O&M team. Where development and operations teams are separate, they may espouse competing priorities, where the development team is focused on delivering initial functionality and performance while the O&M team bears the responsibility of software maintenance. In all environments, it's important for developers to understand O&M priorities and requirements—whether these are the responsibility of the developers or of a separate team—to ensure that software will sufficiently meet performance requirements once released.

Failure of Maintainability

A failure to maintain software leads to its eventual demise, as it may fall victim to defects or errors, a changing environment, or evolution of needs and requirements. On the other hand, a failure to develop maintainable software can increase technical debt and decrease software quality, as SAS practitioners must exert more effort to perform equivalent maintenance. A failure of maintainability thus decreases the efficiency with which developers can provide necessary modifications to software and, in some cases, ultimately contributes to the decision not to perform necessary software maintenance.

When software is not maintained, functional or performance failures are often cited as ramifications of poor maintenance practices. For example, software fails, a cause is determined, examination of the risk register reveals that the specific vulnerability had been identified four months earlier, and suddenly stakeholders are asking "Why wasn't the corrective maintenance performed?" A lack of maintainability in software, conversely, is often highlighted when maintenance activities take forever to complete. If a SAS practitioner must take days to modify software in a dozen different places just to effect a slight change in function or performance, a lack of maintainability (and modularity) may be the culprit. Or if your ETL software fails and it's such a rat's nest of poorly documented code that you have to call a former coworker who built it to beg for guidance—you may have a maintainability crisis on your hands.

Requiring Maintainability

Maintainability is difficult to convey in formal requirements documentation because it represents an aggregate of other static performance characteristics. When documenting technical specifications, it's typically more desirable to reference modularity, readability, or other attributes specifically, with the awareness that they collectively benefit maintainability.

One of the most important aspects of requirements documentation that will influence the inclusion of maintainability principles in software is the intended software lifespan. Ephemeral analytic software that may be run a few times and then discarded will be less likely to require maintenance. However, SAS software with a longer intended lifespan and especially underpinning critical infrastructure will require more cumulative maintenance throughout its lifespan, benefiting SAS practitioners who can more swiftly and painlessly perform that maintenance.

Critical software also often requires higher reliability and availability, as discussed throughout Chapter 4, "Reliability," and high availability requires that software recover quickly when it does fail. When corrective or emergency maintenance is required in the recovery period (during which software functionality is lost), the ease and speed with which maintenance can be performed demonstrates recoverability but is driven by maintainability principles. Thus, when technical specifications require a recovery time objective (RTO) or maximum tolerable downtime (MTD), described in the "Defining Recoverability" section in Chapter 5, "Recoverability," these objectives can be facilitated by incorporating static performance requirements into software.

WHAT'S NEXT?

In many software quality models, static performance attributes are subsumed under a maintainability umbrella, similar to how dynamic performance attributes may be subsumed under reliability. Modularity, more so than all other static performance attributes, contributes to maintainable software by improving software readability, testability, stability, and reusability. In the next chapter, modular software design is demonstrated, which incorporates discrete functionality, loose coupling, and software encapsulation.

NOTES

1. ISO/IEC 25010:2011. *Systems and software engineering—Systems and software Quality Requirements and Evaluation (SQuaRE)—System and software quality models.* Geneva, Switzerland: International Organization for Standardization and Institute of Electrical and Electronics Engineers.
2. ISO/IEC 14764:2006. *Software engineering—Software life cycle processes—Maintenance.* Geneva, Switzerland: International Organization for Standardization and Institute of Electrical and Electronics Engineers.

CHAPTER 14

Modularity

I just dropped a 1,500-year-old Maya ceramic.

It was already mostly broken ... but the sinking feeling in my heart was indescribable. Beneath the giant *palapa* (open thatched hut) at the Martz farm just south of Benque Viejo del Carmen, a carefully choreographed archaeological dance of sorts had been underway for several hours.

Minanhá, Belize. The Late Classic Maya city center situated between Xunantunich and Caracol was little more than unceremonious piles of rubble amid the *milpa* (terraced fields of maize and beans), but the dig site would be the summer home of our expedition.

We surveyed and mapped hillsides, selected dig sites, and forged ephemeral canopies from palm fronds to protect us from the tropical sun and our excavations from the tropical downpours.

By day, the dig—excavating for hours, measuring, photographing, sketching, and slowly excising the hillside. Tiny trowels and pickaxes delicately chipping away at the past. Balancing soil on large screens, we sifted dirt from potsherds, from bone, from Belizean bugs, from bottles. If you ever find an empty bottle of One Barrel rum at the bottom of your dig, you've wasted hours if not days of effort—it's the telltale sign that the site has been excavated and subsequently filled in.

By night, the dance—the giant *palapa* aglow with the flicker of kerosene, a mix of frenetic yet coordinated activity as the day's plunder was brushed, washed, rinsed, photographed, typed, documented, and bagged for transport back to labs in the United States.

Everyone had a role. Several washers whisked dirt from potsherds, dunking these in a series of successively cleaner buckets of water.

In another corner sat the archaeologists, carefully examining and "typing" potsherds and the occasional bone fragments or carved stone.

But the runners facilitated the entire operation, transporting raw materials to the washers, washed potsherds to drying racks, dried pieces to the archaeologists, and a seemingly endless supply of One Barrel rum to glass Coke bottles strewn about the landscape.

■ ■ ■

The choreographed scene at the Martz farm required tremendous communication, cooperation, and coordination, remarkably similar to that required in modular software design. Everyone had a specific function, which produced assembly-line efficiency while reducing confusion. Software modularity similarly prescribes that individual modules be *functionally discrete*—that is, they should do one and only one thing.

Not only were our responsibilities specialized, but our interaction with each other was limited. For example, the runner brought me dirty ceramics (and rum) and transported my cleaned ceramics to drying racks, allowing me to focus on work and preventing me from stumbling into someone else's space. Modular software should also be *loosely coupled,* in that modules are linked to each other through only limited, prescribed methods.

By specializing only in brushing, washing, and rinsing ceramics, I became the master of my domain. And, if the archaeologists wanted to validate or test my performance, they could do so with ease because I had few responsibilities and because those responsibilities were not entangled with those of my peers. Software similarly is significantly easier to test and validate when modules are both functionally discrete and loosely coupled.

Not only were our roles clearly defined, but our communication routes were also specified. The rum runner who gathered bags of debris for us to wash was constantly bounding in and out of the *palapa,* but primarily interacted with washers at the *palapa* entrance. I knew precisely where he would drop the ceramics, precisely where to leave my empty Coke bottle for refills, and precisely where he expected to pick up my cleaned ceramics. Modular software is also improved when data and parameters are passed through precise methods to child processes, and when return codes are precisely passed back to parent processes.

But the dance could all be thrown into confusion when our roles were changed. For example, when I was doubling as a rum runner, I had to learn what artifacts I was carrying to which people, where to place them, and how not to collide with others in the *palapa.* And, when you're stumbling in the wrong place at the wrong time—that's when really old things get dropped and obliterated. Software modularity similarly facilitates security by prescribing rules and modes of communication that guide successful software execution.

DEFINING MODULARITY

Modularity is "the degree to which a system or computer program is composed of discrete components such that a change to one component has minimal impact on other components."[1] Software modules are often said to be functionally discrete and loosely coupled. *Functionally discrete* describes modules that are typically organized by function and that should perform one and only one function. *Loosely coupled* describes modules that should interact with each other and with software in very limited, prescribed ways.

A third attribute often ascribed to modular programming is *encapsulation*, "a software development technique that consists of isolating a system function or a set of data and operations on those data within a module and providing precise specifications for the module."[2] Encapsulation facilitates secure software by ensuring that data and information inside a module do not unintentionally leak beyond the module and that extraneous, external information does not unintentionally contaminate the module. While modularity is much more common in (and beneficial to) object-oriented programming (OOP) third-generation languages (3GLs) that utilize classes and structures, procedural fourth-generation languages (4GLs) like Base SAS can achieve higher quality software when modular design is embraced.

This chapter introduces modularity through SAS software examples that are functionally discrete, loosely coupled, and encapsulated. These core concepts are repeatedly referenced throughout this text, demonstrating the benefits of modular design to various dimensions of quality. Chapter 7, "Execution Efficiency," and Chapter 12, "Automation," demonstrate SAS parallel processing paradigms that can be operationalized only through modular software design.

FROM MONOLITHIC TO MODULAR

As discussed in the "Preface," data analytic software typically serves a very different purpose from more traditional, user-focused software applications. Applications tend to be interactive; the experience is often driven by variability provided through user input. Data analytic software applications are instead driven by data variability, as data transit one or more transformations or analyses. These two very different software objectives have led to a stark divergence in software design between data analytic and other software. While object-oriented, user-focused applications have evolved into increasingly modular design over the past few decades, data analytic software design has remained predominantly monolithic, espousing behemoth programs that are read from the first line to the last like a novel.

Much of the functionality required by data analytic software does require serialized processing that occurs in prescribed, ordered steps. For example, a data set is ingested, transformed, and subsequently analyzed in rigidly ordered steps. However, because software quality entails delivering not only functionality but also performance, modular software design can often be incorporated into performance-focused aspects of data analytic software. For example, a quality control mechanism to detect data set availability and validity—thereby improving performance—could be built modularly and reused by a number of unrelated SAS programs. Developers accustomed to serialized data analytic software, however, often persist in monolithic design habits even when incorporating performance characteristics that could be modularized. Software quality can often be improved when SAS practitioners embrace modular design, as demonstrated throughout this chapter.

Monolithic Software

Monolithic is derived from the Greek words meaning single stone (monos + lithos). Monolithic software describes a behemoth, single-program software product; unfortunately, this is a software design model prevalent in Base SAS and other data analytic development languages. The disadvantages of monolithic software are numerous. Code complexity increases while readability diminishes. Code maintenance is made more difficult and less efficient, because only one developer can modify software at one time. Monolithic software often performs more slowly due to false dependencies, a disadvantage demonstrated in the "False Dependencies" sections in Chapter 7, "Execution Efficiency."

IF A < B < C THEN A < C
IF ingest < transform < analyze THEN ingest < analyze

Figure 14.1 Transitive Property of Data Analytic Software

However, the monolithic nature of SAS software design is not without reason. As data processes typically have prerequisites, dependencies, inputs, and outputs, these elements establish a natural precedence of processes from which program flow cannot deviate. For example, the transitive property states that if A is less than B, and B is less than C, then A is also less than C. Similarly, within extract-transform-load (ETL) software, if transformation must precede analysis, and ingestion must precede transformation, then ingestion also must precede analysis, thus prescribing serialized program flow. As demonstrated in Figure 14.1, the transitive property is the reason that the functionality within data analytic software tends to be serialized.

The program flow in Figure 14.1 is typical of serialized ETL processes that are sequenced by function and can be simulated with the following code:

```
libname perm 'c:\perm';

* ingestion;
data perm.raw;
   length char1 $10;
   char1='burrito';
run;
```

```
* transformation;
data perm.trans;
   set perm.raw;
   char1=upcase(char1);
run;

* analysis;
proc means data=perm.trans;
   by char1;
run;
```

This serialization is critical in data analytic development because it provides necessary security and data integrity. Data pedigree can be easily traced from ingestion through transformation to ultimate analysis, providing confidence in analytic products. For example, data that are analyzed can be trusted in part because they must first pass through the Transformation module before analysis can begin. In essence, in data analytic development, there is no "Advance to Go" card and you don't collect $200 until you've earned it.

Because functional prerequisites and inputs cannot be altered, SAS software is often written as a monolithic program that can be read sequentially, from the first line to the last. I recall when I was young in my SAS career, proudly demonstrating a SAS program to a coworker when it had surpassed 2,000 lines. We actually relished writing those behemoth beauties, because every process was self-contained in one program. Scroll up, scroll down, scroll anywhere and the consummate functionality was described in one place. Although we were seated together and supporting a shared mission, my team and I wrote code in silos, rarely interacting with the software that others had developed. Despite sharing some functional and performance endeavors—such as processes that ingested the same data sets or which performed identical quality assurance tasks—our respective programs shared no actual code in common, an unfortunate but all-too-common consequence of monolithic design in end-user development environments.

While functional components in data analytic software often do have prerequisites and inputs that necessitate sequential program flow, when examined more closely, isolated functional and performance objectives can often be extracted and extrapolated elsewhere. For example, from a distance, an ETL process might seem wholly unique, contributing to monolithic design in which serialized code is neither shared nor reused. However, up close, shared objectives such as validating the existence of a data set (before use) or deleting the contents of the WORK library can be generalized. By modularizing these aspects of software, multiple modules or programs can benefit from reusable functionality and performance. In some cases, modular design can even enable modules to run in parallel, thus facilitating faster software.

Modular Software

Modular software design separates software components into discrete modules and is standard design practice within OOP languages and their software products. At a high level, modular design somewhat mimics the user experience when interacting with traditional user-focused applications. For example, when Microsoft Word is opened, you immediately have thousands of options—start typing in the blank document space; open a saved document; modify the margins, font, or format; or myriad other activities. As the application has evolved over time, not only has the functionality continued to increase, but specific functions can be accessed through numerous modalities.

For example, to bold text, you can highlight the text and use the <CTRL><B> keyboard shortcut, right-click the text and select the **B** icon, select the **B** icon from the ribbon header, click the <Home> tab and click the **B** icon, or type <ALT><H><1>. With so many ways to reach the bold function, it's imperative that each method lead to identical functionality. Modular software design capitalizes on this versatility, by calling the same "bold" function by each of these respective methods.

While the mechanics of modular software design within OOP are significantly more complex than this representation, this high-level view demonstrates the multi-directionality endemic within user-focused applications, which contributes to modular design. Data analytic software will never be able to fully escape the clutches of sequential software design—nor do we want it to. Where functional necessity prescribes process sequencing, this orderly program flow facilitates security and data integrity and should never be condemned. After all, as demonstrated in Figure 14.1, it's impossible to analyze data before you've ingested them. However, even when functionality must be sequenced, the adoption of modular design can ensure that software is more readable, reusable, testable, and ultimately more maintainable. And, where more generalizable functionality and performance are required, modular components can be more readily implemented to support a larger swath of software.

MODULARITY PRINCIPLES

The *SAS® 9.4 Macro Language, Second Edition*, recommends that users "develop your macros using a modular, layered approach" both to reduce software defects and to improve the ease with which defects can be discovered.[3] The documentation goes on to recommend modular development, stating "instead of writing one massive program, develop it piece by piece, test each piece separately, and put the pieces together." This introduction to modularity speaks to its role in facilitating greater software maintainability, readability, and testability. The benefits of modular design, however, are much greater and are described throughout this chapter.

In SAS software development, SAS macros are commonly referred to or conceptualized as modules. While this can be true, not all SAS macros espouse modular software design, and in fact monolithic software can just as easily be developed with SAS macros as without them. Modular software design principles state that software should be:

- *Functionally discrete*—A software module performs one and only one function.
- *Loosely coupled*—A software module directly interacts with other modules in as few ways as possible.
- *Encapsulated*—A barrier protects the software module and the rest of the software from unintended interaction with each other.

While the size of software modules is not typically included as a defining characteristic of modularity, software modules are inherently much smaller because they are functionally discrete, loosely coupled, and encapsulated. Some more anecdotal references to modularity define modules as the quantity of code that can fit on one screen—about 20 to 40 lines, depending on font size, screen size, and resolution. Thus, while the aim of building functionally discrete code is to increase flexibility and reusability, this practice also results in significantly smaller segments of code.

Loose coupling and encapsulation are similar, and some texts even describe them synonymously. The presence of one often does signify the presence of the other, but their objectives and thus methods do differ. The primary goal of loose coupling is to facilitate flexibility, principally by severing unnecessary bonds between parent and child processes. A child will always depend on some parent because it must be invoked from somewhere, but loose coupling seeks to ensure that the child can be invoked from as many parents as possible in as many appropriate ways as possible. For example, a module that capitalizes all text in a data set could be flexibly designed so that the transformation could be performed dynamically on any data set with any quantity of variables. This flexibility is demonstrated in the %GOBIG macro in the following "Functionally Discrete" sections.

On the other hand, the primary goal of encapsulation is to facilitate software security through what is commonly referred to as *black-box design*. In the analogy, a black box is sometimes implemented to protect the *confidentiality* of information, methods, or actions that occur inside a child process, but more often is implemented to protect the *integrity* of the child, its parent, the software, and possibly the software environment. For example, transient information inside a module that is required to produce some result but not required to accompany that result should be locally defined so it cannot interact with other aspects of the software. For example, if a counter is required to produce some result within a macro, but the counter value itself has no importance outside the macro after process completion, then the counter should be locally defined and restricted only to the module. Thus, encapsulation can be conceptualized as the black box placed around a software module while loose coupling is the tiny pinhole in the box through which information can be transmitted through prescribed channels.

Functionally Discrete

That software modules should do one and only one thing runs counter to the common practice of shoving as much functionality as possible into a single DATA step. After all, why do in two DATA steps what you can accomplish in one? In general, DATA steps should not be broken apart, as this unnecessarily decreases both execution speed and efficiency. In part due to the fixed costs of input/output (I/O) operations that read and/or write observations, DATA steps that do more are always more efficient than those that do less. Thus, to be clear, hacking up a DATA step into bits will make it smaller but won't demonstrate modular design. In Base SAS, functionally discrete code typically refers to SAS macro language modules that improve software functionality and performance.

Novel Code

Novice coders write novel code—that is, code that reads cover to cover like a novel, from the first DATA step to the final RUN statement. As mentioned in the "Monolithic Software" section, serialized design is expected and unavoidable when data analytic processes must be sequenced due to prerequisites and inputs. However, SAS practitioners can be so accustomed to designing and developing monolithically that they fail to perceive opportunities where modular design would improve program flow or static performance attributes.

The following code simulates monolithic ETL software saved as a single program but conceptualized as three functional modules:

```
libname perm 'c:\perm';

* ingestion;
data perm.raw;
   length char1 $10 num1 8;
   char1='burrito';
   num1=443;
run;

* analysis 1;
proc means data=perm.raw;
   by char1;
run;

*analysis 2;
proc freq data=perm.raw;
   tables num1;
run;
```

In some cases, this long-windedness is unavoidable and separating data processes into discrete modules—that is, separate programs—would not benefit dynamic or static performance. For example, the following attempt to modularize the ETL code does create a series of macros but provides no improved performance:

```
libname perm 'c:\perm';

* ingestion;
%macro ingest();
data perm.raw;
   length char1 $10 num1 8;
   char1='burrito';
   num1=443;
run;
%mend;

* analysis 1;
%macro anal1();
```

```
proc means data=perm.raw;
   by char1;
run;
%mend;

* analysis 2;
%macro anal2();
proc freq data=perm.raw;
   tables num1;
run;
%mend;

%ingest;
%anal1;
%anal2;
```

All code is still contained inside a single program and that program demonstrates sequential processing of all steps. Nothing has changed except that the complexity of the software has increased due to unnecessary macro definitions and invocations. Faux modularity is often observed in SAS software because certain macro statements, such as %IF–%THEN–%ELSE conditional logic, are viable only inside the confines of a macro. Software may appear to be broken into discrete chunks of functionality, but these delineations are often created only so that macro statements and logic can be utilized.

In this example, both analytic modules could in theory be run in parallel to increase performance. As demonstrated in the "False Dependencies" section in Chapter 7, "Execution Efficiency," because the MEANS and FREQ procedures each require only a shared file lock on the data set, they can be executed simultaneously. The advantages of modularity cannot be realized until monolithic code is conceptualized not as a single entity but rather as discrete functional elements. Through functional decomposition, demonstrated in the "Functional Decomposition" section later in the chapter, this code embarks on its journey toward modularity.

Overly Complex Code

Whereas SAS practitioners should try to cram as much functionality as possible into every DATA step, this data management best practice does not translate to modular design. And that's okay! If you can ingest, clean, and transform your data in a single DATA step, there's no reason to arbitrarily kill one bird with three stones—keep the single DATA step. But certain functionality required in ingesting, cleaning, and transforming data can often be generalized to other software and thus can benefit through modularization and reuse.

The following %GOBIG macro, which appears in the "Flexibility" section in Chapter 18, "Reusability," dynamically capitalizes all character variables in a data set:

```
%macro gobig(dsnin= /* old data set in LIB.DSN or DSN format */,
   dsnout= /* updated data set in LIB.DSN or DSN format */);
%local dsid;
%local vars;
%local vartype;
%local varlist;
%local i;
%let varlist=;
%let dsid=%sysfunc(open(&dsnin,i));
%let vars=%sysfunc(attrn(&dsid, nvars));
%do i=1 %to &vars;
   %let vartype=%sysfunc(vartype(&dsid,&i));
   %if &vartype=C %then %let varlist=&varlist %sysfunc(varname(&dsid,&i));
   %end;
%let close=%sysfunc(close(&dsid));
%put VARLIST: &varlist;
data &dsnout;
   set &dsnin;
   %let i=1;
   %do %while(%length(%scan(&varlist,&i,,S))>1);
      %scan(&varlist,&i,,S)=upcase(%scan(&varlist,&i,,S));
      %let i=%eval(&i+1);
      %end;
run;
%mend;
```

When the following code is executed, a new Uppersample data set is created, and the values sas and rules become SAS and RULES!, respectively:

```
data sample;
   length char1 $10 char2 $10;
   char1='sas';
   char2='rules!';
run;

%gobig(dsnin=sample, dsnout=uppersample);
```

The dynamic macro capitalizes values in all text variables and can be implemented on any data set. However, the code is not functionally discrete and, as demonstrated later in the "Loosely Coupled" sections, it unfortunately is highly coupled.

The single functional requirement of the module is to capitalize the values of all text variables within a data set dynamically. And, in a nontechnical sense, this is what the end user perceives as the functional outcome. For example, when I hit <CTRL><B> in Microsoft Word, the characters I've highlighted are made bold—a single functional outcome. However, behind the scenes, the software must assess whether I'm turning off or turning on the bold and whether I've highlighted text or want to affect the cursor and subsequent text I'm about to type. Thus, despite having a single functional outcome, a functional requirement will often have numerous functional and performance objectives subsumed within it that are inherently required.

The %GOBIG macro could still be segmented more granularly. The first function performed opens a stream to the data set, retrieves the number of variables, and makes a comma-delimited list of character variable names. In modular software design, this functionality alone should be represented in a separate macro saved as a separate SAS program. The next function dynamically generates code that capitalizes the values, but the code is unfortunately wrapped inside a DATA step, decreasing reusability and violating the loose coupling principle of modular design. Thus, while the code may have only one functional requirement, other implicit functionality may be required to achieve that goal. The next section demonstrates functional decomposition, which aims to separate monolithic software and overly complex macros into true discrete functionality.

Functional Decomposition

Functional decomposition describes identifying separate functions that software performs, including both functionality explicitly described in technical requirements and functionality implicitly required to meet those objectives. In some cases, functional decomposition is a theoretical exercise and occurs during planning and design before any code has been written. This thought experiment involves decomposition of explicit software requirements to generate implicit tasks that will be required to achieve the necessary function and performance. In other cases, as described later, functional decomposition is the actual breakup of code into smaller chunks.

For example, the %GOBIG macro previously introduced in the "Overly Complex Code" section has a single functional requirement: create a macro that can dynamically capitalize the values of all character variables in any data set. However, during software planning and design, as SAS practitioners begin to conceptualize how this will occur, they functionally decompose the objective into additional functional steps required for success. If the software demands robustness, a first step might be to validate the existence and availability of the data set. A second step could be to create a space-delimited list of the character variables. A third step might be to exit with a return code if no character variables were found. A fourth step could be to generate lines of code dynamically that would capitalize the values with the UPCASE function.

After the thought experiment, it's clear that %GOBIG demonstrates diverse rather than discrete functionality. Developers might decide to coalesce some functions into a single module, while separating other functions into external macros that can be called from reusable macro programs. Thus, as a theoretical exercise, functional decomposition doesn't require software to be built modularly, but rather is intended only to identify distinct parts of composite software or requirements. Functional decomposition is extremely useful early in the SDLC because it enables future development work to be conceptualized, assigned, and scheduled based on overarching requirements. In project management, *functional decomposition* similarly describes breaking a project into discrete tasks that can be reconstituted to give a fuller perspective of the breadth and depth of the statement of work. Functional decomposition also often exposes unexpected process prerequisites and dependencies that can add complexity to software design—complexity that is much more welcome when discovered early in the SDLC than in the throes of development.

But in software development, functional decomposition does not represent only casual conversation and whiteboard diagramming; more than theory, functional decomposition is the literal cleaving of software into discrete chunks, typically saved as separate programs. For example, the ETL software demonstrated earlier in the "Novel Code" section was broken into three separate macros. However, because the macros were still confined to the same program, no performance gains could be experienced due to this limitation. The macro hinted at reusability, but the code would have had to have been cut-and-pasted into other software—and this only describes code *extension* or *repurposing,* not *reuse.*

During functional decomposition, the second identified implicit task of the %GOBIG requirement was to generate a space-delimited list of all character variables. The following %FINDVARS macro now performs that more specific task:

```
* creates a space-delimited macro variable VARLIST in data set DSN;
%macro findvars(dsn= /* data set in LIB.DSN or DSN format */,
   type= /* ALL, CHAR, or NUM to retrieve those variable types */);
%local dsid;
%local vars;
%local vartype;
%global varlist;
%let varlist=;
%local i;
%let dsid=%sysfunc(open(&dsn,i));
%let vars=%sysfunc(attrn(&dsid, nvars));
%do i=1 %to &vars;
   %let vartype=%sysfunc(vartype(&dsid,&i));
   %if %upcase(&type)=ALL or (&vartype=N and %upcase(&type)=NUM) or
         (&vartype=C and %upcase(&type)=CHAR) %then %do;
      %let varlist=&varlist %sysfunc(varname(&dsid,&i));
      %end;
   %end;
%let close=%sysfunc(close(&dsid));
%mend;
```

But wait, it actually does more! The developer realized that he could add one parameter and alter a single line of code to enable the macro not only to identify all character variable names, but also to identify either all numeric variable names or all variable names in general. With this additional functionality and its modular design, the %FINDVARS macro is a great candidate for inclusion in a reuse library to ensure that other SAS practitioners can benefit from its function and possibly later improve its performance.

For example, while modular, the %FINDVARS macro contains no exception handling, so it is susceptible to several easily identifiable threats. The data set might not exist or might be locked exclusively, both of which would cause the %SYSFUNC(OPEN) function to fail. This failure, however, would not be passed to the parent process because no return code exists. The code might also want to specify a unique return code if the macro is called to compile a list of character variables and none are found. Notwithstanding these vulnerabilities, additional performance can be added over time with ease when modules are managed within reuse libraries and reuse catalogs, discussed in the "Reuse Artifacts" sections in Chapter 18, "Reusability."

The %FINDVARS macro represents only the second of four implicit functions that were generated from the single functional requirement. The "Modularity" section in Chapter 18, "Reusability," demonstrates the entire modular solution that incorporates %FINDVARS into the %GOBIG macro. For a more robust solution, the "Exception Inheritance" section in Chapter 6, "Robustness," additionally demonstrates an exception handling framework that incorporates exception inheritance. In the next section, the second principle of modular design—loose coupling—is introduced and demonstrated.

Loosely Coupled

Loose coupling increases the flexibility of software modules by detaching them from other software elements. Detachment doesn't mean that modules cannot communicate, but requires that this communication must be limited, flexible, and occur only through prescribed channels. The most ubiquitous example of loose coupling (to SAS practitioners) is the structure prescribed through Base SAS procedures. For example, the SORT procedure is powerful and highly versatile but essentially requires parameters to function, which are conveyed as SORT procedure keywords or options.

The DATA parameter reflects the data set being sorted while the optional OUT parameter specifies the creation of an output data set. The BY parameter includes one or more variables by which the data set is to be sorted and other parameters can additionally modify the functionality of the SORT procedure. Thus, the *only* way to interact with SORT is through these prescribed parameters, and any deviation therefrom will produce a runtime error—the return code from the SORT procedure that informs the user that something is rotten in the state of Denmark.

Since software modules are often operationalized in Base SAS through SAS macros, limited coupling is enforced through parameters that are passed from a parent process to a child process. If the macro definition is inside the software that calls it, or if the %INCLUDE statement is used to reference an external macro saved as a SAS program, parameters are passed to the child process through the macro invocation. Or, when child processes represent external batch jobs that are spawned with the SYSTASK statement, all information necessary to execute that module must be passed through the SYSPARM parameter to the &SYSPARM automatic macro variable.

To provide another example, the International Space Station (ISS) was built through incremental, modular production and demonstrates loose coupling. Despite the prodigious size of some compartments, each connects only through designated ports, providing structural integrity while limiting the number of precision connections that must be manufactured on Earth and assembled in space. Because individual compartments were constructed by different countries, the connective ports had to be precise and predetermined so that they could be built separately yet joined seamlessly. This connectivity simulates the precision required when passing parameters to software modules.

Not only do ISS compartments have a limited number of ports through which they can connect, but also those ports had to be able to dock flexibly with both Russian Soyuz spacecraft as well as U.S. space shuttles. This flexibility (loose coupling) increased the usefulness and reuse of the ISS, because it could be accessed from multiple space vehicles and countries. Like the ISS, a software module that operates as child process similarly should be flexibly designed so that it can be called by as many processes and programs as possible that can benefit from its functionality. Loose coupling represents the modularity principle that can facilitate this objective.

Loosely Coupled Macros

Loose coupling is especially salient when modularity is achieved through SAS macro invocations as opposed to through SAS batch jobs that spawn new SAS sessions. When a new session is spawned, a SAS module runs with a clean slate—customized system options, global macro variables, library assignments, and other leftover attributes cannot influence the session, so it is more stable. However, when a module is invoked instead by calling a macro—whether defined in the program or defined externally and called with the %INCLUDE statement, the SAS Autocall Macro Facility, or the SAS Stored Compiled Macro Facility—that macro will always be susceptible to elements and environmental conditions in the current SAS session.

At the far end of the coupling spectrum are macros that are completely coupled. For example, the following %SORTME macro relies exclusively on attributes—the data set name and variable name—that are created in the program itself but not passed through the macro invocation. The macro walls, like a sieve, expect this information to percolate from the DATA step:

```
data mydata;
   length char1 $15;
   char1='chimichanga';
run;

%macro sortme();
%global sortme_RC;
```

```
proc sort data=mydata;
   by char1;
run;
%if &syscc>0 %then %let sortme_RC=FAILURE!;
%mend;

%sortme;
```

To decouple the %SORTME macro from the program, all dynamic attributes must be passed through parameters during the macro invocation:

```
%macro sortme(dsn= /* data set name in LIB.DSN format */,
   var= /* BY variable by which to sort */);
proc sort data=&dsn;
   by &var;
run;
%mend;

%sortme(dsn=mydata, var=char1);
```

The macro can now be called from other SAS programs because it relies on no host-specific information from the program besides the parameters that are passed. However, the original macro also included exception handling that tested the global macro variable &SYSCC for warning or runtime errors, and this logic must also be decoupled. Macros are porous to global macro variables in a SAS session, even those belonging to irrelevant programs. To ensure %SORTME is not reliant on or contaminated by existent values of &SYSCC or &SORTME_RC, both of these macro variables must be initialized inside the macro:

```
%macro sortme(dsn= /* data set name in LIB.DSN format */,
   var= /* BY variable by which to sort */);
%let syscc=0;
%global sortme_RC;
%let sortme_RC=GENERAL FAILURE;
proc sort data=&dsn;
   by &var;
run;
%if &syscc=0 %then %let sortme_RC=;
%mend;

%sortme(dsn=mydata, var=char1);
```

The %SORTME macro is now also decoupled from the global macro variables that it references. All information is passed to the macro through parameters, and from the macro through a single return code &SORTME_RC. The return code is initialized to GENERAL FAILURE so that an abrupt termination of the macro doesn't accidentally report a successful execution, a best practice discussed in the "Default to Failure" section in Chapter 11, "Security."

While the macro appears to be decoupled from the program, modifying the %SORTME macro invocation exposes another weakness. When the parameterized data set is changed from Mydata to PERM.Mydata, this now requires that the PERM library be available inside the macro. While the LIBNAME assignment for PERM may exist in the current SAS session, were the %SORTME macro saved as a SAS program and called as a batch job, the PERM library would not exist unless PERM were already globally defined through SAS metadata, the Autoexec.sas file, a configuration file, or other persistent method. This illustrates the importance of globally defining commonly used artifacts such as persistent SAS libraries and SAS formats.

Loosely Coupled Batch Jobs

Sometimes performance can be improved by replacing serialized program flow with invocations of batch jobs that can facilitate parallel processing and decreased execution time. Because macro variables cannot be passed to a batch job through macro parameters, all necessary information must be passed through the SYSPARM parameter of the SYSTASK statement. For example, the following program (parent process) now executes the Sortme.sas program as a batch job (child process):

```
libname perm 'c:\perm';

data perm.mydata;
   length char1 $15;
   char1='chimichanga';
run;

%let dsn=perm.mydata;
%let var=char1;
systask command """%sysget(SASROOT)\sas.exe"" -noterminal -nosplash
   -nostatuswin -noicon -sysparm ""&dsn &var"" -sysin
   ""c:\perm\sortme.sas"" -log ""c:\perm\sortme.log""" status=rc
   taskname=task;
waitfor _all_;
systask kill _all_;
```

The following child process should be saved as C:\perm\sortme.sas:

```
libname perm 'c:\perm';
%let dsn=%scan(&sysparm,1,,S);
%let var=%scan(&sysparm,2,,S);

%macro sortme(dsn=,var=);
%let syscc=0;
%global sortme_RC;
%let sortme_RC=GENERAL FAILURE;
proc sort data=&dsn;
   by &var;
run;
%if &syscc=0 %then %let sortme_RC=;
%mend;

%sortme(dsn=&dsn, var=&var);
```

The PERM library must be assigned inside the child process so that the PERM.Mydata data set can be accessed. However, enabling the batch job to function with the current SYSTASK invocation doesn't guarantee success for later invocations that might reference different libraries. Recall that loose coupling in part is designed to ensure that a child process can be called from as many parents as possible who can utilize the functionality. Thus, a hardcoded solution like this is never recommended. Moreover, if SAS system options were specifically required in the batch process, this request also could not be passed from parent to child given the current paradigm.

To deliver the necessary functionality, the manner in which SYSPARM is passed and parsed must become more flexible. The following updated SYSPARM value will now be passed from the parent process via SYSTASK—hardcoded here for readability, but passed through SAS macro variables in the actual SYSTASK statement:

```
opt=mprint * lib=perm c:\perm * dsn=perm.mydata * var=char1
```

In this example, the asterisk is utilized to tokenize the SYSPARM parameters into individual parameters (tokens) and, within each token, the equal sign is utilized to further tokenize individual parameters into discrete elements. For example, OPT=MPRINT is first read, after which OPT is identified to represent the command to invoke the SAS OPTIONS statement, while MPRINT is identified as the specific option to invoke.

The updated parent process follows, with only the SYSPARM value modified:

```
libname perm 'c:\perm';

data perm.mydata;
   length char1 $15;
   char1='chimichanga';
run;

%let dsn=perm.mydata;
%let var=char1;
systask command """%sysget(SASROOT)\sas.exe"" -noterminal -nosplash
   -nostatuswin -noicon -sysparm ""opt=mprint * lib=perm c:\perm
   * dsn=&dsn * var=&var"" -sysin ""c:\perm\sortme.sas"" -log
   ""c:\perm\sortme.log""" status=rc taskname=task;
waitfor _all_;
systask kill _all_;
```

The updated child process, saved again as C:\perm\sortme.sas, now includes an additional %PARSE_PARM macro that tokenizes and parses the &SYSPARM automatic macro variable:

```
%macro parse_parm();
%local tok;
%let i=1;
%do %while(%length(%scan(&sysparm,&i,*))>1);
   %let tok=%scan(&sysparm,&i,*);
   %if %upcase(%scan(&tok,1,=))=LIB %then %do;
      libname %scan(%scan(&tok,2,=),1,,S) "%scan(%scan(&tok,2,=),2,,S)";
      %end;
   %else %if %upcase(%scan(&tok,1,=))=OPT %then %do;
      options %scan(&tok,2,=);
      %end;
   %else %if %upcase(%scan(&tok,1,=))=DSN %then %do;
      %global dsn;
      %let dsn=%scan(&tok,2,=);
      %end;
   %else %if %upcase(%scan(&tok,1,=))=VAR %then %do;
      %global var;
      %let var=%scan(&tok,2,=);
      %end;
```

```
   %let i=%eval(&i+1);
   %end;
%mend;

%parse_parm;

%macro sortme(dsn=,var=);
%let syscc=0;
%global sortme_RC;
%let sortme_RC=GENERAL FAILURE;
proc sort data=&dsn;
   by &var;
run;
%if &syscc=0 %then %let sortme_RC=;
%mend;

%sortme(dsn=&dsn, var=&var);
```

When the parent process executes and calls the Sortme.sas batch job, the %PARSE_PARM macro is able to assign one or more SAS libraries dynamically, ensuring not only that this parent process can pass the PERM library invocation, but also that additional, unspecified parents will be able to pass different library definitions. Although redundant (because the SYSTASK statement itself can specify SAS system options), the %PARSE_PARM macro can also recognize SAS system options and specify these with the OPTIONS statement. Finally, the macro variables &DSN and &VAR that represent the data set name and BY variable, respectively, are also still passed via SYSPARM, albeit in a more flexible form.

This type of dynamism may seem gratuitous and, for many intents and purposes, the added code complexity (inasmuch as time spent designing, developing, and testing a solution) may not provide sufficient business value to warrant its inclusion. Notwithstanding, software that truly demands loose coupling to support development and reuse of independent modules will require this extensive decoupling if modules are intended to be saved as separate SAS programs and executed as batch jobs.

To improve readability in this example, no exception handling is provided; however, in production software, all parameters should be validated and, with each successive level of tokenization, additional exception handling should prescribe program flow when exceptional values are encountered. Exception handling techniques are demonstrated throughout Chapter 6, "Robustness," and the validation of macro parameters and return codes is discussed in the "Macro Validation" sections in Chapter 11, "Security."

Passing SAS Global Macro Variables

Passing a global macro variable through a SAS macro invocation may seem uncanny, but it's one of the final indicators of loosely coupled code. For example, the following code creates PERM.Mydata, then prints the data set using a title supplied through the global macro variable &TIT1:

```
libname perm 'c:\perm';
%let tit1=Mmmmm Mexican food;

data perm.mydata;
   length char1 $15;
   char1='chimichanga';
run;

proc print data=perm.mydata;
   title "&tit1";
run;
```

The PRINT procedure simulates a more complex process, so, espousing modular design, a SAS practitioner might wrap this functionality inside a macro module:

```
%macro print(dsn= /* data set name in LIB.DSN format */);
proc print data=&dsn;
   title "&tit1";
run;
%mend;

%print(dsn=perm.mydata);
```

The %PRINT macro can now be both defined and invoked in the same program, or the %PRINT macro could be saved as a separate SAS program so that it can be utilized by other parent processes as well. This second usage, however, exposes the vulnerability that the global macro variable &TIT1, while defined in *this* parent process, is not necessarily defined in *other* parent processes. Thus, as demonstrated in the following code, loose coupling necessitates that even global macro variables be explicitly passed through macro parameters:

```
%macro print(dsn= /* data set name in LIB.DSN format */,
   title1= /* title */);
proc print data=&dsn;
   title "&title1";
run;
%mend;

%print(dsn=perm.mydata, title1=&tit1);
```

Although this additional effort may seem redundant, it is necessary to ensure loose coupling. Because loose coupling represents the tiny pinhole in black-box design through which all interaction with a module must be prescribed and limited, if %PRINT is receiving information directly from global macro variables (in lieu of being appropriately passed parameters), the integrity of the macro is compromised. Moreover, the revised %PRINT macro can now be saved as a SAS program and called by a subsequent, independent parent process which will easily be able to identify the necessity to pass the parameter TITLE1 due to its presence in the macro definition.

Encapsulated

The objectives of loose coupling and encapsulation are similar, as well as many of the techniques used to achieve them. However, encapsulation focuses on facets of security—including confidentiality and integrity, aiming to protect modules as well as the larger body of software. Loose coupling, on the other hand, focuses on software and module flexibility, seeking to ensure that modules can be reused in as many ways and by as many processes as possible. In general, because of the overlap between these two objectives, they are often achieved in tandem.

Every surface has two sides, so in the black-box analogy that represents an encapsulated module, dual roles of encapsulation exist. The interior surface of the box aims to prevent the module from harming the parent process that called it or its software while the exterior surface aims to prevent the parent process and its software environment from harming the module. Because SAS global macro variables can unfortunately pass freely in and out of black boxes, they are often a focus of encapsulation discussions. The encapsulation of macro variables is further discussed and demonstrated in the "Macro Encapsulation" section in Chapter 11, "Security."

Encapsulating Parents

The first test for software encapsulation is to run software from a fresh SAS session in which no code has been executed. Because modified SAS system options, global macro variables, WORK library data sets, open file streams, and locked data sets can all persist until a SAS session is terminated, only a clean slate can ensure a stable, consistent, uncontaminated environment. This freshness is one of the primary benefits of spawning SAS batch jobs, which execute in new SAS sessions.

It's critical that all global macro variables be uniquely named to avoid confusion, collisions, and corruption. Confusion results when software readability is diminished because of similarly named variables or, in some cases, reuse of macro variable names. Collision and corruption can occur when macro variable names are not unique and interact in unexpected ways. For example, the following code initially creates the global macro variable &PATH to represent the SASROOT directory:

```
%let path=%sysget(SASROOT);
%put PATH: &path;
```

However, when a %PATHOLOGY macro is called, it reassigns the &PATH macro, unknowingly overwriting the original value:

```
%macro pathology();
%global path;
%let path=pathogen;
%mend;

%pathology;
%put PATH: &path;
PATH: pathogen
```

To overcome this vulnerability, where global macro variables are intended to be enduring, consider macro names that are unlikely to be reused, even erring on the side of long and unwieldy to ensure that reuse is unlikely. Global macro variables are not the only vulnerability; local macro variables can also contaminate parent processes. The following code is intended to write the ASCII characters 65 through 69 (i.e., A through F) twice:

```
%macro inner(); * child;
%do i=65 %to 69;
   %put -- %sysfunc(byte(&i));
   %end;
%mend;

%macro outer(); * parent;
%do i=1 %to 2;
   %put &i;
   %inner;
   %end;
%mend;

%outer;
```

However, the code iterates only once and produces the following output:

```
1
-- A
-- B
-- C
-- D
-- E
```

In this example, because the macro variable &I in the %INNER macro was not defined as a local macro variable, it overwrites the &I in the %OUTER macro. When %INNER completes the first iteration, the value of &I is 70 and, when this value is assessed by the %DO loop within %OUTER, the %DO loop exits because 70 is greater than the threshold value 2. To remedy this error, %I must at least be defined as local within %INNER, but because %OUTER itself could also be invoked at some point as a child process, both instances of %I should be distinguished as local macro variables:

```
* CHILD;
%macro inner();
%local i;
%do i=65 %to 69;
   %put -- %sysfunc(byte(&i));
   %end;
%mend;

* PARENT;
%macro outer();
%local i;
%do i=1 %to 2;
   %put &i;
   %inner;
   %end;
%mend;

%outer;
```

While these examples have demonstrated the perils of macro variable creation and initialization, child processes can have other unintended effects on parent processes. For example, if a child process fails to unlock a data set (that it has explicitly locked), fails to close a data stream (that it has opened), fails to redirect log output (that it has directed with PRINTTO), or fails to KILL (or WAITFOR) a SYSTASK statement, each of these events can adversely affect functionality and performance once program flow returns to the parent process. In all cases, the best *prevention* of damage to parent processes is to launch child processes as batch jobs while the best *remedy* to limit damage is to prescribe exception handling that identifies exceptions and runtime errors within child processes and communicates these to parent processes.

Batch processing can eliminate many of the security concerns that exist when a child process is called. For example, in the black-box paradigm, the parent provides inputs to and receives outputs and return codes from children, but should be unaware of other child functionality. For example, in calling a child process by invoking a SAS macro, that child process could alter global macro variables or delete the entire WORK library relied upon by the parent process—because both parent and child are running in the same session. However, when a child process is instead initiated as a batch job, the new SAS session cannot access the global macro variables, WORK library, or system options of the parent session, thus providing protection against a number of threats.

In other cases, threats cannot be avoided, but they must be detected. An exception handling framework that detects and handles exceptions and runtime errors in child processes, provides necessary exception inheritance to parent processes, and prescribes a fail-safe path (if process or program termination is necessary) is the best method to minimize damage caused by software failure within child processes. Fail-safe paths are further discussed in the "Fail-Safe Path" section in Chapter 11, "Security."

Encapsulating Children

Nobody puts Baby in a corner, but sometimes you need to put Baby in a big black box. As demonstrated in the previous "Loosely Coupled" section, macros are susceptible to elements of the SAS session, such as global macro variables, system options, and WORK library data sets. Thus, modules should only rely on information passed through SAS macro parameters or the SYSPARM option. In some cases, module functionality will be dependent upon SAS system options, and if user permissions allow certain options to be changed within the SAS session, it is possible that these could have been modified.

The most effective way to protect child processes is to ensure they receive the necessary information from the parent, whether passed through macro parameters, the SYSTASK parameter, or some other modality. Especially where these parameters are dynamically passed, quality control routines within child processes should validate parameters to ensure that sufficient (and valid) information exists to continue. Macro validation is further discussed and demonstrated in the "Macro Validation" sections in Chapter 11, "Security."

BENEFITS OF MODULARITY

Modularity is so central to software development best practices that some static performance attributes not only benefit from but require modular software design. Most other static performance attributes are conceptualized primarily as investments, because they support software maintainability but don't directly influence dynamic performance. Functional decomposition and modular software design, however, are often prerequisites to implementing parallel processing solutions that can increase software speed. In the following sections, the benefits of modularity are discussed relative to other dimensions of software quality.

Maintainability

Software maintainability can be greatly increased where software is divided into discrete modules and saved as separate SAS programs. In addition to being big and cumbersome, monolithic software products represent single programs that cannot be modified efficiently because multiple developers cannot edit the code simultaneously. When software is divided into modules of discrete functionality, several SAS practitioners can tackle different maintenance activities concurrently to achieve faster maintenance and, if software functionality has been interrupted, a faster recovery period.

Readability

Software is inherently more readable when it fits on one screen. A study by Ko et al. recorded software developers' software development and maintenance tasks and found they spent "35 percent of their time with the mechanics of redundant but necessary navigations between relevant code fragments."[4] Tremendous amounts of time are consumed merely navigating within software as developers ferret out dependencies, prerequisites, inputs, outputs, and other information so that they can perform development tasks. The implication is that modular software is much easier to read and thus results in more efficient software maintenance and management activities.

Testability

Unit testing describes testing done on a single unit or module of code and is ideally suited for modular software design. Because functionally discrete code is only intended to do one thing, this simplicity of function translates into simplicity of testing. Furthermore, by reducing module interaction through loose coupling methods that prescribe precision in parameter passing and return code generation, testable components are more easily isolated and their behavior validated. Monolithic code is difficult to test through a formalized test plan with test cases because of the inherent complexity and multifunctionality.

Stability

Code stability is substantially increased due to loose coupling of modules. If one component needs to be modified, in many cases, other aspects of the software can remain intact. For example, the %SORTME macro in the "Loosely Coupled Macros" section dynamically sorts a data set, given parameters for the data set name and sort variable:

```
%macro sortme();
%global sortme_RC;
proc sort data=mydata;
   by char1;
run;
%if &syscc>0 %then %let sortme_RC=FAILURE!;
%mend;
```

However, if maintenance was required to replace the SORT procedure with an equivalent SQL procedure sort or hash object sort, these modifications could be made within the %SORTME macro without disturbing or even accessing its parent process. This degree of stability facilitates software integrity and reduces subsequent software testing following maintenance because relatively smaller chunks of software must be modified with each maintenance activity.

Reusability

Software modules are much more reusable when they are functionally discrete and loosely coupled. Modules that perform several functions can sometimes be bundled together efficiently when those functions are related and commonly used together, such as determining whether a library exists, then whether a data set exists, and finally whether that data set is unlocked and can be accessed. Whenever functionality is bundled, however, the risk exists that some aspect of the functionality will not apply for reuse situations, thus requiring development of a similar but slightly divergent module. In robust software, loose coupling also benefits software reuse because prerequisites, inputs, and outputs can be validated within an exception handling framework.

Execution Efficiency

Functional decomposition of software into discrete elements of functionality is the first step toward critical path analysis that can demonstrate where process bottlenecks lie and where parallel processing could be implemented to increase performance. By conceptualizing software as bits of discrete functionality rather than one composite whole, SAS practitioners can better understand individual requirements, prerequisites, and dependencies of specific functions, thus enabling them to better manage program flow and system resources during software execution.

WHAT'S NEXT?

Over the next few chapters, the benefits of modular software design are demonstrated as they relate to static dimensions of software quality. The next chapter introduces readability, which facilitates code that can be more readily and rapidly comprehended. Readability is often the most evanescent of all performance attributes, disappearing with the addition of other dynamic dimensions of quality that inherently add content and complexity. Thus, the balance of modularity and readability is an important one, because through modular design, readability can often be maximized even as additional functionality and performance are incorporated.

NOTES

1. ISO/IEC 25010:2011. *Systems and software engineering—Systems and software Quality Requirements and Evaluation (SQuaRE)—System and software quality models.* Geneva, Switzerland: International Organization for Standardization and Institute of Electrical and Electronics Engineers.
2. ISO/IEC/IEEE 24765:2010. *Systems and software engineering—Vocabulary.* Geneva, Switzerland: International Organization for Standardization, International Electrotechnical Commission, and Institute of Electrical and Electronics Engineers.
3. SAS Institute. "Chapter 10: Macro Facility Error Messages and Debugging," in *The SAS® 9.4 Macro Language, Second Edition*: 124.
4. A. J. Ko, B. A. Myers, M. J. Coblenz, and H. H. Aung, "An Exploratory Study of How Developers Seek, Relate, and Collect Relevant Information during Software Maintenance Tasks," *IEEE Transactions on Software Engineering* 32, no. 12 (2006, Dec.): 971–987.

CHAPTER 15

Readability

As the airport immigration officer thumbed through my passport for the hundredth time, I couldn't help but wish I'd hitchhiked back to the States on the KC-135 as planned. You know, sometimes they let you operate the boom if you ask nicely …

Anything would have been better than secondary screening but, as my mind wandered, I was comforted that at least the latex gloves hadn't come out yet.

Quito, Ecuador. In country on counternarcotics operations, I had taken advantage of the free ticket to South America to trek around for a week of R&R: climbing the spires of gothic cathedrals, bisecting my body at *Mitad del Mundo* (misplaced equatorial monument), up *Teleférico* (gondola to 13,000 feet), and of course the nightly Gringolandia *discotecas* (clubs). So of course I didn't regret staying, but this was getting ridiculous.

"*Ummm* … those pages don't come apart." I gently reprimanded.

Back to reality, as one officer attempted to pull apart my passport as though some hidden coffer would be revealed. Two guards to his right similarly scoured my military ID and orders, repeatedly passing these between each other. I think that they thought they were trying to validate my credentials somehow, but holding my orders up to the light did not reveal some arcane watermark or mysterious palimpsest.

All this commotion was about a tiny entry stamp that nonresidents receive when immigrating to Ecuador—a tiny stamp I didn't have in my passport, having arrived on a C-130. Their concerns swelled around my mysterious entry into their country.

"No, I didn't sneak into Ecuador. No, I've never been to Colombia. No, I don't have an arrival ticket, either—I don't think the jumpseats were numbered, but I could be wrong …"

At last I was saved by my salsa and merengue skills, as one officer asked where I'd been staying. "Gringolandia," naturally: the tawdry strip of hostels, hookah joints, *discotecas*, and all-night eateries. At that instant, my credentials were suddenly validated and the trio only wanted to discuss their favorite bars on the strip. I was glad to have learned how to evade secondary (which would come in handy during future adventures) and emerged elated, not having missed my flight back to the States.

■ ■ ■

Part of the difficulty that I encountered at the Quito immigration office was due to a lack of readability. However, it was compounded by a general lack of comprehension. Each officer spoke limited English, and with my limited

Spanish abilities and English documentation, communication in general and comprehension of my military orders were difficult.

In software development, it can also seem at times like someone is speaking—or coding—in a foreign language. I remember being introduced to the SAS macro language; although it grew on me quickly, I had many dazed and confused afternoons in the infancy of my macro career.

In addition to the language barrier, military orders in general are not an easy read. Even if you have a military background and are familiar with the jargon, nomenclature, and acronyms, familiarizing yourself with a specific set of orders can take a few moments. Software development is no different—even if you can absorb the intent of individual statements, collective comprehension of software can take some time.

Not only the content but also the presentation of military orders can render them inscrutable—ALL UPPERCASE FONT and as little formatting as possible ensure that no section stands out from another. Even to skilled SAS practitioners, code that lacks structure and standardization can consume more effort to comprehend.

The consternation of the officers was fueled not only by my passport, which lacked the required entry stamp, but also by my orders, which showed an expected departure date a week earlier. In software development, comprehension of software is imparted not only through code, but also through external documentation such as technical requirements.

If requirements specify one thing, but software does another, you can bet that readability will be compromised because of the extra scrutiny required to resolve the discrepancy. Just as the officers repeatedly passed my documentation around, trying to make sense of the discrepancy, SAS practitioners and other stakeholders can waste precious time inspecting code and documentation when software does not match stated functional and performance requirements.

DEFINING READABILITY

Readability is "the ease with which a system's source code can be read and understood, especially at the detailed, statement level."[1] Readability typically describes only the examination of code, although to gain a fuller understanding of software intent and function, developers must often additionally reference external documentation, such as technical requirements, Unified Modeling Language (UML) diagrams, data models, data dictionaries, and quality control constraints. Moreover, descriptive remarks about software are often imbedded in code but also can be included in external knowledge management systems, such as code reuse libraries and reuse catalogs. Thus, while full comprehension of software may require examination of external documentation, readability speaks to deriving understanding from code itself.

In infancy, all software is readable, if for no reason other than its simplicity. But as code grows and matures, software readability often plummets. Especially in production software, higher levels of readability are often incompatible with the inclusion of other dimensions of quality because of the inherent complexities those dimensions bring, as well as increased code size. For example, robust software that aims to identify and respond dynamically to exceptions and errors through an exception handling framework will typically require more lines of code to facilitate this increased performance than lines of code that deliver ultimate software functionality.

This chapter introduces software readability and demonstrates the inherent trade-off between it and many other quality characteristics. Nevertheless, through modular software design, clear syntax, and consistent structure and naming conventions, readability can be maximized despite software complexity. Software readability is a critical component of software maintainability, because code that cannot be comprehended will take longer to modify and test. Especially when maintenance responsibility is shared across a development team, readability best practices can facilitate swift recovery and restoration of business value.

PLAN TO GET HIT BY A BUS

While working as researchers, my friend Thomas and I would continually joke about being hit by buses—in our minds, the measure of good code was the ability to run your coworker's software after he was pummeled by a bus and wasn't around to help decipher it. As much as we'd joke about that Central Ohio Transit Authority (COTA) bus bearing down on us, the continuity of operations mind-set enabled us to write software (and documentation) that was more readily understood by others when we were unexpectedly out of the office.

While bus incursions are a reality few of us will face, more realistic scenarios include vacations, sickness, job transfers, and eventual retirement, each of which temporarily or permanently removes us from the dear SAS software we've developed. Despite this reality, many end-user developers persist in writing software as though they alone are the intended audience. This attitude can tremendously reduce software maintainability and readability, especially when the original developers are no longer available to perform necessary maintenance. Developing the mind-set that your date with a bus might occur at any moment helps enforce the reality that code standardization, commenting, and other documentation, while perhaps perceived as onerous, should be performed as part of software development and not as an afterthought.

SOFTWARE READABILITY

Readability is influenced by naming conventions, text formatting, indentation, vertical and horizontal spacing, commenting, and other rather subjective factors that can clutter or clarify software. While influencing readability, these facets are only briefly enumerated because they speak to aesthetics best selected and perfected at the team or organizational level. This chapter instead focuses on higher-level aspects of readability, including modular code design, SAS macro definitions, code standardization, and other methods that can facilitate defect identification and overall maintainability.

To fully understand software, however, you must first understand its objective, which extends from customer needs and requirements. Especially when reading another developer's code, a project or product vision as well as technical specifications set the mood and provide necessary context to begin to decipher software. While this chapter begins with a focus on internal readability, when reviewing software, requirements documentation should be consulted before code examination. It is far more difficult—and may be impossible—to assess whether software achieves its functional and performance objectives if the required level of quality cannot first be determined.

Syntax and Structure

Clear, unambiguous syntax and clean aesthetics facilitate code that can be read and absorbed more quickly. Because developers typically read and interact with software they have not authored, the extent to which syntax can be standardized within a team, organization, or even a software language facilitates readability. For example, the indentation of code within this text represents standard SAS horizontal spacing convention found throughout literature, which supports readability. But even if an alternative code structure has been selected—whether in a text or an organization—more important is standardization of the cumulative body of code.

One anecdotal tenet of software development states that developers shouldn't add comments just to make code comprehensible—if code can't be understood, it should be simplified. While this mentality itself is an oversimplification, there is truth in the quest for code reduction, and there are times when a decrease in dynamic software performance is warranted if code readability can be improved. Simplification can be accomplished through modular software design, discussed extensively in Chapter 14, "Modularity," as well as by reducing nested loops, nested functionality, complex branching, dead code, and other elements that can lead to confusion.

Thus, like other software performance attributes, readability displays (and is assessed on) a broad spectrum rather than a dichotomous outcome; however, unlike other software performance attributes, readability is often inversely related to its performance counterparts. For example, the following clear syntax is reprised from the "Executing Software Concurrently" section in Chapter 7, "Execution Efficiency." It ingests process metrics from a control table (PERM.Control) and assigns three macro variables (&INGEST, &MEANS, and &FREQ) based on business logic:

```
%let ingest=;
%let means=;
%let freq=;
data control_temp;
   set perm.control;
   if process='INGEST' and not missing(start) then do;
      if missing(stop) then call symput('ingest','IN PROGRESS');
      else call symput('ingest','COMPLETE');
      end;
   else if process='MEANS' and not missing(start) then do;
      if missing(stop) then call symput('means','IN PROGRESS');
      else call symput('means','COMPLETE');
      end;
   else if process='FREQ' and not missing(start) then do;
      if missing(stop) then call symput('freq','IN PROGRESS');
      else call symput('freq','COMPLETE');
      end;
run;
```

The business logic specifies that if a start date is missing, the process (Ingest, Means, or Freq) has not initiated so nothing is updated. If a start date exists but a stop date is missing, the process is currently executing so the respective macro variable is updated to IN PROGRESS. If both start and stop dates exist, the process has completed so the respective macro variable is updated to COMPLETE. The logic is intuitive because it can be read and deciphered easily within the code.

The code, however, can be simplified with use of the IFC function, which evaluates the logical truth of a statement and subsequently assigns one of two possible character values. If a logical statement evaluates to be true, the first character value is assigned and, if false, the second is assigned:

```
data control_temp;
   set perm.control;
   if process='INGEST' and not missing(start) then call symput('ingest',
      ifc(missing(stop),'IN PROGRESS','COMPLETE'));
   else if process='MEANS' and not missing(start) then call
      symput('means',ifc(missing(stop),'IN PROGRESS','COMPLETE'));
   else if process='FREQ' and not missing(start) then call
      symput('freq',ifc(missing(stop),'IN PROGRESS','COMPLETE'));
run;
```

The code is much leaner, with the number of CALL SYMPUT statements halved. Moreover, because CALL SYMPUT statements are combined, the interior IF–THEN logic is not required, thus eliminating code blocks. But in examining the code more closely, it can be reduced even further to deliver the same functionality. By realizing that the values of the PROCESS variables are also used as the respective macro variable names, the 15-line DATA step can be simplified to four.

```
data control_temp;
   set perm.control;
   if process in('INGEST','MEANS','FREQ') and not missing(start)
      then call symput(process,ifc(missing(stop),'IN PROGRESS',
      'COMPLETE'));
run;
```

But does making this code four times smaller make it more readable, or less? Note that the question is not *is this readable* but rather *is this an improvement or a setback to readability*? That question, however, can only be answered at the individual level, or team or organizational level if readability is being prescribed. In compressing the code, a greater body of code can fit on a single screen, which can improve readability and maintainability. But if parsing the nested functionality of the compressed version is less intuitive than the expanded version, this decreases readability. This is not to say that one method is preferred over another—each method is functionally equivalent. But, based on the level of expertise of SAS practitioners as well as the prioritization of code readability, one version may be preferred over the other two, based on individual, team, or organizational priorities and expertise.

In this example, the somewhat subjective interpretation of readability was driven primarily by differences in syntax and structure. In other cases, however, a clearer tradeoff between readability and dynamic performance (such as software speed or efficiency) may exist. For example, should a team modify a SORT-SORT-MERGE process from dual SORT procedures to a more complex but commensurately more efficient SQL procedure? In other words, if dynamic performance is improved but readability decreased, is this a win? Only at the individual and team levels can these decisions be made based on software requirements, including the extent to which software is intended to be modified and maintained throughout its lifespan.

Personal Knowledge

Understanding of code begins with Base SAS syntax comprehension. Before software context is even considered, the functional intent, capabilities, and limitations of a specific procedure, function, statement, option, or operator must be understood. Developers gain this knowledge over time through SAS language documentation, technical literature, blogs and Internet resources, training, and SAS customer support. But because the flexibility of Base SAS allows developers to achieve similar functionality through diverse methods, developers often gravitate naturally to tried-and-true methods to the possible exclusion of other worthy techniques.

For example, the SQL procedure offers an alternative to many DATA steps and some procedures such as SORT, providing similar functionality for many common tasks. For SAS practitioners unfamiliar with the SQL procedure, however, a multistep SQL join may be virtually indecipherable, presenting a tremendous obstacle to code readability and maintainability. This knowledge divide can also occur in development environments that rely upon the SAS macro language heavily if not all SAS practitioners are skilled in macro development. Thus, readable code to some will be inscrutable to others due to each developer's SAS lexicon and comprehension of syntax.

Software readability should also be assessed from the standpoint of stakeholders who must interact with the code. Teams can benefit from espousing practical knowledge standards, such as by requiring that SAS practitioners receive the SAS Certified Base Programmer, SAS Certified Advanced Programmer, or SAS Certified Clinical Trials Programmer certificates to demonstrate their expertise. Attainment of these credentials can set the bar for the level of sophistication that will be expected within SAS software, which in part drives software readability.

Modularity

Software readability tends to decrease as software shifts from functionality-focused to quality-focused, comprising both functional and performance requirements. For example, as software is made more reliable and robust, the code base often doubles, making software more difficult to comprehend. Modular software design, however, can restore some of this readability by simplifying software design, often enabling entire SAS modules to fit on a screen at once. Modular software design improves readability through the following modularity principles:

- *Functionally discrete*—Because each code module is intended to perform one and only one function, modules are more compact and straightforward than monolithic software.
- *Loosely coupled*—Because information required to run modules is passed through designated parameters, SAS practitioners are able to examine a single location—the macro invocation or SYSTASK statement—to understand module inputs.

The degree to which software is modularized will depend on functional complexity, anticipated benefits, and individual preference or team standards. Some SAS practitioners prefer scrolling through thousands of lines of monolithic code and cannot be shaken from this practice more common in procedural languages. They view modularity—which might require a monolithic SAS program to be divided into 10 or 20 individual programs—as a step away from, not toward, software readability.

Other developers with more experience in object-oriented programming (OOP) third-generation languages (3GLs) will naturally gravitate toward modular design because it's more familiar and because they recognize its intrinsic benefits. While the merits of modular design supporting readability won't be debated here—since readability inherently represents one's own comprehension of software—modular design does support more stable, testable, reusable, and maintainable software.

Code Structure

Code structure includes horizontal and vertical spacing and the general arrangement of text. For example, each of the following four DATA steps provides equivalent functionality, but does so with varying horizontal and vertical spacing. While the first solution is the one most commonly observed throughout SAS documentation and literature, if a separate standard is adopted within a team or organization, it should be used consistently by all stakeholders.

```
data final;
   set original;
run;

data final;
set original;
run;

data final;
   set original;
   run;

data final; set original; run;
```

Because Base SAS is agnostic to code structure, unlike some languages like Python that require precise indentation, SAS practitioners have far more options in the appearance of their code, such as:

- Vertical spacing, including optional space before or after declarative statements, macro definitions, macro invocations, and other code
- Horizontal spacing, including code block indentation, inline comment placement, and optional spacing between tokens, operators, variables, and other content
- Maximum line size before continuation on a new line
- Maximum line size before termination with semicolon
- Number of statements per line
- Semicolon placement

Vertical and horizontal spacing are most commonly altered to achieve subjectively more readable software. For example, some SAS practitioners are content to let the characters fall as they may:

```
data words;
   set nouns;
   if noun in ('taco','burrito','chimichanga') then group='food';
```

```
   else if noun in('parrot','mallard') then group='bird';
   else if noun in ('California','Pennsylvania') then group='state';
run;
```

Other SAS practitioners prefer horizontal alignment, believing this improves the readability or professionalism of software:

```
data words;
   set nouns;
   if noun in ('taco','burrito','chimichanga')      then group='food';
   else if noun in('parrot','mallard')              then group='bird';
   else if noun in ('California','Pennsylvania')    then group='state';
run;
```

Both solutions are functionally identical; however, the importance of code structure should never overtake code function or maintainability. In the horizontally aligned example, if this list were to continue to grow in size while the number of items in the IN sets also continued to increase, the structure and alignment could become unwieldy to maintain, forcing developers to constantly move the positioning of the "then" column, not only for new but also for existing lines of code. Thus, while software readability should be valued, a pursuit of aesthetics should not decrease the efficiency with which software can be modified or maintained.

Naming Conventions

Whether explicitly defined or derived from common practice, naming conditions exist in all software development environments. Variables, formats, informats, arrays, macro variables, macro names, SAS data sets, SAS libraries, and the logical structures they represent all must be named. The SAS Language Reference describes actual limitations of Base SAS names, as well as best practices for naming conventions. The following naming convention considerations can be standardized within a development environment:

- Maximum character length to be used (equal to or less than the Base SAS threshold)
- Descriptive versus nondescriptive variable names, including the situations for which each is warranted
- Use of capitalization or the underscore to distinguish words or text
- Descriptor inclusion in name, such as requiring that all arrays begin with ARRAY_

- Identification of temporary information, such as data sets or variables that are intended to be discarded immediately after use, such as by starting temporary data sets with X_
- Enforcement of uniqueness, in that no variable, macro variable, data set, or other name can be reused or repurposed within software

In considering naming conventions, the competing priorities of simplicity, descriptiveness, uniqueness, and others must be assessed. For example, one SAS practitioner might want to name a macro variable &ReturnCodeForMyFavoriteMacro, which, although specific and unique, might decrease code maintainability due to substantially increased line size. If the macro variable relates, however, to a macro named %DOIT rather than %MYFAVORITEMACRO, the uniqueness of the macro variable is decreased, because for other developers the macro variable could ambiguously relate to *their* favorite macros, respectively.

Because global macro variables can unintentionally perseverate from one SAS macro or program to the next, a best practice demonstrated in the "Perseverating Macro Variables" section in Chapter 11, "Security," recommends that macro return codes be named after their macros. For example, because only one macro named %DOIT can exist in a SAS session, only one macro variable &DOIT_RC can also exist in a SAS session, preventing confusion, collisions, and corruption of the macro variable. More than improving the security of software, standardized naming practices such as this can reduce confusion among developers, thus improving readability.

Dead Code

Dead code includes lines or blocks of code that have no possibility of being executed. Most commonly, code doesn't execute when it's "commented out" as demonstrated in the later "Inline Comments" section. During development, debugging, and testing, code is often commented out temporarily to test functionality as code is altered. As part of code hardening before software release, however, commented code should be removed. For example, the first ELSE statement in the following code will never execute, so it should be removed from production software:

```
data words;
   set nouns;
   if noun in ('taco','burrito','chimichanga') then group='food';
/*   else if noun in('parrot','mallard') then group='bird'; */
   else if noun in ('California','Pennsylvania') then group='state';
run;
```

Confusion can erupt when it's unclear whether code was commented out because it was unnecessary, redundant, faulty, a vestigial remnant of past functionality, or a harbinger of future functionality. In some Agile or rapid development environments in which production software is updated weekly or biweekly, elements of functionality that could not be completed and tested in the current iteration are released in subsequent iterations. Despite the temptation to release software containing dead code that is commented out (because it's still under development), dead code ideally should not be included in a production release until it is functional, tested, and accepted by the customer.

Dead code can also result when logical evaluations are definitively false or impossible given data or metadata definitions or quality controls. This code atrophy typically results from latent defects rather than from the commenting of code. For example, the following %BADCODE macro evaluates the expression 1=2, which simulates a conditional logic evaluation that will always evaluate to false, creating a dead zone of code that never runs:

```
%macro badcode();
%put this will print;
%if %eval(1=2) %then %do;
   %put this will never print;
   %end;
%mend;

%badcode;
```

This practice, similar to commenting out code, enables software to run while skipping over segments that will never execute. While it is acceptable during development and debugging, dead code should be removed before testing, acceptance, and release of production software.

Code Comments

In software applications development, comments facilitate development, testing, and maintenance. They have no utility during software operation because software is typically encrypted so the underlying code is inaccessible—users are not intended to modify, maintain, or even understand the inner workings of software. Despite not playing a role in the operational phase, comments generally remain in software during compilation and encryption so that a single, uncompiled version of the software can be maintained. In some languages, comments are removed automatically by the compiler so that they don't unnecessarily increase the size of compiled software executable files.

This paradigm contrasts sharply with data analytic development, in which code transparency is paramount and end users are expected not only to understand but also often to modify production software. Tautological comments that describe straightforward functionality which can be inferred through code inspection should be avoided. But comments that convey software product vision, intent, requirements, deficiencies, or defects can be invaluable during development and maintenance. Thus, comments should be included when they improve readability by elucidating function or intent not found within or easily gleaned from code.

Like other aspects of readability, best practices prescribing the location, type, quantity, and verbosity of comments are rather subjective. Standardization of commenting practices, inasmuch as the content of comments themselves, can improve readability. In some development environments, comments can provide not only an individual remark at the line or program level, but also a quantitative measure of software quality. For example, the "Symbolism in Software" section later in the chapter demonstrates standardized commenting paradigms that can be used to calculate technical debt for a module, program, or software product by automatically parsing and compiling software comments.

Project Notes

Common content for project-level notes includes instructions for running the project (if not automated), inputs and prerequisites that may exist, and a brief enumeration of caveats or vulnerabilities—anything high level that could facilitate software success and militate against its failure. In some environments, historical versioning may also be recorded, demonstrating a brief account of changes made to software throughout its lifespan. In essence, comments whose content would span multiple programs should be included in project-level documentation to eliminate redundancy across multiple program files or other technical documents.

Within SAS Enterprise Guide, both projects and programs can be defined and project-level information can be stored within notes modules visible in the SAS Enterprise Guide workspace. Notes can alert users to vulnerabilities, or provide execution instructions or other other high-level information. The functionality of Enterprise Guide notes is limited, however, enabling only plain, unformatted text to be recorded. Thus, project notes can often encompass too much critical information to relegate to a tiny, nondescript text box.

A further downside of utilizing Enterprise Guide notes is the inability to parse Enterprise Guide project files (.egp) automatically to extract comments or other relevant information. Whereas SAS program files (.sas) are saved in plain text and can be easily parsed to extract program headers, comments, and other information from the programs, project files are compressed .zip files. It is possible to rename an Enterprise Guide project file as a zip file, open the file, interrogate the Project.xml file, and search for all <Element Type="SAS.EG.ProjectElements.Note"> markings in the text to inspect the content of all note elements—but I'll save that capability for a future white paper. Automated parsing of SAS comments within SAS program files is described later in the "Automated Comment Extraction" section.

Program Headers

Program headers introduce developers to software, providing useful information that might not be conveyed through code review alone. Headers often include a mix of static, dynamic, historical, technical, and administrative information, including:

- Program author(s)
- Milestones, such as creation date, testing date, customer acceptance date, or software production release date
- Revision history, including dates and content of revisions, as well as type (e.g., adaptive, corrective, emergency, perfective, preventive)
- Program objectives, including both high-level goals and technical objectives
- Known vulnerabilities and latent defects that contribute toward technical debt
- Program prerequisites, such as required input or programs that must have already executed
- Program dependencies, such as generated output or subsequent programs that rely on program execution

The extent to which the structure and content of program headers can be standardized across all software within a development environment can substantially improve readability and maintainability. Especially where header formats are standardized, automated routines can be developed that extract and catalog metadata from headers for inclusion in external software databases and documentation, as described in the "Automated Comment Extraction" section later in the chapter.

Macro Comments

SAS macros are a common method to facilitate software modularity through functional decomposition of monolithic programs into smaller chunks of code. Just as program headers can provide information to developers to promote readability and maintainability, macro headers are important to ensure that macros are invoked with the correct parameters and that all prerequisites have been met. Macro comments should also include known limitations, caveats, or other vulnerabilities that could lead to accidental misuse or unintentional results. Of course, if this information is already conveyed separately through a risk register or other database, the duplication is unnecessary.

The following macro definition includes a single comment that describes its objective: to produce and print a space-delimited list of variables occurring in a data set. The DSN and TYPE parameters are each commented inline, which not only standardizes the comment placement but also enables the macro invocation popup window within SAS Enterprise Guide. For example, with the macro definition inside a SAS Enterprise Guide program file, typing %PRINT (will invoke a popup box demonstrating all macro parameters and showing the DSN definition "DSN: data set name in LIB.dataset or dataset format":

```
*<DESC> creates a space-delimited list of variables in a data set;
***** does not validate existence of the data set;
***** assumes a shared lock can be obtained on the data set;
***** does not validate the TYPE parameter;
***** creates no return code to demonstrate macro success or failure;
%macro varlist(dsn= /* data set name in LIB.DSN or DSN format */,
   type= /* ALL, NUM, or CHAR - the type of variables returned */);
%local vars;
%local dsid;
%local i;
%global varlist;
%let varlist=;
%let dsid=%sysfunc(open(&dsn, i));
%let vars=%sysfunc(attrn(&dsid, nvars));
```

```
%do i=1 %to &vars;
   %let vartype=%sysfunc(vartype(&dsid,&i));
   %if %upcase(&type)=ALL or (&vartype=N and %upcase(&type)=NUM) or
         (&vartype=C and %upcase(&type)=CHAR) %then %do;
      %let varlist=&varlist %sysfunc(varname(&dsid,&i));
      %end;
   %end;
%let close=%sysfunc(close(&dsid));
&varlist;
%mend;

proc means data=&dsn;
   var %varlist(dsn=mydata, type=NUM); * creates list of numeric
      variables;
run;
```

Several vulnerabilities are also listed in the macro header, demonstrated with the ***** symbolism. These don't necessarily denote errors or defects in the code, but rather intrinsic weaknesses that should be understood and that could be mitigated or eliminated in the future if technical specifications were modified to require higher performance. The use of standardized commenting conventions, such as *<DESC> to denote a macro description or ***** to denote vulnerabilities, can be harnessed by automated systems that extract comments from SAS program files for inclusion in external documentation, as demonstrated later in the "Automated Comment Extraction" section.

In a production environment that espouses modular software design, the %VARLIST macro would be saved as a separate SAS program and called by the MEANS procedure, having been referenced via the %INCLUDE statement or SAS Autocall Macro Facility. Because reusable macros are isolated from other code modules and because they may be called on weeks, months, or years later to perform some task, macro headers and comments are an integral part of ensuring continued and appropriate use of macros.

Slashterisk versus Asterisk

Inline comments appear with code and are often imbedded in SAS statements. They are useful for describing aspects of performance or functionality that might not be intuitive, such as complex business rules operationalized through nested, conditional logic. Inline comments are also a preferred method for documenting code defects and known vulnerabilities, which in turn should relate to entries in a risk register if maintained. Inline comments are often used to segment sections and improve readability.

SAS provides three methods to implement commenting within code, the asterisk (*), the slashterisk (/*), and the macro comment (%*). Each has pros and cons. The asterisk denotes a comment at the end of a line (or filling an entire line) and must be followed by a semicolon:

```
data test;
   set somedata; * this is a comment;
   * also a comment;
run;
```

The asterisk is the preferred method to include manual comments, which aid the readability of code and are intended to persist in software through development and into production. These comments cannot contain unmatched quotation marks or semicolons, but this is a minor restriction.

The slashterisk, by comparison, is the preferred method to convert single lines or large sections of code automatically to commented text for development and testing purposes:

```
data test;
   set somedata;
run;

/*proc sort data=somedata;*/
/*   by char1;*/
/*run;*/
```

For example, the SORT procedure was initially valid code but has been automatically commented out so that it will not execute. Automatic commenting is an efficient way to develop, debug, and test software. To comment code automatically in the SAS interface, highlight the text and type <CTRL></>. To remove the comments automatically, highlight the commented code again and type <CTRL><SHIFT></> in the SAS Display Manager or SAS Enterprise Guide, or <CTRL></> in the SAS University Edition.

The slashterisk can also be used to insert comments into code manually:

```
proc means data=&dsn;
   var %varlist(dsn=&tab, type=NUM); /* comment */
run;
```

The automatic commenting functions in SAS work well with manual comments that have used the asterisk method, but fail when a slashterisk comment already exists and is encountered. For example, if a SAS practitioner wanted to quickly comment the preceding MEANS procedure, adding automatic commenting in the SAS University Edition produces the following code:

```
/* proc means data=&dsn; */
/*    var %varlist(dsn=&tab, type=NUM); /* comment */
/* run; */
```

However, if the automatic commenting is subsequently removed with the <CTRL></> shortcut, the SAS University Edition fails to restore the code to its original format and instead produces invalid code that will fail because the closing */ is missing:

```
proc means data=&dsn;
   var %varlist(dsn=&tab, type=NUM); /* comment
run;
```

A different failure occurs when automatic commenting is attempted on commented text in the SAS Display Manager or SAS Enterprise Guide. Note the extra */ that follows the manual comment, as contrasted with the SAS University Edition:

```
/* proc means data=&dsn; */
/*    var %varlist(dsn=&tab, type=NUM); /* comment */*/
/* run; */
```

Because this code now contains duplicate comment delimiters, any code that follows the RUN statement will appear to the SAS compiler to still be commented; thus, it will not execute. Because of all these issues, the asterisk is generally recommended for manual commenting over the slashterisk. One exception noted in SAS documentation states that the slashterisk is the preferred method for comments occurring inside a macro definition because the comments are not stored in the compiled macro, whereas asterisk comments are stored.[2] Notwithstanding this recommendation, in macros that are modular and well-written, the dearth of comments occurring inside a macro definition should not be so great as to deter a developer from using whichever method—asterisk, slashterisk, or macro comment—is desired.

Commenting inside macro definitions is the only required use of the manual slashterisk method. For example, the following %SORTME macro definition includes two parameters, both of which are defined in the %MACRO statement:

```
* sorts a data set;
%macro sortme(dsn= /* in LIB.DSN or DSN format */,
   var= /* variable by which to be sorted */);
proc sort data=&dsn;
```

```
   by &var;
run;
%mend;

%sortme(dsn=mydata, var=char1);
```

As demonstrated previously in the "Macro Comments" section, use of the slashterisk in macro parameter definitions facilitates conformity and, when executed within SAS Enterprise Guide, displays a popup window that lists each parameter and its respective definition. This use of the slashterisk is always recommended in macro construction.

Unintuitive Code

It's sometimes said that code should speak for itself; if readability is compromised, then code should be made simpler. Often in complex logic, however, even clear and concise code can be baffling at first glance. As a general rule, if perfecting a convoluted solution to finagle some bit of functionality took several attempts to get it right, or required consulting SAS or other technical documentation, that section of code probably warrants a comment.

For example, the following code randomly generates a capital letter from A to Z using the RAND and BYTE functions. Unless you've memorized the ASCII character table, however, it might take a minute of legwork to establish this, so the comment facilitates immediate comprehension:

```
data mydata;
   length char1 $2;
   char1=byte(int(rand('uniform')*26)+65); *A through Z;
run;
```

Automated Comment Extraction

Comment extraction involves automatic parsing of SAS program files to extract metadata such as program header information or comments about software capabilities or vulnerabilities. Because SAS programs are saved as plain text, parsing can be accomplished with any language, including Base SAS. To be effectively parsed, SAS headers (and other comments) should be standardized to the extent possible, allowing the parser to tokenize delimiters easily to extract meaningful information.

Element definitions can be flexibly defined as needed, but a standardized methodology at the team or organizational level facilitates greater readability of software as well as automation of comment extraction. The following four elements are defined for use in this chapter:

- *<DESC>—description of program, module, or macro
- ***—refactoring opportunity
- ****—functional or performance deficit identified
- *****—vulnerability or defect identified that could cause functional or performance failure

The following macro, %PARSE_SAS, represents a rudimentary comment extractor for a Windows environment. It iterates through a parameterized folder, opening each SAS program and extracting metadata from comments based on the previous metadata element definitions:

```
%macro parse_sas(dir=);
%local i;
%local fil;
filename f pipe "dir /b &dir*.sas";
data _null_;
   length filelist $10000;
   infile f truncover;
   input filename $100.;
   filelist=catx('',filelist,filename);
   call symput('files',filelist);
   retain filelist;
run;
data comments;
   length program $200 type $20 text $200;
run;
%let i=1;
%do %while(%length(%scan(&files,&i,,S))>1);
   %let fil=%scan(&files,&i,,S);
   filename ff "&dir&fil";
   data ingest;
      length program $200 type $20 text $200;
      infile ff truncover;
      input text $200.;
      if find(text,'*<DESC>','i')>0 then type='description';
      else if find(text,'*****')>0 then type='defect';
      else if find(text,'****')>0 then type='deficit';
      else if find(text,'***')>0 then type='refactor';
      if not missing(type) then do;
```

```
          program="&dir&fil";
          output;
          end;
     run;
     proc append base=comments data=ingest;
     run;
     %let i=%eval(&i+1);
     %end;
%mend;

%parse_sas(dir=c:\perm\);
```

When the %PARSE_SAS macro is executed, the macro header and additional comments from each SAS program in the C:\perm folder are ingested into the Comments data set. This data set in turn could be printed through the REPORT procedure or saved as a PDF or other file format to enable automated documentation for an entire library of software. The benefits of this type of standardization and structure are numerous. For example, all defects, vulnerabilities, deficits, or elements that should be improved through refactoring can be automatically amassed and displayed. In an instant, this can provide a team with an understanding of unresolved issues within software, as well as opportunities to improve software through refactoring.

A different development team might define additional comment elements that include the primary and ancillary authors of each program, which the parser could use to generate a matrix demonstrating who should be involved in maintenance activities when required. Yet another use of standardized comment parsing could include defining additional comment elements that represent historical software versioning. This information could thereafter be extracted to create external documentation chronicling software functionality and performance.

Automatic comment extraction is a powerful tool for constructing software reuse catalogs that describe the intent, structure, format, and relative level of quality of software modules. Reuse libraries and reuse catalogs are described in the "Reuse Artifacts" sections in Chapter 18, "Reusability," and can improve the effectiveness and efficiency with which existing software can be reused or repurposed. For example, when commenting protocols are standardized within an organization, a SAS practitioner can in seconds automatically parse the cumulative body of SAS software, create a report that defines and describes all programs and modules, and review this catalog to determine whether any existing modules can benefit a new software project.

Symbolism in Software

Symbolism in software comments can succinctly convey meaning while not cluttering software with significant amounts of code or descriptive text. Some examples of common symbols found in comments include designators describing logic errors, defects, vulnerabilities, or other caveats. For example, I use five asterisks (*****) followed by a succinct description to demonstrate a defect or vulnerability in the code that could produce functional failure. Software readability and maintainability are improved because stakeholders are immediately alerted to this symbol and can easily search for it. Furthermore, through standardization of software symbolism, automated comment extraction (described in the previous section) can be facilitated through routines that parse software and extract symbols and meaning.

Refactoring Opportunity

Refactoring involves code modification to improve performance without effect to functionality. For example, if SAS software is running slowly, developers could refactor the code to make it go faster, while leaving the outcome or output unaffected. Especially in Agile and other rapid-development environments, the initial priority is often to generate a working prototype on which additional performance enhancements can be made. Because the highest performing software solution is not often the first developed, refactoring facilitates improving performance throughout development and the software lifespan.

For example, the following initial version of code from the "Clear Syntax" section was hastily typed, but the SAS practitioner may realize that a faster, more efficient, or more readable solution might exist by implementing the IFC function. Because the initial code meets functional intent, it might be released into production initially. Nevertheless, two refactoring notes aim to remind developers that a better method might exist and should be investigated when time permits:

```
data control_temp; *** could be simplified with the IFC function;
   set perm.control; *** could make processes dynamic;
   if process='INGEST' and not missing(start) then do;
      if missing(stop) then call symput('ingest','IN PROGRESS');
      else call symput('ingest','COMPLETE');
      end;
   else if process='MEANS' and not missing(start) then do;
      if missing(stop) then call symput('means','IN PROGRESS');
```

```
      else call symput('means','COMPLETE');
      end;
   else if process='FREQ' and not missing(start) then do;
      if missing(stop) then call symput('freq','IN PROGRESS');
      else call symput('freq','COMPLETE');
      end;
run;
```

The three-asterisk (***) symbol can be used to denote code that functions correctly but which could be refactored to increase performance. This provides a straightforward method to locate areas for future software improvement. The symbolism that is selected is irrelevant so long as it can be consistently applied to software products within a development environment.

Refactoring is often performed to alleviate a performance deficit, in which software meets functional requirements but fails to meet performance requirements. For example, this most commonly occurs when software produces valid results but does so too slowly or fails too often and thus is deemed unreliable. Refactoring in other cases, however, is undertaken to improve software performance just because developers have identified a faster, better, more efficient, or more reliable way of doing something. Where refactoring is undertaken and doesn't aim to achieve stated performance requirements, teams must steer clear of gold-plating (the delivery of unneeded, and often unvalued, performance), as discussed in the "Avoid Gold-Plating" section in Chapter 2, "Quality."

Deficit Identification

This text focuses on performance alone and, in all examples, assumes that functional intent of software is achieved. But in reality, a decrease in software quality can be spurred just as easily from failing to meet functional requirements as from failing to meet performance requirements. For example, functional requirements of data analytic software might state that it should upload metrics to a dashboard each morning. The team might have delivered an initial solution that instead produces a static report as a PDF each morning, thus failing to meet functional requirements. Whether citing intentional or unintentional reasons that motivate this deviation, the rationale is irrelevant because technical debt is incurred regardless. And, until the functional objective is met, the debt remains.

The four-asterisk symbol (****) can be utilized to demonstrate functional or performance deficits—not defects—that exist in software, clearly indicating to SAS practitioners and other stakeholders that failure to meet objectives is noted. As discussed later in the "CYA" section, it is far better to point out your own mistakes, weaknesses, or unfinished code within software than to have a customer or coworker point them out for you. Deficit identification does this by highlighting a gap between requirements and software execution, tacitly stating, "Hey, we realize this doesn't fully meet software intent, but we're on it."

Defect Identification

Whereas refactoring aims solely to improve the performance of software, and deficit identification demonstrates areas that still need improvement, defect resolution identifies and eliminates vulnerabilities that pose risks to software functionality. Production software is often released with latent defects and known vulnerabilities so long as they individually and collectively pose an acceptable risk to stakeholders. These defects and vulnerabilities, however, can be parsed and ingested directly into a risk register.

As stated previously, I use the five-asterisk symbol (*****) to denote defects or vulnerabilities in software, alerting other developers to software issues that, when presented with the right threat, will cause functional or performance failure. For example, the following code from the "Environmental Variables" section in Chapter 10, "Portability," dynamically assigns the macro variable &PATH based on the SAS automatic macro variable &SYSSCP, which demonstrates the OS environment (e.g., Windows, UNIX) in which SAS is executing:

```
***** could produce invalid PATH value if run outside of Display
   Manager, Enterprise Guide, or SAS University Edition;
%macro getcurrentpath();
%global path;
%let path=;
%if %symexist(_clientprojectpath) %then %do;
   %let path=%sysfunc(dequote(&_clientprojectpath));
   %let path=%substr(&path,1,%length(&path)-%length(%scan(&path,-1,\)));
   %end;
%else %if &SYSSCP=WIN %then %do;
   %let path=%sysget(SAS_EXECFILEPATH);
   %let path=%substr(&path,1,%length(&path)-%length(%scan(&path,-1,\)));
   %end;
%else %if &_CLIENTAPP=SAS Studio %then %do;
   %let pathfil=&_SASPROGRAMFILE;
```

```
   %let pathno=%index(%sysfunc(reverse("&pathfil")),/);
   %let path=%substr(&pathfil,1,%eval(%length(&pathfil)-&pathno+1));
   %end;
%else %put Environment Not Defined!; ***** does not provide return code
   if other OS;
%mend;

%getcurrentpath;
```

The software has been tested and validated in the SAS Display Manager (for Windows), SAS Enterprise Guide (for Windows), and the SAS University Edition; however it makes no claims about performance in other SAS-recognized OSs such as z/OS or VMI. As a result, if the software is executed in an environment other than Windows or UNIX, the &PATH macro variable could remain empty if none of the three conditional paths are followed. Worse yet, the software might produce unexpected, invalid results if it did follow a conditional path erroneously. Thus, the use of ***** alerts SAS practitioners to these risks in operational software.

In an environment that operates the SAS application only on Windows machines, this latent defect poses no risk. But because the software as written has only been tested with three specific SAS interfaces, some description of the vulnerability could be warranted if there is a chance that the code could appear in an untested environment. Thus, although the decision to comment this vulnerability will depend on portability requirements, the use of established symbolism in general can more readily alert stakeholders to risk.

Standardizing the identification of defects within SAS software is important not only to support manual readability of code but also to facilitate automated comment extraction. Because technical debt can comprise unresolved defects, vulnerabilities, performance deficits, and functional deficits, automatic parsing of SAS programs to identify these distinct issues through standardized symbolism can automatically build effective quality control reporting. This automation is demonstrated previously in the "Automated Comment Extraction" section.

CYA

Cover Your Analyst. Just do it. Software is rarely perfect the first time; even if it meets full functional intent, it can often be improved—made faster, more efficient, more readable, or more something—through continued refactoring. In other cases, software contains latent defects when released, but because time is money, an earlier release date is prioritized over higher-performing or more functional software. Whatever the rationale for releasing less-than-perfect software into production, make the investment to document software issues, especially those that pose operational risks.

Simple comments such as those demonstrated in the prior three sections not only alert developers to vulnerabilities or weaknesses in their own code but also alert other SAS practitioners who may be less familiar with the code. Moreover, to developers, customers, or other stakeholders who are reviewing code, documentation not only demonstrates the developer's technical capabilities (which might be in question if the code is considered to be buggy), but also shows clear commitment to risk management principles. Even without reviewing technical requirements or understanding the relative level of quality that software was intended to demonstrate, stakeholders can view code and perceive a developer's clear identification of threats, vulnerabilities, defects, and other issues that may affect software. With this knowledge, stakeholders can more easily distinguish the SAS practitioner who identified, calculated, and communicated risks from the SAS practitioner who unknowingly or negligently wrote faulty software.

EXTERNAL READABILITY

While readability describes the ease with which code itself can be read and understood, true software comprehension often requires knowledge of software intent and objectives. Requirements documentation can demonstrate the overarching goals and requirements of software, while design documentation can include program flow information and diagrams describing how those objectives will be technically achieved. Because data analytic software can inherit a substantial degree of variability from data injects, data documentation (e.g., entity relationship diagrams, data dictionaries, data integrity constraints) can be beneficial—if not necessary—to gaining a thorough understanding of software.

Requirements Documentation

My first question before reviewing code is always "Where are the specs?" In other words, what are the business needs that breathed this software into existence, and what technical requirements will it be measured against? Without this basic information, you may be able to decipher technically what software does, but you'll have no idea whether it's doing what it's supposed to be doing.

Even when formalized technical requirements are lacking, high-level goals can help refine the intent of software and provide context essential to deciphering code and assessing its quality.

Design Documentation

The *software design description (SDD)* is "a representation of software created to facilitate analysis, planning, implementation, and decision-making. The software design description is used as a medium for communicating software design information, and may be thought of as a blueprint or model of the system."[3] The SDD outlines how the software needs and requirements will be met. Often, it can include Unified Modeling Language (UML) diagrams, data process flows, and other representations of the software to be designed. SDDs are typically reserved for larger software projects that require an attack plan that organizes the way in which specific technical requirements will be addressed.

Data Documentation

Because SAS data analytic software inherently relies on data sources as dynamic injects, it's typically necessary to understand software as well as the data it processes. For example, a DATA step referencing the variable CHAR1 could be valid, but if that variable doesn't exist in the input data set, it could also represent an error. Common data-centric documentation includes data dictionaries, data integrity constraints, and entity resolution diagrams (ERDs). Data dictionaries describe data sets or larger data collections, including data objects, structure, format, external relationships, and often aspects of data quality. Data integrity constraints can be defined in documentation (such as data dictionaries) or programmatically through quality controls that enforce data quality standards. ERDs demonstrate relationships among data sets and other data objects, enabling SAS practitioners to model, transform, and represent data.

WHAT'S NEXT?

Readability facilitates faster and more complete understanding of software through static analysis of code. As software is made more modular and clear, software modules can be more easily and thoroughly tested to support increased reliability and accuracy. The next chapter introduces testability, which describes the ease with which software can be tested. Because software testing is a critical phase of the SDLC through which software function and performance are demonstrated and validated, the extent to which software can be made more testable improves its quality.

NOTES

1. ISO/IEC/IEEE 24765:2010. *Systems and software engineering—Vocabulary*. Geneva, Switzerland: International Organization for Standardization, International Electrotechnical Commission, and Institute of Electrical and Electronics Engineers.
2. "%* Macro Comment Statement." *SAS(R) 9.4 Macro Language: Reference, Fourth Edition*. Retrieved from http://support.sas.com/documentation/cdl/en/mcrolref/67912/HTML/default/viewer.htm #n17rxjs5x93mghn1mdxesvg78drx.htm.
3. IEEE Std 1012-2004. *IEEE standard for software verification and validation*. Geneva, Switzerland: Institute of Electrical and Electronics Engineers.

CHAPTER 16
Testability

I didn't know it yet, but I'd torn my right meniscus to shreds.

Four days, 26 miles, gear in tow, summiting a 13,000-foot peak—the Inca Trail is no marathon, but it's also no joke.

We had booked the trek through SAS Travel—naturally—a stellar tour company albeit with no connection to Cary, North Carolina. Despite a leisurely backpacker's lifestyle for a few months, I thought I had trained adequately for the jaunt.

While in Guatemala, I'd joined the Antigua Gym, a converted 17th-century Spanish Colonial house. Even in the open courtyard, tarps had been pitched and staked to a central mast like the big top of an itinerant circus, the now-quiet, centuries-old plaster fountain oddly the centerpiece of modern fitness equipment.

Despite my high-energy start, more than 4,000 miles spent on buses from Guatemala to Peru had tempered my fitness regimen. I knew I was in trouble when in Quito, Ecuador, I was out of breath just walking around town at 12,000 feet.

The trek itself was indescribable, up one mountain and down the next, winding along cliffs, bridges, worn stone steps, and muddy paths rutted from centuries of use. Hypnotized by the trail, placing one foot steadfastly ahead of the other like our pack mule companions, we plodded along, pausing only for photos, snacks, and jungle restrooms.

We wore rain ponchos at first, believing somehow that the ephemeral plastic would keep us dry—yet even during the few, brief rainless respites, the sempiternal fog cloaked us in moisture. By the second day, I'd given up even wearing a shirt.

And the porters wore less, often bounding up and down the stepped landscape barefoot carrying packs two-thirds their size, the pitter-patter of their feet signaling weary gringos to briefly step aside.

And finally, day four: the end of the line. Emerging into Machu Picchu, the landscape is so overwhelming you don't even know where to point your camera. And when you do, the images don't just look surreal—they look fake.

Although my knees were screaming, one final summit remained—Huayna Picchu—the jagged peak jutting behind Machu Picchu in every classic photo. There would be a week of painkillers and ice to come—and surgeries to follow—but, undaunted, we set out and conquered Huayna.

■ ■ ■

In planning for the Inca Trail, I thought I'd adequately trained through jogging, stair climbers, and a pretty extensive daily workout routine. Training not only strengthens your body but also enables you to better understand your weaknesses—testing those vulnerabilities that might more easily be exploited through fatigue, stress, or injury.

Software development is no different and, while software doesn't get stronger simply by being executed, it can garner increased trust of stakeholders when testing demonstrates reliability. Moreover, testing can expose vulnerabilities that pose operational risks. If your legs are sore after a mile jog uphill, there's a pretty good chance you're not going to survive hiking seven miles a day.

An important component of any training (or testing) regimen is to ensure it's done in a realistic environment. While jogging and stair climbers had been necessary cardiovascular training, the actual terrain I faced was tremendously varied—there was nothing routinized about the grade, texture, or speed of the trail. Not having trained for that variability, my knees were not happy.

In other cases, risks are identified that are accepted. For example, in Quito I realized that I would have some issues with the altitude of the trek but, rather than strapping on an elevation mask for hypoxic training while chicken busing, I chose to deal with the elevation when I encountered it, thus accepting the risk. Similarly, in software development there may be components that are too difficult to test or for which the costs outweigh the benefits, in which case those risks are accepted.

In addition to variability, there was the sheer duration of the hike. An hour at the gym—even if a machine could somehow replicate the arduous climbing environment—still would only have been an hour. Muscle training and strength training, yes; endurance training that prepares a body to climb up and down mountains all day, no. Data analytic testing, similarly, must account for not only the variability of data but also the size. The degree to which you can replicate the anticipated operational environment will make testing that much more effective.

At least I'd thoroughly tested my equipment, having worn my backpack for months. The weight was perfectly distributed, straps snug, and I was barely cognizant I was wearing it. Others, you could tell, either had just purchased their packs or were not accustomed to them, and were constantly stopping to tug straps, readjust, and redistribute items.

In software development you also have to test the equipment you build. Does a module perform well individually? Does it perform well when integrated with other software? The hikers who initially struggled with their packs didn't have poor-quality packs—but they had never tested the packs on their backs as a combined unit.

DEFINING TESTABILITY

Testability is the "degree of effectiveness and efficiency with which test criteria can be established for a system, product, or component and tests can be performed to determine whether those criteria have been met."[1] Testability defines the ease with which functional testing, performance testing, load testing, stress testing, unit testing, regression testing, and other testing can be implemented successfully to demonstrate technical requirements. In many cases, both positive and negative testing are required, in that testing must demonstrate not only that software is doing what it *should* do, but also that software is not doing what it *shouldn't* do.

Test plans are commonly created during software planning and design in conjunction with technical requirements that are being generated. Through the inclusion of test cases, stakeholders effectively enumerate not only the functional and performance requirements that must be met for software acceptance, but also specifically how those requirements will be unambiguously demonstrated and measured before software release. Testability principles such as modularity and readability facilitate testing clarity and can yield tests that more effectively demonstrate software success and adherence to test plan and test case requirements.

Because software testing—and especially testing done through formalized test plans—is egregiously absent from many end-user development environments, this chapter describes, differentiates, and demonstrates multiple software testing modalities. It further describes the benefits of formalized test plans, including their critical role in software quality assurance plans, as well as their usefulness in creating a battery of test cases that can collectively demonstrate software success. While *testability* remains the quality characteristic that high-performing software should demonstrate, a culture of software *testing* must first exist to ensure benefits will be derived from delivering more testable software.

SOFTWARE TESTING

Software testing is "the dynamic verification of the behavior of a program on a finite set of test cases, suitably selected from the usually infinite executions domain, against the expected behavior."[2] Testing does two things: it demonstrates correct function and performance, and it identifies software vulnerabilities and the threats that could exploit them. Because of the critical role of testing in software development, the SDLC always contains a testing phase that validates software against requirements.

The primary role of testing is to validate whether software meets functional and performance objectives. Testing doesn't include the refinement and refactoring that occur naturally during development as SAS practitioners strive to correct defects and improve performance to meet software objectives. For example, a SAS practitioner might first write a process that relies on the SORT procedure to order a data set but, after "testing" the process, might refactor it to order the data with a data set index instead. This type of empirical testing can be integral to the development process, but it does not represent the testing phase of the SDLC. Testing instead examines software or software modules that are believed to be complete and validates them.

The secondary role of testing is to ensure that software is robust against sources of failure, thus testing often requires both normal and exceptional injects and environmental characteristics. In other words, testing aims to uncover all vulnerabilities, including defects, software errors, and errors in technical requirements, that pose a risk to software. It's not enough to know that software is doing the right thing—it's equally important to be confident it's not doing the wrong thing.

Testers

Too frequently in software development literature, software testing seems to be performed by some elite warrior class of testers who operate as part of an independent quality assurance team. While these environments do exist, in reality, much software testing is performed by developers with concomitant software testing responsibilities. This cross-functionality, a commonly sought objective within Agile development, can effectively increase the skillset of developers and the team as a whole.

To be effective testers, however, developers must espouse a new perspective. Rather than focusing on developing quality software, their focus must turn to ways to undermine, antagonize, and otherwise threaten that software. When a team completes software, they should believe that they've produced a masterpiece—an amalgam of perfectly poised statements that will thrill the customer and meet all needs and requirements for the expected lifespan of the software. If latent defects or vulnerabilities do exist when the software is released, the developers should have documented these within the code or in a risk register, and the respective risks they pose should be known to and accepted by the customer.

Dedicated software testers, on the other hand, receive code, review software requirements, and seek to ensure that the former fulfills the latter, doing nothing more and nothing less. Demonstration of positive (or correct) behavior can be straightforward given adequate technical specifications, but testers must also identify negative behavior. In doing so, they must attempt to break software—assume that vulnerabilities exist, find them, exploit them, document them, and possibly suggest remedies. Rather than trying to create something beautiful, testers are trying to tear it apart. But their role and unique attitude are absolutely essential.

Developers can make excellent testers, but not necessarily for the software they have developed. Because testing immediately after development would require SAS practitioners to attack a masterpiece they had just created, a cooling-off period between development and testing can facilitate more objectivity. However, because this delay can thwart business value and deadlines, peer code reviews offer one form of testing in which developers swap code (and respective requirements documentation) for a more immediate quality assurance review.

Within end-user development environments, SAS practitioners themselves are both software developers and users. These environments often diverge sharply from SDLC best practices, for example, in teams for which no formal software testing phase or protocols exist, whose developers often "fix it on the fly." Moreover, in end-user development environments, software integrity is more likely to be conveyed to customers and other stakeholders through faith in and friendship with developers than through formalized testing and test documentation. Even within end-user development environments, however, SAS practitioners can espouse and benefit from formalized software testing plans and methods. Thus, in addition to wearing the hats of developer and user simultaneously, end-user developers should additionally incorporate *software tester* into the litany of roles they perform.

Test Plans

A *test plan* is a "document describing the scope, approach, resources, and schedule of intended testing activities."[3] A test plan is beneficial because it formalizes the importance of testing activities within the SDLC, demonstrating a commitment to quality to and from all stakeholders. While technical requirements convey the objectives that software must achieve, a formalized test plan conveys the methods and metrics to determine whether those objectives were achieved.

Comprehensive test plans should reference specific software requirements. For example, a requirement might state that a software module should be able to process 10 million observations efficiently while not exceeding a threshold level of system resources. In development, the module would have been executed, but not necessarily with the volume or variability of data throughput expected in software operation. Specifying within a test plan that load testing is required with a realistic volume of 12 million observations allows testers to demonstrate to stakeholders that software will sufficiently scale as required.

Many test plans require automated tests, which are essentially additional SAS programs that test and validate the actual software being developed. Automated testing is preferred because results represent a repeatable, defensible product that can be submitted to the customer with the final software product. Where test plans are required to be submitted to a customer or saved for posterity, test data and test cases should also be archived with the automated testing software.

Test Cases

Test cases are a "set of inputs, execution conditions, and expected results developed for a particular objective, such as to exercise a particular program path or to verify compliance with a specific requirement."[4] Test cases are used to test the infinite with the finite. Therefore, while the aggregate variability facing software may be multifaceted and fairly limitless (when environmental factors and data variability are included), a finite set of test cases should be extracted that, when cumulatively executed, adequately ensures that all aspects of the software have been tested.

Test cases allow developers to maintain autonomy over the design and development of software, but provide an enumeration of standards against which that software will ultimately be judged. A test plan often comprises an enumeration of test cases that must be demonstrated for the software to be validated and accepted. For example, because SAS comments can begin with an asterisk, the following test plan excerpt demonstrates test cases that can be checked off as software developers or testers complete the test plan:

```
* TEST: demonstrate an existent data set succeeds with ALL type;
* TEST: demonstrate an existent data set succeeds with CHAR type;
* TEST: demonstrate an existent data set succeeds with NUM type;
* TEST: demonstrate an existent data set succeeds with the type missing;
* TEST: demonstrate an existent data set fails with GIBBERISH type;
* TEST: demonstrate a missing data set fails;
```

In the following "Unit Testing" section, these test cases are expanded with test code that demonstrates success or failure of each test. Some formalized test plans additionally include specific return codes that should be created for successful and failed conditions.

Test Data

While test *cases* represent positive and negative conditions that software must successfully navigate to validate technical requirements, test *data* are the data that software evaluates to validate its function and performance during testing. In user-focused software applications, these data often represent user input as well as other information passed inside software. Thus, parameters passed to a SAS macro module or return codes from a child process also represent test data that could be used to validate test cases. For example, to validate a macro with one parameter, several test data might be required, each representing different parameter conditions—a correct value, an incorrect value, an invalid value, and a missing parameter.

In data analytic development environments, test data more commonly represent data sets or input that are ingested and processed. In some environments, especially those with low-volume data, actual data may be used as test data. This is common, for example, in clinical trials environments that have relatively small sample sizes. One risk of using real data to test software is the likelihood that real data, especially in smaller quantities, will not demonstrate the full variability of valid data described through data models or other business rules. For example, business logic might prescribe that SAS software take certain actions when an observation is encountered for a patient with epilepsy. However, if no epileptic patients exist in a data set, then the data are insufficient to validate test cases.

A second risk of using real data to test software is the likelihood that sufficient invalid data (i.e., those representing negative test cases) will not exist. For example, if a business rule specifies that a patient's sex must be either M or F in a data set, and if data constraints enforce this rule through conditional SAS logic, that logic can only be fully tested when valid (M and F) and invalid (H and J) test data are processed. Although actual data may be a good substrate on which to build test data sets, additional data must often be added to ensure that all positive and negative test cases are sufficiently represented. Caution should also be exercised when real and fictitious data are commingled to ensure that test data are never confused or integrated with actual data.

While Base SAS does not directly support data relationships indicative of relational databases, data sets are often constructed to represent relational database tables. Thus, integrity constraints and other business rules may effectively create an equivalent relational structure among SAS flat data sets. Where complex relationships do exist among data sets, test data will be commensurately complex. For example, if a one-to-many relationship exists between a primary key and foreign key in two data sets, test data should not only represent valid and invalid values but also valid and invalid relationships between the respective data sets to ensure that quality controls appropriately flag, expunge, delete, or modify the invalid values.

Test data sets tend to accrete over time as additional variability is discovered or incorporated within operational data sets. For example, a development team might not consider during initial planning and design how to handle duplicate (i.e., invalid) values in a particular data set. When a duplicate value is first encountered, the error is detected, stakeholders establish a business rule to prescribe error handling, a test case is created, and test data should be modified to include the exceptional data so that updated software can demonstrate that it identifies and handles duplicate data. Because test cases and test data tend to increase over time—even after software release—it can be beneficial to archive test code, test cases, test data, and test results with software whenever production software is validated. This versioning enables stakeholders, if necessary during a review or audit, to recreate historical testing requirements, conditions, and results.

Testing Modalities

A multitude of software testing modalities exist, many of which overlap each other in intent, function, and technique. The majority of testing activities understandably occur within the testing phase of the SDLC. Some testing also continues into software operation, especially where development and test environments can't fully replicate the production environment. As defined by the International Organization for Standardization (ISO), common types of software testing include:

- *Informal Testing*—"Testing conducted in accordance with test plans and procedures that have not been reviewed and approved by a customer, user, or designated level of management."[5]
- *Formal Testing*—"Testing conducted in accordance with test plans and procedures that have been reviewed and approved by a customer, user, or designated level of management."[6]

- *Automated Testing*—Any type of software testing performed in which tests are conducted by repeatable programs that validate or invalidate functionality or performance of the original software.
- *Unit Test*—"1. Testing of individual routines and modules by the developer or an independent tester. 2. A test of individual programs or modules in order to ensure that there are no analysis or programming errors."[7]
- *Branch (Path) Testing*—"Testing designed to execute each outcome of each decision point in a computer program."[8]
- *Functional Testing*—"Testing that ignores the internal mechanism of a system or component and focuses solely on the outputs generated in response to selected inputs and execution conditions; testing conducted to evaluate the compliance of a system or component with specified functional requirements."[9]
- *Performance Testing*—"Testing conducted to evaluate the compliance of a system or component with specified performance requirements."[10]
- *Load Testing*—Testing to evaluate a component at the anticipated load, demand, or usage level. While the ISO definition of stress testing incorporates load testing, many software texts differentiate load testing as testing at but not exceeding the anticipated volume of data, demand, users, or other attributes.
- *Stress Testing*—"Testing conducted to evaluate a system or component at or beyond the limits of its specified requirements."[11]
- *Integration Testing*—"Testing in which software components, hardware components, or both are combined and tested to evaluate the interaction among them."[12]
- *Regression Testing*—"Selective retesting of a system or component to verify that modifications have not caused unintended effects and that the system or component still complies with its specified requirements; testing required to determine that a change to a system component has not adversely affected functionality, reliability, or performance and has not introduced additional defects."[13]
- *Acceptance (Operational) Testing*—"Testing conducted to determine whether a system satisfies its acceptance criteria and to enable the customer to determine whether to accept the system."[14]

Informal Testing

The ISO definition of informal testing differs from formal testing only in that informal test plans and testing procedures have not been "reviewed and approved by a customer, user, or designated level of management."[15] But this supposes that test plans and procedures do exist, which may not be the case in some environments. Informal testing, especially in end-user development environments, can include developers reviewing the output of some process under ideal conditions and with ideal data injects. This type of manual inspection essentially provides face validity but not integrity. That is, the software prima facie appears to perform as it should and is believed to be correct—until demonstrated otherwise.

Informal testing is adequate for some purposes and environments, but its convenience underlies its vulnerabilities and detractors. When informal testing is not documented, developers, customers, and other stakeholders must essentially take someone's word that the code "looks right" or "ran correctly," but this informal process is neither repeatable nor defensible. For example, as demonstrated in the "Plan to Get Hit by a Bus" section in Chapter 15, "Readability," when poor Thomas fails to show up to work one day, because his test plans and test results are documented nowhere, his quality assurance efforts cannot be demonstrated and testing must redundantly be performed again. When no documentation of testing exists, stakeholders are essentially placing their trust only in the developer performing the testing rather than appropriately in the integrity of the SDLC and formal testing procedures.

The ISO distinction between informal and formal testing is the review and approval process of testing procedures. The specification of technical requirements during software planning is intended to ensure that all stakeholders have a clear, unambiguous vision of required functionality and performance. However, if a performance requirement specifies that software must efficiently scale to handle 10 million observations and be portable between the SAS Display Manager and SAS Studio environments, then a test plan should exist that details how those requirements will be demonstrated to the customer. Test plans do not have to be lengthy documents; in this example, a single statement could indicate that the software would be required to produce identical results and performance in the two SAS environments. In formal testing environments, test cases would be required to state specifically how equivalent function and performance would be both defined and measured.

Formal Testing

Formal testing specifies not only that a test plan and procedures exist, but also that the plan has been accepted by stakeholders. Where software technical requirements exist and customer acceptance of software is required, formal software testing can facilitate the acceptance process by providing concrete methods and metrics against which performance is tested. Moreover, those performing the testing will understand that their work is part of an organized, empirical process as they carry out the test plan. Thus, the formal testing process ensures that developers, customers, and other stakeholders agree on the methodology that will evaluate whether software is complete—often referred to anecdotally as the *definition of done*.

Automated Testing

Automated testing (or test automation) describes software that tests other software. This includes third-party open source or commercial-off-the-shelf (COTS) test software, as well as test software developed internally by developers and testers themselves. The benefits of automated testing are numerous, and within many development environments, "testing" is synonymous with "automated testing" in that all tests must be automated. Automated testing can be run with little effort and provides structured, repeatable, defensible results that can validate software or uncover defects or vulnerabilities. Moreover, automated tests can be archived with code and test data so that test conditions (and results) can be replicated if necessary for an audit.

Automated testing is the standard for third-generation languages (3GLs) such as Java or Python. Despite its prevalence in other languages, and in part because third-party testing software is not widely available for Base SAS, automated testing is rarely described in SAS literature and likely to be implemented on a limited scale in few environments. The quality of SAS software can be substantially improved through automated testing, however, in part because of the vulnerabilities that tests can uncover, and in part because automated testing encourages SAS practitioners to embrace a testability mind-set throughout planning, design, and development. SAS practitioners who develop within environments that espouse automated testing typically know what tests their software will have to pass even before they begin developing software, which helps frame design and development activities.

Two of the most common automated tests are unit testing and functional testing. Unit testing inspects an individual module of software, typically irrespective of other software elements or functionality. For example, a unit test for a SAS macro might validate how the module handled valid and invalid data and parameter input. Especially where the unit being tested represents a SAS macro, unit tests can be conceptualized as fake parent processes that call the SAS macro to validate specific test cases. Functional testing conversely examines only the big picture—software inputs and outputs—irrespective of what steps occur in the middle. In a sense, functional testing describes a unit test at the software level. Automated unit testing and functional testing are demonstrated later in the "Unit Testing" and "Functional Testing" sections, respectively.

Automated tests can be written before software is developed, as is central in test-first development and test-driven development (TDD) environments. Automated tests can also be written in tandem with or immediately after code within the development phase, or after software is completed in a separate testing phase. Especially in Agile and other rapid-development environments, writing automated tests often occurs with or immediately following development because software components must be designed, developed, tested, accepted, and released all within a time-boxed iteration. This proximity of development and automated testing contrasts with phase-gate Waterfall environments in which automated tests may be written weeks or months after software development has occurred.

Regardless of the manner in which automated testing is implemented, all automated tests should be functional when software is validated and accepted. In environments that utilize automated testing, the degree to which software morphs throughout the SDLC will require that automated tests are also refined to ensure they accurately represent software behavior before software acceptance. For example, if the functional intent of a SAS macro changes after production or the macro later incorporates additional parameters, these changes should be reflected in commensurately updated automated unit testing.

Unit Testing

Unit testing tests the smallest functional module of software. Under the principles of software modularity, modules should be functionally discrete, loosely coupled, and encapsulated. Thus, testing a unit in theory is straightforward, because it tests a small portion of code that should lack complexity. Unit testing is often referred to as *white-box testing* because the inner mechanics of modules are exposed and validated, as contrasted with functional testing, referred to as *black-box testing*. To be clear, however, *white-box testing* can also refer to other invasive species of testing, such as branch testing, integration testing, or regression testing.

In SAS development, unit testing typically refers to testing a single macro, DATA step, or batch job, so long as modular design principles were espoused. If your macro is 150 lines, however, this may include composite—not discrete—functionality, more closely representing monolithic than modular design. While it is possible to *functionally test* monolithic code, it is definitively impossible to *unit test* a complex software module with composite functionality. Just as modular software is expected to be loosely coupled, unit tests should also have as few interdependencies with other unit tests as possible. Modular software design and functional decomposition to identify and isolate discrete functionality are discussed throughout Chapter 14, "Modularity."

Consider the %VARLIST macro, demonstrated in the "Macro Comments" section in Chapter 15, "Readability":

```
* creates a space-delimited list of all variables within a data set;
***** does not validate existence of the data set;
***** assumes a shared lock can be obtained on the data set;
***** does not validate the TYPE parameter;
%macro varlist(dsn= /* data set name in LIB.dataset or dataset format */,
   type= /* ALL, NUM, or CHAR, the type of variables returned */);
%local vars;
%local dsid;
%local i;
%global varlist;
%let varlist=;
%let dsid=%sysfunc(open(&dsn, i));
%let vars=%sysfunc(attrn(&dsid, nvars));
%do i=1 %to &vars;
   %let vartype=%sysfunc(vartype(&dsid,&i));
   %if %upcase(&type)=ALL or (&vartype=N and %upcase(&type)=NUM) or
         (&vartype=C and %upcase(&type)=CHAR) %then %do;
      %let varlist=&varlist %sysfunc(varname(&dsid,&i));
      %end;
   %end;
%let close=%sysfunc(close(&dsid));
%mend;
```

Unit testing specifies that the smallest module should be tested, but it doesn't specify the type of test that is being performed. For example, if conditional logic branching occurs within a unit, branch testing might be performed as part of a unit test. Functional testing is also typically included because units should have some function or purpose, even if they are only called as a child process. Unit testing, like many other types of testing, overlaps significantly with testing modalities but specifies that each module should be tested thoroughly and severally.

The following code demonstrates three unit tests that show the expected output from the %VARLIST macro:

```
* TEST GROUP: pass existent data set with valid and invalid data types;
libname test 'c:\perm';
%macro test_varlist;
data test.test_dsn;
   length char1 $10 char2 $10 num1 8 num2 8;
run;
* TEST: demonstrate an existent data set with ALL type;
%varlist(dsn=test.test_dsn,type=ALL);
%if &varlist=char1 char2 num1 num2 %then %put ALL: PASS;
%else %put ALL: FAIL;

* TEST: demonstrate an existent data set with NUM type;
%varlist(dsn=test.test_dsn,type=NUM);
%if &varlist=num1 num2 %then %put NUM: PASS;
%else %put NUM: FAIL;

* TEST: demonstrate an existent data set with missing type;
%varlist(dsn=test.test_dsn,type=);
%if %length(&varlist)=0 %then %put Missing TYPE: PASS;
%else %put CHAR: FAIL;
%mend;

%test_varlist;
```

If any of the tests produce a FAIL, this indicates an error in the module. In this example, three test cases are demonstrated. The first two occur under normal operation (i.e., positive test cases) while the third represents an exceptional event (i.e., negative test case) in which the TYPE parameter was omitted. Other unit tests would be required to test this module fully, but this represents a start in the right direction.

Unit testing is invaluable because it tests discrete modules of software functionality and, in theory, unit tests should be sufficiently decoupled from each other to enable them to follow the code they validate. For example, if a modular SAS macro is developed and is flexible enough that it warrants inclusion in a reuse library—discussed in the "Reuse Library" section in Chapter 18, "Reusability"—not only the macro but also its separate unit testing code and logs can be included in the reuse library to benefit future reuse and repurposing. Unit testing code and unit testing results are sometimes provided (with completed software) to customers and other stakeholders to validate functionality during software acceptance.

Branch Testing

Branch or path testing tests the sometimes-complex reticulations of conditional logic statements. In theory, all possible branches should be tested, but this can be difficult where branches represent environmental attributes rather than parameters that are passed. For example, if one branch of a program executes only when the software detects it is running in a UNIX environment, this can be difficult to fake when testing is occurring within Windows.

To demonstrate branch testing for the %VARLIST macro described in the "Unit Testing" section, two additional unit tests can be added to demonstrate program flow when the TYPE parameter is either missing or invalid:

```
* TEST: demonstrate an existent data set with the type missing;
%varlist(dsn=test.test_dsn,type=);
%if %length(&varlist)=0 %then %put Missing: PASS;
%else %put Missing: FAIL;

* TEST: demonstrate an existent data set with an invalid type;
%varlist(dsn=test.test_dsn,type=BLAH);
%if %length(&varlist)=0 %then %put Missing: PASS;
%else %put Missing: FAIL;
```

When these additional tests are run within the structure of the %TEST_VARLIST macro, they demonstrate that either an empty TYPE parameter or an invalid TYPE parameter will produce a global macro variable &VARLIST that is empty. Coupled with the prior three unit tests, these five tests collectively demonstrate that the branch testing has passed.

One of the most critical uses for branch testing can be to test the fail-safe path to ensure that software fails gracefully in a secure manner. The fail-safe path is discussed in the "Fail-Safe Path" sections in Chapter 11, "Security," and is required when robust software encounters a runtime error or other exception from which it cannot recover. For example, when software fails, data sets may be explicitly locked, file streams may be open, or temporary or invalid data sets may have been created—each of which can require some action in the fail-safe path to return software to a secure state. Branch testing the fail-safe path is the only way to ensure that when software does fail, it does so as prescribed through business rules.

Functional Testing

Functional testing takes into account software inputs and outputs but does not test what occurs between these two extremes. For this reason, functional testing is often referred to as *black-box testing* because software (or, in theory, even developers) cannot view the module interior—only inputs and outputs are observable. The test cases in both the "Unit Testing" and "Branch Testing" sections are examples of functional testing at the module level. In general, however, functional testing refers to testing significantly larger software products having complex, composite functionality.

In the following example, the %GET_MEANS macro runs the MEANS procedure and saves the mean value of one or more variables to a data set. The macro calls the %VARLIST macro, which provides input for the VAR statement in the MEANS procedure, after which output is generated to the Means_temp data set:

```
data final;
   length num1 8 num2 8 char $10;
   char="n/a";
   do num1=1 to 10;
      num2=num1+10;
      output;
      end;
run;

* saves mean values of one or more variables to a data set;
%macro get_means(dsn=);
%varlist(dsn=&dsn,type=num);
proc means data=&dsn noprint;
   var &varlist;
   output out=means_temp;
run;
%mend;

%get_means(dsn=final);
```

To functionally test the %GET_MEANS macro, it's not necessary to directly test the inputs, outputs, or any other aspect of the %VARLIST procedure. The functional test for %GET_MEANS should only evaluate the input (Final data set) and output (Means_temp data set) for the macro itself. The %VARLIST macro and other functionality inside %GET_MEANS essentially occur within the black box, as demonstrated in the following functional test:

```
* TEST GET_MEANS;
* show means in output data set are accurate;
proc means data=final noprint;
   var num1 num2;
   output out=test_get_means;
run;
%get_means(dsn=final);
proc compare base=test_get_means compare=means_temp;
run;
```

The COMPARE procedure validates that the MEANS procedure produces identical results to the %GET_MEANS macro. As demonstrated, when the functionality being tested is simple, such as that in the %GET_MEANS macro, the test itself may be as long as or longer than the code being tested. As the complexity of larger programs increases, however, functional testing code will remain relatively small because it relies only on primary inputs and outputs and does not reference or utilize secondary (or intermediary) inputs or outputs.

Some functions specified through technical requirements can only be tested and validated by inspecting software or its products. In data analytic development, for example, a data product such as a SAS report can be examined in part through automated testing if the REPORT procedure utilizes the OUT statement to create a testable data set. However, other functional requirements of the report such as formatting and aesthetic attributes can only be validated through manual inspection.

Performance Testing

Performance testing tests performance rather than function, and includes load testing, stress testing, and typically all tests of speed, efficiency, and scalability. Performance testing is especially useful in data analytic development due to data variability, including fundamental differences that can exist between test data and production data. Load testing and stress testing are described later in their respective sections while methods to facilitate greater speed, efficiency, and scalability are described in Chapter 7, "Execution Efficiency," Chapter 8, "Efficiency," and Chapter 9, "Scalability."

Performance testing often involves testing software against a *boundary value*—that is, a value "that corresponds to a minimum or maximum input, internal, or output value specified for a system or component."[16] Boundary values can be used to facilitate secure software, for example, by detecting erroneous inputs that could lead to buffer overflows or other failures. More relevant to data analytic development, boundary values are used in technical requirements to specify data volume or velocity capabilities required by software. Load testing is distinguished from stress testing in that the former performs testing *at* boundary limits while the latter performs testing *beyond* boundary limits. In other words, successful load testing should be demonstrated before software acceptance, whereas stress testing places software in situations or environments to elicit when likely or inevitable failure will occur.

Performance testing is sometimes conceptualized to include reliability testing, although this lies in a grey zone that spans functionality and performance testing. If software functions but does so unreliably, its functionality will also be diminished when availability of the data product, output, or solution is compromised. For example, the %GET_MEANS macro demonstrated in the earlier "Functional Testing" section fails if the data set is exclusively locked by another user or process, which threatens the robustness and reliability of the software.

To detect the exception—a locked data set—an additional module (%DSN_AVAILABLE) is developed that tests the existence and availability of a data set. The OPEN function returns a 0 value for the &DSID macro variable if the data set does not exist or is exclusively locked:

```
* tests existence and shared lock availability of a data set;
%macro dsn_available(dsn= /* data set in LIB.DSN or DSN format */);
%global dsn_available_RC;
%let dsn_available_RC=;
%local dsid;
%local close;
%let dsid=%sysfunc(open(&dsn));
%if &dsid^=0 %then %do;
   %let dsn_available_RC=TRUE;
   %let close=%sysfunc(close(&dsid));
   %end;
%else %let dsn_available_RC=FALSE;
%mend;
```

By implementing the macro within the %GET_MEANS program and testing the value of the &DSN_AVAILABLE_RC return code immediately before attempting the MEANS procedure, robustness is improved because the MEANS procedure cannot be invoked on a missing or locked data set. Nevertheless, software reliability really is not improved because the software still terminates and business value is lost when a data set is locked. While software has been made robust to one type of exception, to be made more reliable, additional exception handling would be required to wait for the data set to be unlocked, provide a duplicate data set to failover to, or provide some other mechanism through which business value could still be achieved.

The following revised %GET_MEANS macro includes the %DSN_AVAILABLE quality control:

```
* saves mean values of one or more variables to a data set;
%macro get_means(dsn= /* data set in LIB.DSN or DSN format */);
%dsn_available(dsn=&dsn);
%if &dsn_available_RC=TRUE %then %do;
%varlist(dsn=&dsn,type=num);
   proc means data=&dsn noprint;
      var &varlist;
      output out=means_temp;
   run;
   %end;
%else %put DATA SET MISSING OR LOCKED;
%mend;

%get_means(dsn=final);
```

After integrating the new module into the %GET_MEANS macro, integration testing should ensure the software functions as intended. This is demonstrated in the "Integration Testing" section later in the chapter, which—spoiler alert—uncovers a logic error in the %GET_MEANS macro.

Load Testing

Load testing describes testing software against data volume or demand (usage) described in software requirements. While load (demand) testing describes testing a software application to ensure it functions effectively with 50 simultaneous users (if this represents the expected user demand), for purposes of this chapter, only load testing that tests the predicted volume or velocity of data throughput is discussed. For example, if an extract-transform-load (ETL) system is being developed and is required to process a maximum of 20 million observations per day, at some point the software would need to be load tested against this predicted volume. During development and early testing, however, it's common to run and test software with less voluminous data to facilitate faster performance and test results.

Load testing should not, however, be misconstrued to represent testing at only the expected data levels specified in requirements. For example, if 20 million observations is the expected data volume, load testing might test the software at 125 percent of this value to demonstrate how software would handle this variability. In other cases, load testing is utilized if the predicted load is expected to increase over time. For example, other requirements might not only specify an initial volume of 20 million observations per day but also specify that an additional 5 million observations should be expected per year thereafter. In this second scenario, load testing might be conducted to the expected five-year mark—that is, 45 million observations—to determine how the software will handle the eventual load. This additional load testing beyond current load volume enables stakeholders to plan programmatic and nonprogrammatic solutions for the future if test results demonstrate that software will not meet performance requirements as data attempt to scale.

Load testing often demonstrates the ability of software to scale—with respect to data volume and velocity—from a test environment to a production environment, and possibly to an additional theoretical future environment. By ISO definition, load testing is subsumed under stress testing and not differentiated. However, because this differentiation is commonly made throughout software development literature, it is made throughout this text. As stated in the "Test Data" section, test data should also be sufficiently variable to demonstrate the breadth of both positive and negative test cases; however, variability is typically not the focus of load testing.

Stress Testing

Stress testing demonstrates the performance of software not only within expected boundaries but also beyond. Where load testing and stress testing are differentiated (as throughout this text), however, stress testing refers more selectively to testing software *beyond* its operational capacity, typically with the intent to expose vulnerabilities, performance failure, or functional failure. In some cases, stress testing is conducted up to a statistical threshold determined from requirements specifications. For example, if requirements state that software must process 20 million observations per day, stress testing test cases might specify testing at 500 percent of that value.

In other cases, stress testing denotes incremental testing until the software experiences functional or performance failure. For example, stress testing might create test data sets of increasingly larger size until some component of the program runs out of memory or produces other runtime errors. This equates to functional stress testing, because it tests the software product as a whole against one or more stressors such as data volume. In Chapter 8, "Efficiency," the "Memory" section demonstrates stress testing of the SORT procedure until it begins to perform inefficiently and fails. And, in the "Inefficiency Elbow" section in Chapter 9, "Scalability," the use of FULLSTIMER performance metrics (to aid in performance and functional failure prediction) is demonstrated.

In applications development, stress testing is a critical component of facilitating software security because malicious parties can attack and potentially exploit software by stressing it, for example, by overwriting software with code injection or buffer overflow attacks. While software attacks are beyond the scope of this text, stress testing remains a critical step toward ensuring and validating that software will not fail when grossly gratuitous inputs, injects, or data throughput are processed. Stress testing in part aims to test the limits of data validation and other quality controls to ensure they are adequate.

Environmental stress testing changes aspects of the environment to see how software responds. In hardware stress testing, this could involve testing how equipment performs in extreme heat, cold, moisture, or in the face of some other physical threat. In software stress testing, however, more common environmental threats include the decrease of system resources, such as low-memory conditions. For example, stress testing might halve the amount of memory available to the SAS application to test how software performs in a low-memory environment. A complete stress test would continue decreasing available memory until functional or performance failure resulted, thus generating a threshold that could be used to deter and detect future failures of this type.

True stress testing stands alone as the one type of test that software should often *not* be able to pass. Especially in cases where stress testing is designed to elicit software failure, it's understood that failure will be evident, and testing is intended to document the boundary at which failure occurs. In some cases, where stress tests reveal that failure thresholds are too proximal to predicted data loads, additional refactoring may be required to ensure that software does not fail under normal utilization. In other cases, stress testing will yield thresholds so high (or low) that they will never be achieved in software operation and thus present no risk.

In some cases, developers are actually testing the limits of the software language itself. For example, SAS documentation describes the SYSPARM command line option but fails to specify any upper limits, such as the maximum length of the string that can be passed. This might be crucial to understand for peace of mind, so the following child process can be saved as C:\perm\stress.sas, which assesses the length of the &SYSPARM macro variable after it is passed from a parent process:

```
libname perm 'c:\perm';

%macro test;
%if %sysfunc(exist(perm.stress))=0 %then %do;
   data perm.stress;
      length text $10000 len 8;
   run;
   %end;
data perm.stress;
   set perm.stress end=eof;
   output;
   if eof then do;
      text=strip("&sysparm");
      len=length(text);
      output;
      end;
run;
%mend;

%test;
```

When the following parent process is executed, it tests the lengths of SYSPARM iteratively from 1 to 100 characters:

```
%macro sysparm_stress_test();
%local var;
%do i=1 %to 100;
   %put &i;
   %let var=&var.1;
   systask command """%sysget(SASROOT)\sas.exe"" -noterminal -nosplash
      -sysparm ""&var"" -sysin ""&perm\stress.sas"" -log
      ""&perm\stress.log"" -print ""&perm\stress.lst"""
      status=stress_status taskname=stress_task;
   waitfor _all_ stress_task;
   %end;
%mend;
```

The partial results demonstrate that at the 83rd iteration, one threshold of the SYSTASK statement itself (rather than the SYSPARM option, which was the intended test subject) was reached, as the length of the quotation used by the SYSTASK COMMAND statement exceeded the BASE SAS threshold:

```
82
NOTE: Task "stress_task" produced no LOG/Output.
83
WARNING: The quoted string currently being processed has become more
than 262 characters long.
         You might have unbalanced quotation marks.
NOTE: Task "stress_task" produced no LOG/Output.
```

Despite the warning message, examination of the PERM.Stress data set reveals that the SYSTASK statement continued to execute and the SYSPARM length continued to increment, so this warning poses no actual limitation or threat. Further stress testing (not demonstrated) reveals that at a length of 8,076 characters, the SYSTASK statement itself begins to fail, producing a STATUS return code of 104 but no warnings or runtime errors in the log. This demonstrates the value not only in stress testing but also in exception handling to validate the success of critical processes. The SYSTASK statement and SYSPARM option are discussed further throughout Chapter 12, "Automation." Armed with this new information (and system boundary) achieved through stress testing, SAS practitioners can more confidently implement SYSTASK and the SYSPARM parameter.

Integration Testing

Once modules of code have been independently tested and integrated into a larger body of software, it's critical to understand how all the pieces fit together. Integration testing accomplishes this by demonstrating that modules work together cohesively and do not create additional vulnerabilities in their collective implementation. The %GET_MEANS macro demonstrated earlier in the "Performance Testing" section includes a logic error that makes it vulnerable to either missing or locked data sets—the very two vulnerabilities that the additional exception handling sought to alleviate!

While the OPEN function can demonstrate that a data set exists and is not exclusively locked, the %DSN_AVAILABLE macro subsequently closes the data set with the CLOSE function. Once closed, a separate process running in another session of SAS could sneak in and either delete or exclusively lock the data set. Therefore, although the &DSN_AVAILABLE_RC return code would indicate TRUE, by the time the MEANS procedure would have been executed, the data set could have been locked or deleted, causing the MEANS procedure to fail. This demonstrates the importance of integration testing, because at face value, the %DSN_AVAILABLE macro does determine data set existence and availability, its only two objectives. However, when implemented within this macro, it fails to eliminate the risk.

To solve this problem, the I/O stream to the data set must remain open while the MEANS procedure executes, thus preventing a concurrent session of SAS from deleting or exclusively locking the data set. Once the MEANS procedure terminates, the CLOSE function should be executed from within the %GET_MEANS module rather than from within the %DSN_AVAILABLE macro. This modification also requires that the &DSID macro variable be changed from a local to a global macro variable so that it can be read by the %GET_MEANS macro:

```
* tests existence and shared lock availability of a data set;
%macro dsn_available(dsn= /* data set in LIB.DSN or DSN format */);
%global dsn_available_RC;
%let dsn_available_RC=;
%global dsid;
%let dsid=%sysfunc(open(&dsn));
%if &dsid^=0 %then %let dsn_available_RC=TRUE;
%else %let dsn_available_RC=FALSE;
%mend;

* saves mean values of one or more variables to a data set;
%macro get_means(dsn= /* data set in LIB.DSN or DSN format */);
%dsn_available(dsn=&dsn);
%if &dsn_available_RC=TRUE %then %do;
%varlist(dsn=&dsn,type=num);
   proc means data=&dsn noprint;
      var &varlist;
      output out=means_temp;
   run;
   %let close=%sysfunc(close(&dsid));
   %end;
%else %put DATA SET MISSING OR LOCKED;
%mend;
```

While the updated code now achieves the functional intent and is robust against the possibility of a missing or exclusively locked data set, software modularity has unfortunately been decreased. Having to open the file stream inside the child process but close it in the parent process diminishes loose coupling and encapsulation. The child process performs a risky task—opening a file stream that requires a later %SYSFUNC(CLOSE) function—but has no way to guarantee that the file stream is closed by the parent process. Nevertheless, this tradeoff—the loss of static performance (modularity) to gain dynamic performance (robustness)—is warranted if the process needs to be protected from incursions from external SAS sessions. Loose coupling and encapsulation are discussed within Chapter 14, "Modularity."

Regression Testing

Integration testing adds a module to a larger body of code and then seeks to ensure that the composite software still functions as required. As demonstrated in the "Integration Testing" section, however, this can require subtle or not-so-subtle changes to modules, such when the %DSN_AVAILABLE macro had to be overhauled to correct a logic error. Regression testing, on the other hand, occurs after modules have been modified to ensure that current functionality or performance is not diminished. For example, in stable production software, a reusable module such as the %DSN_AVAILABLE macro might have been saved in the SAS Autocall Macro Facility or to an external SAS program and referenced with an %INCLUDE statement. This stability of a tested module facilitates its reuse and generalizability to other software and solutions.

However, because the macro had to be modified, all other software that referenced it would need to be tested to ensure that the macro still functioned in those diverse use cases. And, because the CLOSE statement had to be removed, in all likelihood, the modifications to the %DSN_AVAILABLE macro would have caused it to fail within software that referenced its original version. Regression testing thus seeks to demonstrate backward compatibility of modules or software after modification to ensure that past uses are not corrupted or invalidated.

A more secure (and backward compatible) solution instead modifies the %DSN_AVAILABLE macro for use in %GET_MEANS while not sacrificing its original design (that includes the CLOSE statement). This more robust modification passes regression testing by enabling backward compatibility:

```
* tests existence and shared lock availability of a data set;
%macro dsn_available(dsn= /* data set in LIB.DSN or DSN format */,
   close=YES /* default closes the data set, NO keeps the stream
   open */);
```

```
%global dsn_available_RC;
%let dsn_available_RC=;
%global dsid;
%let dsid=%sysfunc(open(&dsn));
%if &dsid^=0 %then %do;
   %let dsn_available_RC=TRUE;
   %if &close=YES %then %let close=%sysfunc(close(&dsid));
   %end;
%else %let dsn_available_RC=FALSE;
%mend;
```

The function is now backward compatible because it *overloads* the macro invocation with an extra parameter, CLOSE, that represents whether %DSN_AVAILABLE should close the data stream. Older code can still call the macro with the following invocation, which will close the I/O stream before the macro exits (by defaulting to CLOSE=YES):

```
%dsn_available(dsn=&dsn);
```

Or, as is required by the %GET_MEANS macro, the %DSN_AVAILABLE macro can be overloaded by including the CLOSE=NO parameter, which will keep the I/O stream open so that it can be closed by the parent process (rather than the child) after the MEANS procedure has completed:

```
* saves mean values of one or more variables to a data set;
%macro get_means(dsn=);
%dsn_available(dsn=&dsn, close=NO);
%if &dsn_available_RC=TRUE %then %do;
%varlist(dsn=&dsn,type=num);
   proc means data=&dsn noprint;
      var &varlist;
      output out=means_temp;
   run;
   %let close=%sysfunc(close(&dsid));
   %end;
%else %put DATA SET MISSING OR LOCKED;
%mend;

%get_means(dsn=final);
```

Especially where libraries of stable code exist that are used throughout production software by a development team or an organization, every effort should be made to ensure that modules are backward compatible when modified so that existent calls to and uses of those modules do not also have to be individually modified. Regression testing is an important aspect of software maintainability because it effectively reduces the scope of system-wide modifications that are necessary when only small components of software must be modified.

Regression Testing of Batch Jobs

Batch jobs are automated SAS programs spawned in self-contained sessions of SAS, launched from the command prompt, the SAS interactive environment, or other batch jobs. As demonstrated in the "Passing Parameters with SYSPARM" section in Chapter 12, "Automation," use of SYSPARM enables one or more parameters to be passed from the parent process (or OS environment) to the batch job. Inside the batch job, the &SYSPARM automatic macro variable receives the SYSPARM parameter, which can be parsed into one or multiple parameters or values.

A batch job typically represents the culmination of SAS production software because it has been tested, validated, automated, and often scheduled to ensure reliable execution. Even the most reliable batch jobs, however, must be maintained and occasionally modified to ensure they remain relevant to shifting needs, requirements, or variability in the environment. Furthermore, because development and testing occur in an interactive environment rather than the batch environment, batch jobs should ensure that they retain backward compatibility to the interactive environment, which can be facilitated through performing regression testing by running a batch job manually from the interactive environment.

To demonstrate a lack of backward compatibility, the following plain text batch file is saved as C:\perm\freq.bat:

```
"c:\program files\sashome\sasfoundation\9.4\sas.exe" -noterminal
   -nosplash -sysin c:\perm\freq.sas -log C:\perm\freq.log -print
   c:\perm\freq.lst -sysparm "dsn=perm.final, table_var=char1"
```

The batch file calls the SAS batch job C:\perm\freq.sas, which runs the FREQ procedure on the data set specified in the SYSPARM internal DSN parameter. To ensure the data set is available (only in this simulation), the LIBNAME statement assigns the PERM library while the DATA step generates the PERM.Final data set. The following code should be saved as C:\perm\freq.sas:

```
libname perm 'c:\perm';
data perm.final; * produces required data set;
   length char1 $10;
```

```
   do i=1 to 10;
      char1="obs" || strip(put(i,8.));
      output;
      end;
run;

* accepts a comma-delimited list of parameters in VAR1=parameter
   one, VAR2=parameter two format;
%macro getparm();
%local i;
%let i=1;
%if %length(&sysparm)=0 %then %return;
%do %while(%length(%scan(%quote(&sysparm),&i,','))>1);
   %let var=%scan(%scan(%quote(&sysparm),&i,','),1,=);
   %let val=%scan(%scan(%quote(&sysparm),&i,','),2,=);
   %global &var;
   %let &var=&val;
   %let i=%eval(&i+1);
   %end;
%mend;

* retrieves the parameters DSN (data set name) and TABLE_VAR
   (variable name to be analyzed);
%getparm;
proc freq data=&dsn;
   tables &table_var;
run;
```

This Freq.sas program runs from batch but, when executed interactively—including from either the SAS Display Manager or SAS Enterprise Guide—the software fails because the &SYSPARM macro variable is empty, not having been passed a SYSPARM parameter. To ensure backward compatibility, the software should be able to be run manually from the interactive mode as well as from batch. The following inserted code tests the &SYSPARM macro variable and, if it is empty, assigns default values to the global macro variables that otherwise would have been created with the %GETPARM macro:

```
* REMOVE IN PRODUCTION! * initializes SYSPARM for testing environment;
%macro testparm();
%if %length(&sysparm)=0 %then %let sysparm=dsn=perm.final,
   table_var=char1;
%mend;

%testparm; * REMOVE IN PRODUCTION!;
```

Inserting this code between the %MEND statement and %GETPARM invocation assigns &SYSPARM to a default value, thus allowing the %GETPARM macro to be executed interactively. The inserted code does have to be removed in production because it could mask errors that occurred, for example, if the batch job were called erroneously without the SYSPARM parameter. However, by inserting this succinct test code, it allows developers to run, modify, and test the program within an interactive environment. This discussion continues in the "Batch Backward Compatibility" section in Chapter 12, "Automation."

Acceptance Testing

Acceptance testing in many environments is less actual software testing and more a dog-and-pony show that demonstrates software to the customer, users, or other stakeholders. Acceptance testing primarily demonstrates software function and performance and, during formal acceptance testing, stakeholders may scrutinize requirements documentation in conjunction with software to validate that software has demonstrated all technical specifications. By validating the enumerated list of test cases, the customer or other stakeholders can be confident not only that software meets functional and performance requirements, but also that it meets formal testing requirements.

But in data analytic development environments, business value is typically conveyed not through software products (or the archetypal "releasable code" espoused in Agile literature), but through resultant data products. Moreover, unlike user-focused applications that may include a graphical user interface (GUI) that begs stakeholder interaction, Base SAS software products derive value by being accurate and fast, not flashy. In fact, because a common end-goal is to automate SAS production software as batch jobs that unobtrusively run in the background, in many cases there won't even be a SAS log that a customer can pretend to view as it scrolls by at an indecipherable pace.

The fundamental differences between software applications development and data analytic development may manifest in such a stark contrast that customers have little to no interest in reviewing software functional and performance requirements during acceptance testing. In cases in which business value is conveyed solely through a derivative data product—such as a report that must be written, derived from data generated through SAS software—some customers may not even acknowledge the milestone of software completion because software alone provides no business value. For example, after four long days of work, I once remarked to a customer, "I finished the ETL software for the new data stream," to signal that he could review the software product. "Let me know when you finish the report," was his only response, because only the derivative data product conveyed value to him.

In other cases, customers may recognize the intrinsic value in software that has been created, but only show interest in functional objectives that were met. Still other customers may be interested in function and *dynamic* performance, yet have no interest in *static* performance attributes. Regardless of the environment in which software is developed and from where business value is derived, if software performance and quality are important enough to discuss during planning and design for inclusion in software requirements, then they should be commensurately important enough to acknowledge and validate when software is completed as part of software acceptance.

When Testing Fails

Tests will fail. Sometimes tests fail expectedly because some functional or performance element was not specified in software requirements or was intentionally omitted. For example, if software requirements say nothing about testing for data set existence (before referencing the data set), some degree of robustness is lost, but this may be acceptable in software in which it is unlikely that the data set would be missing or in which the missing data set would pose little risk. Thus, it would be a waste of time to create an automated test to demonstrate what occurs when a data set is missing in software not required to demonstrate this type of robustness. When specific risks are accepted by stakeholders, there is no need to develop tests to demonstrate those failures—they can be chronicled in the failure log if necessary.

Sometimes tests fail unexpectedly when an unforeseen vulnerability is detected through software testing. After all, the intent of software testing is to identify both known and unknown vulnerabilities. The identified error or vulnerability should be discussed but, if it requires extensive modification of code, and depending on the software development environment and methodology espoused, the required maintenance might need to be authorized by a customer or other stakeholders. Where new risks are identified, they can be eliminated by correcting software defects, but customers also commonly accept the results of failed tests because the exposed vulnerabilities pose little overall risk to the software, or because additional functionality or the software release schedule are prioritized over maintenance.

While in many cases adequate software testing will identify previously unknown defects in software that must be remedied through corrective maintenance, in other cases the defects will be recorded in a risk register, accepted, and effectively ignored thereafter. The quality assurance aspect of testing is not intended to force unnecessary performance into software, but rather to identify vulnerabilities that exist and to ensure that software is being released with the desired level of functionality and performance and an acceptable level of risk.

TESTABILITY

The shift toward embracing testability first and foremost requires a commitment to software testing. Where software testing is not a priority within a team or organization and the testing phase of the SDLC is absent, testability principles may still benefit software in other ways but won't suddenly spur stakeholders to implement testing. The decision to implement formalized testing in software will only be made when its benefits are understood and its costs not prohibitive. Thus, where high-performing software is demanded, and especially where software audits are likely or expected, the embracing of testability principles will facilitate higher quality software.

Testability Principles

Testability principles facilitate higher quality software because software more readily and reliably can be shown to be functional and free of defects. The previous enumeration of formalized testing methods may be foreign to some SAS development environments that rely on less formal, manual code review to facilitate software validation. Notwithstanding the breadth of differences in intensity or formality with which testing can be approached, all software testing—from formalized test plans to informal code review—are benefited by testability principles.

SAS data analytic development environments are less likely to implement formalized software testing in part due to the practice of building monolithic software—large, single-program software products. This design practice in turn encourages programs that are more complex and less comprehensible to SAS practitioners, and thus not only more error-prone but also more difficult to debug, test, and validate. By deciphering smaller chunks of software, not only is understanding improved, but software function and performance can more easily be compared against technical requirements.

Technical Requirements

All testing begins with the definition and documentation of technical requirements that ideally should specify not only what software *should* do but also what it *shouldn't* do. With this groundwork in place, whether a formal test plan is created or not, subsequent tests will be able to be measured against those technical specifications. In software development environments that lack technical requirements, it is virtually impossible to test software because no functional or performance expectations exist, so no definition of done or software acceptance criteria can be validated.

Modularity

As demonstrated earlier in the "Unit Testing" section, even a small code module that is functionally discrete and relatively simple can have multiple vulnerabilities that should be tested, documented, and potentially resolved. While testing alone will not mitigate or eliminate vulnerabilities, it can uncover them and raise their attention to developers and stakeholders to ensure they are documented in a risk register and prioritized for possible future corrective maintenance. Without modular software design, testing will be significantly more difficult due to its inherent complexity while unit testing will be all but impossible where composite functionality exists and test cases cannot be disentangled.

Readability

Software readability improves the ability to test software efficiently and effectively. Readable software, described in Chapter 15, "Readability," references primarily the ease with which code can be comprehended. But where requirements documentation, project documentation, data documentation, or other relevant information can more fully describe software, these artifacts can also further comprehension and may be required in data analytic software. When software requirements convey the objectives that software should achieve, and code is clear and concise, software testing will be facilitated because test cases often follow naturally from requirements or software use cases. By eliminating confusion or ambiguity from both requirements and software, the assessment of how that software meets those requirements can be made much more readily.

WHAT'S NEXT?

The goal of software testing is to demonstrate the validity of software as assessed against technical requirements. Once tested, validated, and accepted, software can be released into production and, at that point, software should be sufficiently stable that it can perform without frequent modifications or maintenance. In the next chapter, the benefits of software stability are described and demonstrated, which can in turn facilitate greater code reuse and repurposing.

NOTES

1. ISO/IEC 25010:2011. *Systems and software engineering — Systems and software Quality Requirements and Evaluation (SQuaRE)—System and software quality models*. Geneva, Switzerland: International Organization for Standardization and Institute of Electrical and Electronics Engineers.
2. ISO/IEC/IEEE 24765:2010. *Systems and software engineering—Vocabulary*. Geneva, Switzerland: International Organization for Standardization, International Electrotechnical Commission, and Institute of Electrical and Electronics Engineers.
3. ISO/IEC 25051:2014. *Software engineering—Systems and software Quality Requirements and Evaluation (SQuaRE)—Requirements for quality of Ready to Use Software Product (RUSP) and instructions for testing*. Geneva, Switzerland: International Organization for Standardization.
4. Id.
5. ISO/IEC/IEEE 24765:2010.
6. Id.
7. Id.
8. Id.
9. Id.
10. Id.
11. Id.
12. Id.
13. Id.
14. Id.
15. Id.
16. Id.

CHAPTER 17

Stability

Backpacking 8,000 miles throughout Central and South America, you learn to take very little for granted except that change is constant.

Sleeping in so-called *primera clase* (first class) buses seemed cramped and unforgiving—that is, until you've spent the night on the cement floor of an actual bus terminal. However, it wasn't the occasional squalid conditions but rather the constant variability that at times grew exhausting.

While El Salvador and Ecuador use exclusively U.S. currency, the first stop in other countries was always to exchange or withdraw local money. Upon departing a country, I often had so little currency (or such small denominations) that it wasn't worth exchanging the remainder, so I began amassing wads of small bills and bags of coins that I'd squirrel away.

In many cities I was only able to spend a single night—the hostel little more than a secure place to sleep. There is the frisson of arrival, not knowing whom you'll encounter and what thrilling adventures you'll share for a night, but never a sense of stability.

Private rooms, shared rooms, dorm rooms—often a surprise until you arrive, as the act of reserving a bed sometimes carries more wishful intention than actual binding contract. Most hostels did have Wi-Fi, although often only dial-up service on which 50 guests were simultaneously trying to Skype to parts unknown.

Nearly all hostels had wooden boxes or metal gym lockers in which valuables could be securely stowed, but each time I was convinced that I'd finally purchased a lock that universally fit all lockers, I was surprised to find one that didn't. But of course, for only US$5—sometimes half the price of the room—any hostel will sell you a lock that fits their brand of lockers.

Cell phone carriers and cell service were just as diverse—Tigo in Guatemala and El Salvador, Claro in Colombia, Movistar in Peru. I was switching out SIM cards and cell phones as fast as a narcotrafficker. And because charging equipment always matched the country's power supply, it was not uncommon to see three and four converters and adapters creatively connected to charge a single phone.

When I finally arrived in Buenos Aires, having looped through Patagonia and trekked across the Pampas, I checked into a hostel and didn't leave the city for a week until my flight to the States. While I may have missed a rugged adventure or two, I was content to be confined to the Paris of South America, and the one currency, one bed, one shower, one Wi-Fi, one locker, and one cell phone were refreshing. The lack of change ... was a welcome change.

■ ■ ■

By the time I had reached Buenos Aires, I had amassed five locks, four cell phones, several adapters and converters, two pounds of coinage, and more than 100 Wi-Fi passwords saved in my laptop. While these all had fond memories associated with them, they collectively represented 8,000 miles of variability, inefficiency, and risk.

I've since purchased a universal adapter/converter and global cell phone coverage, and I travel with an array of locks. A goal in software development should also be to seek code stability. While some variability is inevitable—like changing money at border crossings—software maintenance can often be avoided through flexible design, development, and testing.

One cost of variability is the associated inefficiency. When it's 10 PM and you're roaming the streets trying to find an open *tienda* (store) to purchase a lock, this isn't an efficient use of time, and it's financially inefficient when the *tienda* can only sell you a five-pack of locks. Software that's not stable often requires corrective and emergency maintenance at the worst possible times, derailing other development priorities and inefficiently diverting personnel.

Another cost of variability is risk—plug an unknown adapter into an unknown power source in a dark dorm room in a new country, and wait for the blue arc to light up the night. We're instinctively drawn to the familiar because it's safe and, just as I grew to trust my surroundings more in Buenos Aires during my last week of travels, software stakeholders grow to trust software that has reliably executed for some period without need for maintenance.

Because we loathe risk, we take actions to mitigate it. In hostels, you can't immediately trust the 19 BFFs with whom you're sharing quarters, so you buy a lock, stow your pack, tuck your passport between your legs, and crash for the night with one eye open. Software testing and validation also act to mitigate risk by exposing vulnerabilities in software and ensuring that software ultimately meets needs and requirements. If software is modified after testing and validation, and thus is unstable, risk returns until retesting and revalidation can occur.

DEFINING STABILITY

Stability can be defined as the degree to which software can remain static, resisting modification or maintenance while continuing to exhibit undiminishing quality. Thus, stability describes the continuing ability of software to meet needs and requirements but does not represent neglected software that is not maintained and whose relative performance and quality decrease over time. To resist preventative and corrective maintenance, stable software should be free of errors and defects. And to resist adaptive maintenance, stable software should flexibly adapt to its environment and other sources of variability to the extent possible. Somewhat paradoxically, the more flexible software can be made, the more stable it can remain over time.

Software stability is so integral to traditional software development that it's often an unspoken expectation. Where developers and users are separate and distinct, software modification requires reentry into the software development life cycle (SDLC) to test, validate, and release each software patch or update to users. Because of this level of effort, cost, and complexity, software stability is required and modifications are typically made only through planned, periodic updates. Within end-user development environments, however, software can be modified daily or hourly as desired by end-user developers because they typically have unfettered access to the underlying code. While this flexibility has certain advantages, continual modification tremendously diminishes the benefits of testing and validation within the SDLC and reduces the integrity of software.

In many ways, stability represents the gateway from software development to software operation; once software is stable, it can be automated, scheduled, and left to perform reliably without further interaction. Moreover, because stable software first requires testing to demonstrate and validate its functionality and performance, the resultant integrity encourages code reuse and repurposing in other software. This chapter demonstrates methods that can be used to implement flexibility within SAS software, including the SAS macro language, Autoexec.sas file, configuration files, and external text files and control tables, all of which can dynamically drive performance, affording users the ability to customize software operation without modification to its code.

ACHIEVING STABILITY

Stability defines resistance to change. One definition within literature references software stability as *invariability in software output or results*. In this sense, stability represents a facet of reliability that describes consistent functionality and thus dynamic rather than static performance. This sense of stability is described in the "Consistency" section in Chapter 4, "Reliability," but is not further discussed in this chapter. Thus, the term *stability*, as used throughout this text, describes static code that resists the need for modification or maintenance while continuing to meet needs and requirements.

As described in Chapter 1, "Introduction," stable software requires stable requirements, defect-free code, and adaptive flexibility. The lack of any one of these elements can result in unstable software that inefficiently requires frequent maintenance. If maintenance is not performed on unstable software, reliability can plummet as software either begins to fail or fails to meet established requirements. This unfortunate relationship too often pits reliability against stability as mutually exclusive or opposing pursuits; however, with appropriate planning, design, development, and testing, both dimensions of quality can be achieved.

STABLE REQUIREMENTS

Stable technical requirements are a luxury afforded very few software development projects, especially within environments that espouse Agile development methods and principles. Even in Waterfall development environments that favor heavy initial planning and design to militate against later design changes, requirements are often modified once software is in production, either to improve performance or deliver additional functionality. For example, a SAS analytic module might be developed to be run on one data stream but, after two months of successful performance, customers might require that it be modified to support additional data streams.

The developers might not have been able to predict this repurposing, so the requirements change doesn't represent a defect. Once the requirement is implemented, however, adaptive maintenance must repurpose the software to encompass its expanded objective. Thus, the extent to which requirements can be finalized during the design phase of the SDLC will facilitate stability in software and efficiency in the software development environment.

Software stability is sometimes conceptualized as being more closely aligned with Waterfall development methodologies that espouse big design up front (BDUF). After all, the Agile Manifesto does value "responding to change over following a plan."[1] Despite a commitment to flexibility, Agile methodologies still expect software requirements to remain stable for the duration of each iteration. For example, while the road ahead may be tortuous and evolving based on new objectives and business opportunity, developers always have a clear, stable vision of their current workload in the present iteration.

Formalizing written requirements aids stability because it memorializes the intent, objective, and technical specifications of software. Requirements provide developers not only with functional intent but also the degree of quality to which software should be developed. Moreover, once development has been completed, stakeholders can compare the software product against technical requirements and, when implemented, a formalized test plan. Without formalized requirements, software projects can suffer from *scope creep* in which software never seems to be complete because new functionality or performance requirements are continually added.

DEFECT-FREE CODE

Defect-free code arguably can never be achieved but, where high-performing software is required, a formalized implementation of the SDLC can improve software performance. The extent to which software can be intentionally designed, developed, and tested in a planned rather than ad hoc, frenetic fashion will always support fewer defects and errors. Software testing is key to facilitating code accuracy and, where a formalized test plan exists, to further validating software against testing requirements.

Because a formalized test plan describes testable elements that have been agreed upon by developers, customers, and possibly other stakeholders, successful *test plan completion* can unambiguously demonstrate *software completion*. But even during software planning and design, discussion about test plans and test cases can elucidate ambiguity in needs and requirements that could have led to errors in later software development. Thus, when metrics are proposed that will later be utilized to measure the success of technical specifications, customers can ensure early in the SDLC whether those requirements will be able to discriminate adequately between software success and failure.

Software upgrades and updates are common components of the SDLC operations and maintenance (O&M) phase. Users of more traditional software applications are accustomed to patches and other downloads that regularly install behind the scenes to correct defects, improve security, or provide additional performance or functionality. However, the extent to which software updates are required for corrective maintenance diminishes time that developers could have invested in other development activities. By adequately testing software and reducing corrective maintenance, greater code stability can be achieved.

DYNAMIC FLEXIBILITY

That flexible, dynamic software facilitates software stability may at first appear to be a misnomer but in fact rigid code is highly unstable. In SAS software, flexibility is often achieved through the SAS macro language, which dynamically writes Base SAS code, enabling software to be more efficiently developed. Most notably, exception handling that utilizes SAS macros can flexibly respond to changes in the environment, in data throughput, and to other shifting needs and requirements.

The implementation of flexibility through SAS macros typically occurs on a continuum. At one end, rigid software is hardcoded without flexibility in mind. For example, the following DATA step and FREQ procedure are hardcoded:

```
data temp;
   length char1 $10;
   char1='pupusas';
run;

proc freq data=temp;
   tables char1;
run;
```

Flexibility can be implemented by representing dynamic aspects of software as malleable macro variables. This middle-of-the-road approach to flexibility is common in SAS software because it enables SAS practitioners to modify functionality (and possibly performance) through the assignment of macro variables, often conveniently located in software headers or parameterized through macro invocations:

```
%macro doit(dsn=, var=);
data &dsn;
   length &var $10;
   &var='pupusas';
run;

proc freq data=&dsn;
   tables &var;
run;
%mend;

%doit(dsn=temp, var=char1);
```

For many software purposes this degree of flexibility is sufficient; however, where software must truly remain static, additional flexibility can be gained by removing parameter assignments from code and passing them externally. The following code can be saved as C:\perm\doit.sas and invoked through a batch job that utilizes the SYSPARM option to pass parameters for the data set name and variable name:

```
%let dsn=%scan(&sysparm,1,,S);
%let var=%scan(&sysparm,2,,S);

%macro doit(dsn=, var=);
data &dsn;
   length &var $10;
   &var='pupusas';
run;

proc freq data=&dsn;
   tables &var;
run;
%mend;

%doit(dsn=&dsn, var=&var);
```

Now, because all sources of variability occur outside the module, it can be saved in a software reuse library in this stable format. Various users and programs can call the batch job and, through dynamic invocation with the SYSPARM parameter, pass any values to represent the data set name and variable name without software modification. One rationale for this more dynamic code would be cases in which a checksum value is being generated after software testing to validate that software is not changed in the future, as demonstrated in the "Toward Software Stability" sections in Chapter 11, "Security." The assignment of macro variables through external files is further demonstrated later in this chapter in the "Custom Configuration Files" section.

Modularity

The remainder of this chapter demonstrates the role that modularity plays in supporting software stability. Modularity decomposes software into discrete chunks that can be developed and tested independently but which ultimately must unite to form a cohesive whole. Because modules are functionally discrete and perform only one or a limited number of actions, they are individually more likely to resist change. For example, if software is decomposed from a single, monolithic program into five separate modules saved as distinct programs, in many cases, maintenance can be performed on one module while leaving the others stable.

This approach dramatically increases stability because four of the five modules can remain unchanged. Additionally, though functional or unit testing would be required for the modified module, and integration testing would be necessary to ensure that module continued to function correctly with the other four modules, those unchanged modules wouldn't need to be tested individually because their integrity would be intact. This overall increased stability facilitates more rapid modification and more efficient maintenance.

STABILITY AND BEYOND

The following sections demonstrate the role that software modularity plays in supporting stability and in turn reusability and extensibility. The %DEL_VARS macro removes all character variables from a data set and simulates a typical child process that would be called from a parent:

```
* deletes all character variables *;
%macro del_vars(dsn= /* data set name in LIB.DSN or DSN format */);
%local vars;
%local dsid;
%local i;
%local vartype;
%local close;
%local varlist;
%let dsid=%sysfunc(open(&dsn, i));
%let vars=%sysfunc(attrn(&dsid, nvars));
%do i=1 %to &vars;
   %if %sysfunc(vartype(&dsid,&i))=N %then %do;
      %let varlist=&varlist %sysfunc(varname(&dsid,&i));
      %end;
   %end;
%let close=%sysfunc(close(&dsid));
data &dsn (keep=&varlist);
   set &dsn;
run;
%mend;
```

To demonstrate the %DEL_VARS macro, the following DATA step and macro invocation simulates use of %DEL_VARS in a parent process, %ETL. In the second invocation of the PRINT procedure, the character variables CHAR1 and CHAR2 have been removed from the Final data set:

```
data final;
   length char1 $10 char2 $10 num1 8 num2 8;
   char1='lower';
   char2='case';
   num1=5;
   num2=0;
   output;
run;

%macro etl(dsn=);
proc print data=&dsn;
run;
%del_vars(dsn=&dsn);
proc print data=&dsn;
run;
%mend;

%etl(dsn=final);
```

In this scenario, the %ETL macro is intended to simulate more extensive production software that has been tested, validated, automated through batch processing, and scheduled to run recurrently. When the %ETL macro is executed, the Final data set is modified so that only numeric variables are retained. These capabilities are discussed and expanded in the following sections.

Stability Promoting Reusability

Because the %DEL_VARS macro in the "Stability and Beyond" section has been tested and validated, it could be removed from the body of the %ETL program and saved to a shared location so that all SAS practitioners within a team or organization can access and utilize it. SAS software can later reference the %DEL_VARS macro with the SAS Autocall Macro Facility or with the %INCLUDE statement.

If SAS practitioners later needed to create a quantitative analysis process that relied on only numeric variables, they could implement the %DEL_VARS macro to reduce input/output (I/O) resources wasted on processing unneeded character variables. Because %DEL_VARS provides this functionality and represents stable, tested code, %DEL_VARS could be implemented with ease. The following code simulates the analytical process with the MEANS procedure.

```
data final;
   length char1 $10 char2 $10 num1 8 num2 8;
   char1='lower';
   char2='case';
   num1=5;
   num2=0;
   output;
run;

%macro analysis(dsn=);
%del_vars(dsn=&dsn);
proc means data=&dsn;
run;
%mend;

%analysis(dsn=final);
```

Note that in this example, the child process %DEL_VARS did not need to be modified, so no unit testing was required, only integration testing to determine how the macro interacts with the parent process. Like the child process %DEL_VARS, SAS macros and modules will be more likely to be reused when they represent independent, modular, tested, stable components.

Stability Promoting Extensibility

Continuing the scenario depicted in the "Stability Promoting Reusability" section, SAS practitioners have been successfully implementing the %DEL_VARS module in various software products for a few months when a developer notes that its functionality could be expanded to retain only numeric variables as well. This is a common example of software extension in which added functionality is desired and, rather than creating new software from scratch, existing code is expanded.

Whenever possible, it's important to maintain backward compatibility to current uses of the module, so design and testing should attempt to leave the default functionality unchanged. In this case, the module should still be able to remove all character variables from a data set. *Software extensibility* speaks to repurposing software or building something new from something old. Where extensibility can also incorporate backward compatibility, however, this benefits software development because new functionality can be added while maintaining a single module or code base.

Extensibility and backward compatibility are achieved and demonstrated in the upgraded %DEL_VARS macro:

```
* deletes either all character or all numeric variables *;
%macro del_vars(dsn= /* data set name in LIB.DSN or DSN format */,
   type=C /* C to drop character vars, N to drop numeric */);
%local vars;
%local dsid;
%local i;
%local vartype;
%local close;
%local varlist;
%let dsid=%sysfunc(open(&dsn, i));
%let vars=%sysfunc(attrn(&dsid, nvars));
%do i=1 %to &vars;
   %if %sysfunc(vartype(&dsid,&i))=%upcase(&type) %then %do;
      %let varlist=&varlist %sysfunc(varname(&dsid,&i));
      %end;
   %end;
%let close=%sysfunc(close(&dsid));
data &dsn (drop=&varlist);
   set &dsn;
run;
%mend;

data final;
   length char1 $10 char2 $10 num1 8 num2 8;
   char1='lower';
   char2='case';
   num1=5;
   num2=0;
   output;
run;

%macro etl(dsn=);
proc print data=&dsn;
run;
%del_vars(dsn=&dsn, type=N);
```

```
proc print data=&dsn;
run;
%mend;

%etl(dsn=final);
```

The new %DEL_VARS macro is now *overloaded,* in that it additionally accepts the optional TYPE parameter—denoted by C or N—to drop only character or numeric variables. Thus, new invocations of %DEL_VARS can take advantage of the additional functionality of the TYPE parameter, while existent invocations that don't include TYPE—such as the example demonstrated in the "Stability Promoting Reusability" section—will default to deleting only character variables when TYPE is absent, thus maintaining past functionality. Overloading is demonstrated in the "Regression Testing" section in Chapter 16, "Testability," and often facilitates backward compatibility in macros that increase in scope over time, thus promoting software stability.

If the %DEL_VARS macro functionality could not have been extended successfully while ensuring backward compatibility (as was accomplished in this scenario), an important decision would have had to have been made. The original version of the module could have remained intact, thus supporting all current uses in various programs, while a new macro could have been created as a new module to meet the new functional requirement. Or, the macro still could have been modified to meet the new functional requirement, but this would have invalidated its other current uses.

Thus, where backward compatibility cannot be achieved, developers must essentially choose between code stability and maintainability. They can modify two separate (but similar) code bases, which incurs future inefficiency every time maintenance is required, or they can modify all current programs that reference the changed module, albeit being able to maintain a single software module going forward. Each method will be warranted in certain circumstances so the costs and benefits must be weighed.

Stability Promoting Longevity

Software longevity is introduced in the "Longevity" section in Chapter 4, "Reliability," and even the most meager software requirements should state an expected software lifespan. Intended lifespan enables developers to understand whether they are designing some ephemeral, fly-by-night software to solve a tactical problem or building an enduring, strategic solution. When software is intended to be enduring, the degree to which software can be maintained with minimal modifications can be facilitated with stable, reusable code.

Continuing the scenario from the "Stability Promoting Extensibility" section, the revised %DEL_VARS module has now been reused multiple times within ETL software, and its functionality has been extended so that either character or numeric variables can be retained. Because the module is stable and has gained significant popularity, it is more likely to be appropriately maintained because effort invested in its maintenance benefits numerous dependent software products or projects. If developers instead had built a new macro (to delete only numeric variables), this macro in addition to %DEL_VARS would have had to have been separately maintained, thus effectively decreasing the value of each. Because maintenance is critical to enduring software, the extent to which new SAS software can be built using existing, tested, trusted, stable modules will facilitate software longevity of the modules being reused.

Stability Promoting Efficiency

Another benefit of stable code is that when advancements are made to a software module, those changes can made in one location yet benefit all programs that call the module. For example, the %DEL_VARS macro as defined is inefficient because regardless of whether any variables are removed, the data set still is read and rewritten, requiring I/O resources. But what occurs if the code attempts to eliminate character variables when none exist? The module unnecessarily recreates the original data set, retaining all variables.

A more efficient solution would instead identify similar cases in which all variables were being maintained—because the original data sets were either all numeric or all character—and skip the gratuitous DATA step. This logic is conveyed in a subsequent version of %DEL_VARS:

```
* deletes either all character or all numeric variables *;
%macro del_vars(dsn= /* data set name in LIB.DSN or DSN format */,
   type=C /* C to drop character vars, N to drop numeric */);
%local vars;
%local dsid;
%local i;
%local vartype;
%local close;
%local varlist;
```

```
%local empty;
%let empty=YES;
%let dsid=%sysfunc(open(&dsn, i));
%let vars=%sysfunc(attrn(&dsid, nvars));
%do i=1 %to &vars;
   %if %sysfunc(vartype(&dsid,&i))=%upcase(&type) %then %do;
      %let varlist=&varlist %sysfunc(varname(&dsid,&i));
      %let empty=NO;
      %end;
   %end;
%let close=%sysfunc(close(&dsid));
%if &empty=NO %then %do;
   data &dsn (drop=&varlist);
      set &dsn;
   run;
   %end;
%mend;

data final;
   length char1 $10 char2 $10 num1 8 num2 8;
   char1='lower';
   char2='case';
   num1=5;
   num2=0;
   output;
run;
%macro etl(dsn=);
proc print data=&dsn;
run;
%del_vars(dsn=&dsn, type=C);
proc print data=&dsn;
run;
%mend;

%etl(dsn=final);
```

Had the module not been stable and tested, the decision to correct the inefficiency through perfective maintenance might not have been prioritized; however, because the module was reused in this and other software, the decision to maintain the module and make it more efficient was more readily supported. Stable, reusable code will generally be more valued (and thus more likely to be refactored) because its value is construed not only from one software product but also from multiple programs that implement it.

But one more glaring inefficiency exists in the software, which results from violation of modular software design. Because the %DEL_VARS macro not only generates the space-delimited list of variables to remove (&VARLIST) but also implements the DATA step, it is neither functionally discrete nor loosely coupled. The DATA step inside the %DEL_VARS macro is gratuitous because &VARLIST could be implemented directly within the MEANS procedure. This inefficiency is demonstrated in the "Stability Promoting Reusability" section:

```
%macro analysis(dsn=);
%del_vars(dsn=&dsn);
proc means data=&dsn;
run;
%mend;
```

To remedy this inefficiency, while ensuring that the code remains backward compatible, the following %DEL_VARS macro requires three modifications. First, the DEL parameter should be added to the macro definition to demonstrate that the macro can optionally generate only the &VARLIST macro variable without deleting those variables within the gratuitous DATA step. The second change requires that &VARLIST be changed from a local to a global macro variable to facilitate access by the parent process. And third, the conditional logic that evaluates whether to execute the gratuitous DATA step should examine the value of the DEL parameter. These changes are finalized in the following code:

```
* deletes either all character or all numeric variables *;
%macro del_vars(dsn= /* data set name in LIB.DSN or DSN format */,
   type=C /* C to drop character vars, N to drop numeric */,
   del=Y /* Y to delete from data set, N to generate VARLIST only */);
%local vars;
%local dsid;
%local i;
%local vartype;
%local close;
%global varlist;
%local empty;
%let empty=YES;
%let dsid=%sysfunc(open(&dsn, i));
%let vars=%sysfunc(attrn(&dsid, nvars));
```

```
%do i=1 %to &vars;
   %if %sysfunc(vartype(&dsid,&i))=%upcase(&type) %then %do;
      %let varlist=&varlist %sysfunc(varname(&dsid,&i));
         %let empty=NO;
      %end;
   %end;
%let close=%sysfunc(close(&dsid));
%if &empty=NO and &del=Y %then %do;
   data &dsn (drop=&varlist);
      set &dsn;
   run;
   %end;
%mend;
```

Executing the following statements now again removes all character variables from the second invocation of PRINT. But because the DROP statement occurs within the PRINT procedure rather than in the %DEL_VARS macro, this demonstrates discrete functionality while avoiding the gratuitous DATA step in newer programs that reference this macro:

```
data final;
   length char1 $10 char2 $10 num1 8 num2 8;
   char1='lower';
   char2='case';
   num1=5;
   num2=0;
   output;
run;

%macro etl(dsn=);
proc print data=&dsn;
run;
%del_vars(dsn=&dsn, type=C, del=N);
proc print data=&dsn (drop=&varlist);
run;
%mend;

%etl(dsn=final);
```

Through overloading, the revised %DEL_VARS macro is now more efficient but is also backward compatible to two former manifestations—the original usage with only the DSN parameter and the second usage with DSN and TYPE parameters. This can be an effective way to build software while supporting stability; however, original modular design should be incorporated for this type of extensibility to be effective. Because the gratuitous DATA step was trapped in the original %DEL_VARS macro, it remains as long as backward compatibility to the original intent of this macro is maintained. While stability should be an objective, the quest for stability should never outweigh common sense. Thus, when truly archaic, inefficient, or insecure former manifestations of software modules exist, their deprecation is often favored in lieu of continuing to support awkward or endless backward compatibility.

Stability Promoting Security

The "I" in the CIA information security triad (confidentiality, integrity, and availability) demonstrates the importance of integrity. Information should be accurate, and where possible, a data pedigree should be maintained that can trace data to its origin, much like the forensic documentation of crime scene evidence that maintains chain of custody. Data integrity is a common objective in data analytic development because business value is derived from data products, thus their accuracy is critical.

However, because *information* describes not only data but also the software that undergirds, manipulates, and analyzes it, information security policies should also be considered for software. In applications development, security policies help to ensure that software is not maliciously modified, such as through computer viruses, code injections, or other attacks. Because data analytic software products are more likely to be used internally, their attack surfaces are significantly reduced. In other words, most SAS practitioners don't write software with the forethought that it will be intentionally attacked. They instead rely on systems security, authentication, physical controls, and other methods to defend against malicious threats.

Even in data analytic software, however, a realistic threat to software integrity exists, not from malice, but from unintentional code modification or from intentional code modification that causes unintended ill effects. Software versioning exists in part for this reason—to ensure that the correct copy of software is being modified, tested, integrated, and executed. When a new version is created, past versions are archived and should remain stable unless they are required for recovery purposes. Archival stability is critical because it timestamps and provides a historical record of software. For example, if developers wanted to turn back time and run software with the functionality and performance that it exhibited two months prior, the versioned software—if archived with the necessary data and other files—would provide the integrity and stability to accomplish this.

In many environments that utilize software versioning, the integrity of the current software version is conveyed through a brief examination of the code or a timestamp representing when it was last saved. Far greater integrity can be achieved, however, by generating checksums on stable SAS program files after software testing and validation. Checksums utilize cryptographic hash functions to perform bit-level analysis of SAS programs (or any files) to create a hexadecimal string that uniquely represents each file. By obtaining a checksum on software before it executes and by validating this checksum against its baseline value, stakeholders can be assured that the software executing is secure and has not been modified since testing and validation. This technique is demonstrated in the "Checksums" section in Chapter 11, "Security."

MODULARIZING MORE THAN MACROS

While modularity is a hallmark of stable software, the examples thus far have demonstrated using modules only within a functional sense. Thus, a parent process calls a child process—a SAS macro—to perform some function after which program flow returns to the parent. While this is a common use of modular design that supports stability, it only scratches the surface of ways in which modularity can facilitate stability.

For example, the following sections simulate an ETL process that ingests, cleans, transforms, and displays data. The initial code will be improved upon throughout the next several sections to demonstrate ways in which modularity and flexibility can be infused into software to remove dynamic elements from code and thereby increase software stability. To improve readability, no exception handling is demonstrated; however, in robust production software, numerous vulnerabilities would need to be rectified through an exception handling framework.

The following three sections of code configure options, load data, and parse those data, and are saved as C:\perm\candy.sas. In the first section of code, SAS system options are defined and two SAS libraries are initialized, PERM and OUTPUT. If an option needs to be modified or added, however, the software must be changed. Furthermore, if the logical location of either library needs to be modified—for example, from C:\perm to C:\permenant—the software must also be changed. More dynamic initialization methods are demonstrated in the later "Autoexec.sas" section:

```
options dlcreatedir nosymbolgen nomprint nomlogic;
libname perm 'c:\perm';
%let outputloc=c:\perm\output;
libname output "&outputloc";
```

The next module simulates data ingestion in ETL software and creates the PERM.Raw_candy data set. Intentional spelling variations of the CANDY values are included:

```
data perm.raw_candy;
   infile datalines delimiter=',';
   length candy $20;
   input candy $;
   format candy $20.;
   datalines;
Almond Joy
Almond Joys
5th Avenue
Fifth avenue
Reese's
Reeses
Reeces
KitKat
Kitkat
Kit Kat
Snickers
Twixx
Twix
;
```

The third module, %CANDY_CLEANER, simulates a quality control step that cleans and standardizes data:

```
* ingests candy data from data set;
%macro candy_cleaner(dsn= /* data set in LIB or LIB.DSN format */);
data perm.clean_candy;
   length clean_candy $20;
   set &dsn;
   * standardize spelling variations;
   if strip(lowcase(candy)) in ("almond joy","almond joys") then
      clean_candy="Almond Joy";
```

```
   else if strip(lowcase(candy)) in ("5th avenue","fifth avenue") then
      clean_candy="5th Avenue";
   else if strip(lowcase(candy)) in ("reese's","reeses","reeces") then
      clean_candy="Reese's";
   else if strip(lowcase(candy)) in ("kitkat","kit kat") then
      clean_candy="KitKat";
   else if strip(lowcase(candy)) in ("snickers") then
      clean_candy="Snickers";
   else if strip(lowcase(candy)) in ("twix","twixx") then
      clean_candy="Twix";
   else clean_candy="UNK";
run;
%mend;
```

It's obvious that with more diverse or voluminous data, this quality control method could grow untenably large, complex, and error-prone. More importantly, however, as new values are encountered in the data and must be binned into the correct, standardized format, the code must be modified each time. Even if a more flexible quality control method (such as PERL regular expressions) were implemented within SAS, those expressions would still need to be maintained in the software itself. Because of these limitations, this module is made more dynamic in the "Business Logic" section later in this chapter.

The FORMAT procedure creates a format that categorizes the candy based on primary ingredients, but it's clear that the raw values are statically tied to the previous quality control conditional logic, so changes in that logic likely would cause the FORMAT procedure to fail outright or produce invalid results. And because the FORMAT procedure is hardcoded in the software itself, any modifications to the format logic require software modification. For example, if an analyst wanted to classify candy by texture rather than ingredients, how would this be accomplished? Moreover, what if two competing candy models were required to be maintained at the same time to satisfy the needs of distinct stakeholders? This functionality is made significantly more flexible later in the "SAS Formatting" section:

```
proc format;
   value $ candy_class
   "Almond Joy" = "chocolate"
   "5th Avenue" = "chocolate and caramel"
   "Reese's" = "chocolate and peanut butter"
   "KitKat" = "chocolate"
   "Snickers" = "chocolate and caramel"
   "Twix" = "chocolate and caramel";
run;
```

The %CANDY_DISPLAY module organizes candy by its classification and name and displays the quantity of candy by name in an HTML report. To highlight the most commonly occurring candy, candies having a frequency greater than two are highlighted in red through stoplight reporting within the REPORT procedure. But what if an analyst wanted to use a different color for highlighting, to add a title, or to save the report to a different location? Each of these customizations would require software modification:

```
* creates an HTML report that organizes and counts candy by
   classification;
%macro candy_display(dsn= /* data set in LIB or LIB.DSN format */);
ods listing close;
ods noproctitle;
ods html path="&outputloc" file="candy.htm" style=sketch;
proc report data=&dsn style(report) = [foreground=black
   backgroundcolor=black] style(header) = [font_size=2
   backgroundcolor=white foreground=black];
   title;
   column clean_candy=candy_class clean_candy n dummy;
   define candy_class / group "Classification" format=$candy_class.;
   define clean_candy / group "Candy";
   define n / "Count";
   define dummy / computed noprint "";
   compute dummy;
      if _c3_>2 then call define("_c3_","style",
         "style=[backgroundcolor=light red]");
      endcomp;
run;
ods html close;
ods listing;
%mend;

%candy_cleaner(dsn=perm.raw_candy);
%candy_display(dsn=perm.clean_candy);
```

As demonstrated, myriad subtle customizations would require software to be updated, and in some instances, competing customizations would inefficiently require multiple instances of the software to be maintained. Thus, by predicting ways in which software may need to be customized while in operation, developers can code dynamism and flexibility into SAS software. The following sections demonstrate incremental changes to this code that further modularize this single program into multiple files that can better support software stability. While maintenance or customization still will require modifications, those changes will be made only to ancillary files rather than the core SAS software itself, which should remain stable and secure to the extent possible.

Autoexec.sas

The Autoexec.sas file is an optional file that can be utilized to initialize a SAS session, specifying system options or executing SAS code before any programs execute. Autoexec.sas is described in detail in the *SAS 9.4 Companion for Windows*, and is introduced here to demonstrate its importance in stabilizing software.[2] In SAS Display Manager for Windows, the default location for Autoexec.sas is C:\Program Files\SASHome\SASFoundation\9.4, but the default can vary by environment, SAS interface, and server configuration. In many environments, this directory will be restricted to specific users, so the SAS application also searches other locations for the Autoexec.sas file. The AUTOEXEC option can be invoked from the command prompt to start a batch session and specify the location of a specific Autoexec.sas file, thus overcoming this restriction.

To demonstrate this technique, the following customized Autoexec.sas file can be saved to C:\perm\autoexec.sas:

```
libname perm 'c:\perm';
options dlcreatedir nosymbolgen nomprint nomlogic;
```

The primary Autoexec.sas file utilized by SAS is unchanged; however, when a batch invocation specifies the AUTOEXEC option and references C:\perm\autoexec.sas, the new file is invoked rather than the standard SAS Autoexec.sas file, ensuring that necessary libraries are assigned and system options initialized. SAS system options that cannot be modified during an active SAS session can be set within the Autoexec.sas file.

The SAS program Candy.sas now no longer requires the LIBNAME or OPTIONS statements because they appear in the Autoexec.sas file. As a golden rule, customization or initialization meant to be applied to all SAS software within an environment can be done in the primary Autoexec.sas file, while options specific to one program should be done via a separate Autoexec.sas file or configuration file, as demonstrated in the following batch invocation:

```
"c:\program files\sashome\sasfoundation\9.4\sas.exe" -sysin
   c:\perm\candy.sas -autoexec c:\perm\autoexec.sas
   -noterminal -nosplash -log c:\perm\candy.log -dlcreatedir
```

Removing the LIBNAME statement from the software and placing it in Autoexec.sas allows the code to be more flexible in the event that the logical location changes. For example, the PERM library might reference a SAS server IP address, and if this address is changed, it can be modified in a single location within Autoexec.sas, allowing all SAS software to remain stable while still referencing the new location. In environments where metadata are enforced, it is always a best practice to define SAS libraries permanently within metadata structures so they are available for use in all software.

Business Logic

The standardization of data is a common theme in SAS data analytic development, for example, where quality control techniques eliminate abbreviations, acronyms, spelling errors, and other variations that can occur in categorical data. Too often, however, this is accomplished through unmaintainable, hardcoded logic that requires modification each time the data model must be updated.

An objective in modular software design should be to remove business rules (or data models) from the software so that changes to one don't necessitate changes to the other. In addition to cleaning and standardizing values, this logic can additionally include organization of values into a matrix or hierarchy. To remove logic from software, it should be operationalized into a structure that can be parsed, often a SAS data set, Excel spreadsheet, XML file, or plain text file. The following comma-delimited text file, saved to C:\perm\candy_cleaning.txt, lists the standardized value in the first position with additional spelling variations following to the right. Text files are straightforward and arguably the most easily maintained representations of simple data models:

```
Almond Joy,almond joy,almond joys
5th Avenue,5th avenue,fifth avenue
Reese's,reese's,reeses,reeces
KitKat,kitkat,kit kat
Snickers,snickers
Twix,twix,twixx
```

The %CANDY_CLEANER macro next utilizes I/O functions to open the text file within the DATA step and place spelling variations into an array VARS which is compared to actual values found in the Raw_candy data set. This code could have been operationalized in numerous ways, including %SYSFUNC calls to I/O functions or a multidimensional array to capture all variations for all candy types:

```
* ingests candy data from data set;
%macro candy_cleaner(dsn= /* data set in LIB or LIB.DSN format */);
filename ref "c:\perm\candy_cleaning.txt";
data perm.clean_candy (keep=candy clean_candy);
   length clean_candy $20;
   set &dsn;
   * standardize spelling variations;
   length var $100 var2 $30;
   f=fopen("ref");
   rc1=fsep(f,",");
   do while(fread(f)=0);
      rc3=fget(f,var);
      array vars{5} $30 _temporary_;
      do i=1 to dim(vars);
         vars{i}="";
         end;
      i=0;
      do while(fget(f,var2)=0);
         i=i+1;
         vars{i}=var2;
         end;
      if strip(lowcase(candy)) in vars then clean_candy=var;
      end;
   if missing(clean_candy) then clean_candy="UNK";
run;
%mend;

%candy_cleaner(dsn=perm.raw_candy);
```

One advantage of using strictly I/O functions to read external data models is that because values are never converted to macro variables, they can include special characters and complex quoting without having to specially handle these values in SAS. And, while IF–THEN conditional logic has been utilized to improve code readability, software performance could likely be gained through the FORMAT procedure or a hash object lookup table.

This pursuit of excellence is exactly what underscores the importance separating business logic from the software engine itself. SAS practitioners desiring to refactor this software and implement a more efficient hash object lookup to clean and standardize data could do so by modifying only the %CANDY_CLEANER module, without ever touching the data model, thus promoting its stability and security. On the other hand, analysts who need to modify only the data model and its logic but do not want to modify the software could do so without the fear of accidentally modifying %CANDY_CLEANER. The two elements are separate and distinct.

Moreover, if analysts maintain different or competing data models, an endless number of models can be swapped in and out of the software simply by changing the SAS FILENAME reference. Thus, an even more modular and flexible solution would require the data model (currently hardcoded as C:\perm\candy_cleaning.txt in the software) to be passed as a parameter so that data models could be changed without touching the code. Especially within empirical and research environments in which data models are expected to be agile, this modular method supports this flexibility objective without compromising the integrity of the underlying software. In differentiated teams in which developers specialize in software and analysts focus on data analysis, this design model allows developers to separately maintain software while analysts separately maintain data models, thus facilitating the stability and security of each.

SAS Formatting

SAS formats represent a common method to clean and standardize data values. The following FORMAT procedure can largely be copied from the original business rules within the %CANDY_CLEANER module. When it is saved to a separate SAS program and referenced from %CANDY_CLEANER with the %INCLUDE statement, the formatting also represents a modular solution that can drive business rules from beyond the code itself:

```
proc format;
   value $ clean_candy
   "almond joy","almond joys" = "Almond Joy"
   "5th avenue","fifth avenue" = "5th Avenue"
   "reese's","reeses","reeces" = "Reese's"
   "kitkat","kit kat" = "KitKat"
   "snickers" = "Snickers"
   "twix","twixx" = "Twix"
   other = "UNK";
run;
```

Because of the necessary quotations, the maintenance of formats is slightly higher than that of raw text files but, as demonstrated later, the implementation can be dramatically more straightforward. The following macro represents an alternative %CANDY_CLEANER module to the one presented earlier in the "Business Logic" section. Note that when formatting is intended to permanently clean or correct data (as opposed to changing its format temporarily for display purposes), a separate variable should often be created to ensure that the format is not accidentally later removed or modified. This also facilitates comparison between the original (unformatted) and transformed (formatted) variables:

```
%macro candy_cleaner(dsn= /* data set in LIB or LIB.DSN format */);
data perm.clean_candy;
   set &dsn;
   length clean_candy $30;
   clean_candy=put(lowcase(candy),$clean_candy.);
run;
%mend;
```

In other cases, formatting is intended to change the representation of data only temporarily for display. For this purpose, the FORMAT function can be applied to the variables directly in REPORT, PRINT, MEANS, FREQ, or other procedures, only modifying the variable display for the duration of the procedure.

The CNTLIN option in the FORMAT procedure allows a control table (i.e., a SAS data set) to be used to populate a SAS format but, unfortunately, is quite restrictive in its implementation. For example, the CNTLIN option cannot accommodate a format such as CANDY_CLASS that aims to group multiple character strings into a single value. To remedy this, a modular, reusable solution can be implemented that transforms comma-delimited text files into SAS formats. This is similar to the methodology demonstrated in the "Business Logic" section but, rather than dynamically creating conditional logic, it dynamically creates a SAS format.

To begin, the business logic is saved to the comma-delimited text file C:\perm\format_candy_ordering.txt. As previously, the first item represents the categorical value to which subsequent values will be transformed. Thus, when Almond Joy or KitKat appears in the data, it will be classified as chocolate in the HTML report when the CANDY_CLASS format is invoked:

```
chocolate, Almond Joy, KitKat
chocolate and caramel, 5th Avenue, Snickers, Twix
chocolate and peanut butter, Reese's
```

Now, rather than hardcoding the REPORT procedure as before, this is done dynamically in a reusable macro, %INGEST_FORMAT, that requires parameters for the control table and the format name:

```
* creates a SAS format that categorizes or bins character data;
%macro ingest_format(fil= /* location of comma-delimited format file */,
   form= /* SAS character format being created */);
filename zzzzzzzz "&fil";
data ingest_format (keep=FMTNAME START LABEL);
   length FMTNAME $32 START $50 LABEL $50;
   fmtname="&form";
   f=fopen("zzzzzzzz");
   rc1=fsep(f,",");
   do while(fread(f)=0);
      rc3=fget(f,label);
      label=strip(label);
      do while(fget(f,start)=0);
         start=strip(start);
         output;
         end;
      end;
run;
%mend;
```

The CANDY_CLASS format that is created is identical to the one created earlier in the “Modularizing More Than Macros” section, with the sole exception that the OTHER option does not specify a default value when the base value cannot be found in the format. If the OTHER option is desired, this can be added with one additional line in the Format_candy_ordering.txt file and logic to parse this to create an OTHER option in the %INGEST_FORMAT macro. Each of these two different methodologies removes business rules and data models from SAS code, thus facilitating stability of SAS production software. In the next section, additional customization will be delivered through a configuration file.

Custom Configuration Files

Custom configuration files are one of the strongest advocates to flexible code that facilitates software stability. For example, in the original Candy.sas program, the location for the HTML report is hardcoded; to change this, modification of the code is necessary. To instead assign this library and logical location dynamically, the &OUTPUTLOC macro variable should be assigned in the configuration file. Thus, the &OUTPUTLOC assignment is removed from the code and the LIBNAME statement now stands alone:

```
libname output "&outputloc";
```

A second hardcoded element is the report name. Because the report is always saved as Candy.htm, each software execution overwrites previous reports. In the future, SAS practitioners might instead want to incrementally name these reports so newer reports don't overwrite older ones, or separate analysts might want to run separate copies of the software concurrently. Thus, the second modification is to assign the report file name dynamically through the software invocation:

```
ods html path="&outputloc" file="&outputfile" style=sketch;
```

A third hardcoded attribute that could benefit from flexibility is the stoplight threshold and the stoplight color inside the COMPUTE statement of the REPORT procedure. In making these elements dynamic, analysts will be able to subtly modify the style and intent of the report without having to modify any of the underlying SAS code—just the associated configuration file:

```
if _c3_>&thresh then call define("_c3_","style",
   "style=[backgroundcolor=&highlight]");
```

To represent these customizations, the text configuration file C:\perm\config_candy.txt is created:

```
This configuration file is required to execute the candy.sas program.
Last updated 12-21 by TMH


Note that the OUTPUT_FILE should NOT include the file path.


<OUTPUT_LIBRARY>
c:\perm\output
<OUTPUT_FILE>
candy_output.html
<REPORT_STYLE>
thresh: 2
highlight: light brown
```

Three configuration categories (e.g., <OUTPUT_LIBRARY>) are specified, depicted in uppercase, and bracketed. Subsumed under each category are either raw data to be transferred to a macro variable or tokenized values to be further parsed. For example, under <REPORT_STYLE>, the values for &THRESH and &HIGHLIGHT are included and can be read in any order.

The configuration file is read via the %READ_CONFIG macro, which parses the data, initializes the OUTPUT library, and assigns global macro variables &OUTPUTLOC, &THRESH, and &HIGHLIGHT to be used in Candy.sas. The %READ_CONFIG macro should be launched initially in the software, just as the SAS application initially reads its own configuration file at startup:

```
* reads a configuration file necessary for Candy.sas execution;
%macro read_config(fil= /* logical location of configuration file */);
%global outputloc;
%let outputloc=;
%global outputfile;
%let outputfile=;
%global thresh;
%let thresh=;
%global highlight;
%let highlight=;
data _null_;
   length tab $100 category $8;
   infile "&fil" truncover;
   input tab $100.;
   if strip(upcase(tab))="<OUTPUT_LIBRARY>" then category="lib";
   else if strip(upcase(tab))="<OUTPUT_FILE>" then category="fil";
   else if upcase(tab)="<REPORT_STYLE>" then category="rpt";
   else do;
      if category="lib" then call execute('libname output
         "' || strip(tab) || '";');
      else if category="fil" then call symput("outputfile",strip(tab));
      else if category="rpt" then do;
         if strip(lowcase(scan(tab,1,":")))="thresh" then call
            symput("thresh",strip(scan(tab,2,":")));
         if strip(lowcase(scan(tab,1,":")))="highlight" then call
            symput("highlight",strip(scan(tab,2,":")));
         end;
      end;
   retain category;
run;
%mend;

%read_config(fil=c:\perm\config_candy.txt);
```

The previous demonstration intentionally contains no exception handling to promote readability. However, in actual configuration files, exception handling is critical to capture input errors that may have been made when creating or modifying the file. One substantial benefit of configuration files is the simplicity with which they can be built, understood, and modified. However, even with appropriate documentation inside configuration files, their strength is also their weakness, due to the vulnerability of improper modification that can cause corresponding software to crash.

For example, if the <REPORT_STYLE> header were omitted, or either the THRESH or HIGHLIGHT parameters were omitted, these parameters would not be read, causing null values to be assigned to their respective global macro variables. To overcome this vulnerability, the following code can be inserted immediately before the %MEND statement, thus testing the length of the macros &THRESH and &HIGHLIGHT and assigning default values if the macros were not assigned:

```
%if %length(&thresh)=0 %then %let thresh=2;
%if %length(&highlight)=0 %then %let highlight=light red;
```

While overcoming these two vulnerabilities, multiple other vulnerabilities exist that would need to be mitigated in production software. For example, what if the value for the output location does not exist? What if a character value rather than numeric is given for the THRESH parameter? What if the HIGHLIGHT parameter includes a non-color value? And, finally, what happens when special characters are detected within a parameter?

Once these vulnerabilities are overcome, however, the software will be more robust and stable while enabling modification through the configuration file alone. With this added stability, both flexibility and security are increased: analysts can make minor, stylistic changes to data products without ever having to open or modify the production software.

Putting It All Together

While the resultant code is longer, more complex, and less readable, it offers flexibility that enables SAS practitioners to customize execution and output without modifying their software. In environments that require stable code—for example, those which have authoritarian change control policies, formalized testing, and validation—removing dynamic elements from software and extrapolating them to external Autoexec.sas files, configuration files, control tables, and data models will facilitate software stability.

The following code represents the updated software which also requires the external files C:\perm\autoexec.sas, C:\perm\config_candy.txt, C:\perm\candy_cleaning.txt, and C:\perm\format_candy_ordering.txt:

```
%include "c:\perm\autoexec.sas";

data perm.raw_candy;
   infile datalines delimiter=',';
   length candy $20;
   input candy $;
   format candy $20.;
   datalines;
Almond Joy
Almond Joys
5th Avenue
Fifth avenue
Reese's
Reeses
Reeces
KitKat
Kitkat
Kit Kat
Snickers
Twixx
Twix
;

* reads a configuration file necessary for Candy.sas execution;
%macro read_config(fil= /* logical location of configuration file */);
%global outputloc;
%let outputloc=;
%global outputfile;
%let outputfile=;
%global thresh;
%let thresh=;
%global highlight;
%let highlight=;
data _null_;
   length tab $100 category $8;
   infile "&fil" truncover;
   input tab $100.;
```

```
   if strip(upcase(tab))="<OUTPUT_LIBRARY>" then category="lib";
   else if strip(upcase(tab))="<OUTPUT_FILE>" then category="fil";
   else if upcase(tab)="<REPORT_STYLE>" then category="rpt";
   else do;
      if category="lib" then call execute('libname output
         "' || strip(tab) || '";');
      else if category="fil" then call symput("outputfile",strip(tab));
      else if category="rpt" then do;
         if strip(lowcase(scan(tab,1,":")))="thresh" then call
            symput("thresh",strip(scan(tab,2,":")));
         if strip(lowcase(scan(tab,1,":")))="highlight" then call
            symput("highlight",strip(scan(tab,2,":")));
         end;
      end;
   retain category;
run;
%if %length(&thresh)=0 %then %let thresh=2;
%if %length(&highlight)=0 %then %let highlight=light red;
%mend;

* ingests candy data from data set;
%macro candy_cleaner(dsn= /* data set in LIB or LIB.DSN format */);
filename ref "c:\perm\candy_cleaning.txt";
data perm.clean_candy (keep=candy clean_candy);
   length clean_candy $20;
   set &dsn;
   * standardize spelling variations;
   length var $100 var2 $30;
   f=fopen("ref");
   rc1=fsep(f,",");
   do while(fread(f)=0);
      rc3=fget(f,var);
      array vars{5} $30 _temporary_;
      do i=1 to dim(vars);
         vars{i}="";
         end;
      i=0;
      do while(fget(f,var2)=0);
         i=i+1;
         vars{i}=var2;
         end;
      if strip(lowcase(candy)) in vars then clean_candy=var;
```

```
      end;
   if missing(clean_candy) then clean_candy="UNK";
run;
%mend;

* creates a SAS format that categorizes or bins character data;
* must include at least two comma-delimited items;
* the first of which is the new format and the second (or additional);
* which will be converted to the first value;
%macro ingest_format(fil= /* logical location of format file */,
   form= /* SAS character format being created */);
filename zzzzzzzz "&fil";
data ingest_format (keep=FMTNAME START LABEL);
   length FMTNAME $32 START $50 LABEL $50;
   fmtname="&form";
   f=fopen("zzzzzzzz");
   rc1=fsep(f,",");
   do while(fread(f)=0);
      rc3=fget(f,label);
      label=strip(label);
      do while(fget(f,start)=0);
         start=strip(start);
         output;
         end;
      end;
run;
proc format library=work cntlin=ingest_format;
run;
%mend;

* creates an HTML report that organizes and counts candy by
   classification;
%macro candy_printer(dsn= /* data set in LIB or LIB.DSN format */);
ods listing close;
ods noproctitle;
ods html path="&outputloc" file="&outputfile" style=sketch;
proc report data=&dsn style(report) = [foreground=black
   backgroundcolor=black] style(header) = [font_size=2
   backgroundcolor=white foreground=black];
   title;
```

```
    column clean_candy=candy_class clean_candy n dummy;
    define candy_class / group "Classification" format=$candy_class.;
    define clean_candy / group "Candy";
    define n / "Count";
    define dummy / computed noprint "";
    compute dummy;
       if _c3_>&thresh then call define("_c3_","style",
          "style=[backgroundcolor=&highlight]");
       endcomp;
run;
ods html close;
ods listing;
%mend;

%read_config(fil=c:\perm\config_candy.txt);
%ingest_format(fil=c:\perm\format_candy_ordering.txt, form=$candy_class);
%candy_cleaner(dsn=perm.raw_candy);
%candy_printer(dsn=perm.clean_candy);
```

WHAT'S NEXT?

Software stability represents the gateway toward software automation and scheduling because stable, tested software can be reliably executed while defending against unnecessary maintenance. Another benefit of increased stability is the likelihood that software modules will be reused and repurposed in the future. The next chapter introduces software reuse and repurposing and describes reusability principles and artifacts that can more effectively promote software reuse within a team and organization.

NOTES

1. The Agile Alliance, *The Manifesto for Agile Software Development*. The Agile Alliance. Retrieved from www.agilealliance.org/the-alliance/the-agile-manifesto.
2. *SAS® 9.4 Companion for Windows, Fourth Edition*. Retrieved from http://support.sas.com/documentation/cdl/en/hostwin/67962/HTML/default/viewer.htm#titlepage.htm.

CHAPTER 18
Reusability

Roasting marshmallows in flowing lava ... bucket list—CHECK!

Antigua, Guatemala. Back in my home away from home, my friend Lisa and I had already toured every cathedral in town, so we decided to take a day trip to the Pacaya volcano, the most accessible active volcano to the city.

Pacaya last gained fame when it erupted in March 2014, sending plumes of fire and smoke thousands of feet in the air and raining down ash throughout central Guatemala. But when not erupting, it's a great place for a hike and picnic.

We bought our bus tickets and, while doing some research online waiting for the bus, I read an offhanded blog post from someone claiming to have roasted marshmallows in one of Pacaya's accessible lava flows. *Hmm ...*

This seemed outlandish—certainly nothing the U.S. Park Service had allowed me to do in Hawaii—but on the off chance, we darted over to the *supermercado* (supermarket) and found not only marshmallows, but also graham crackers and chocolate bars. Sulfur s'mores, here we come!

The school bus (a chicken bus in training) took an hour and a half to wind its way toward the base of the mountain, where we disembarked and set out on the gravelly path. All signs of life and vegetation gave way to rough, volcanic terrain, seemingly smooth in the distance, but painfully unforgiving to the touch. And then, over a crest, distinct patches of red—lava!

There must have been a warning about melting the soles of your feet, that lava can be sharp as glass, that sulfur clouds can kill you, but I was too busy playing with the dog that had bounded up beside me, no doubt because he thought I was pulling treats out of my pack. I was—but the treats weren't for him.

Finding a stick for the marshmallows proved easy, but standing a couple feet from an active lava flow was a feat of endurance. I would wonder later—looking at the photos while I still nursed my blistered knuckles—why I hadn't thought to wrap my shirt (tied around my waist) around my hand, which was in danger of spontaneously combusting.

I burned enough marshmallows for Lisa and me and, as the only s'more representatives of the day, distributed the remaining supplies to the gaggle of tourists.

With a mix of disappointment and horror, we watched as every other tourist paid unaccompanied local children to go burn their hands (rather than roasting marshmallows themselves)—but the kids did get their fill of s'mores, and all survived.

■ ■ ■

Reuse comes in all shapes and sizes, requiring various levels of effort and input to implement. I only reused a crazy notion from a blog—taking s'more fixin's to an active volcano—to achieve a goal I didn't realize was even possible. In software development, code reuse often similarly occurs where technical methods or capabilities are abstracted—if not actually copied—from external sources including texts, white papers, and SAS documentation.

Other reuse can require less or no effort at all. The lazy tourists benefited from the s'more fixin's that Lisa and I had purchased and hauled to Pacaya. Moreover, in a classic build-versus-buy (or build-versus-burn) decision, the tourists further outsourced marshmallow roasting to seven-year-olds. Especially where reusable software modules are maintained within an organization, the objective of software reuse is to facilitate painless implementation of past functionality into future software.

Although a couple tourists did seem discombobulated by the notion of squishing mallow, chocolate, and graham into unwholesome goodness, most immediately began salivating at the sight of the ingredients—you know s'mores are incredible because you've tasted (tested) them your entire life. Software testing also greatly encourages software reuse because software integrity and popularity increase as stakeholders understand and expect its strengths and weaknesses. And s'mores have no weaknesses!

Reuse is also benefited by modularity. I hadn't brought s'mores to the party—I had brought s'more *fixin's*, the primordial ingredients of mallow, chocolate, and graham. So when one guy only wanted a roasted marshmallow, and a woman just wanted to chomp on raw chocolate (rather anticlimactically, given the opportunity at hand), these independent needs could also be fulfilled. Modular software design similarly increases reusability because smaller, functionally discrete components are more likely to be reused and repurposed.

As code reuse becomes more common within an organization and the clear benefits of reuse more recognizable, incorporation of reusability principles into software design begins to occur naturally. In my travels, I've similarly learned always to consult travel blogs for other unique experiences that I can borrow. Here's a hint—smuggle bananas into Machu Picchu if you want a herd of llamas to befriend and follow you relentlessly! Llamas love bananas!

DEFINING REUSABILITY

Reusability is the "degree to which an asset can be used in more than one system, or in building other assets."[1] The International Organization for Standardization (ISO) goes on to define *reusable software* as "a software product developed for one use but having other uses, or one developed specifically to be usable on multiple projects or in multiple roles on one project." Primary software reuse incorporates internally developed code into future software products by the same developer, team, or organization. Secondary reuse further encompasses adoption and use of external software, including code and technical methods derived from textbooks, white papers, software documentation, and external teams and organizations. Therefore, software that exhibits reusability principles will encourage primary reuse by its own development team or organization, and secondary reuse by external teams or organizations.

Software reuse is a commonly sought objective in software development because it can substantially improve the speed and efficiency of the development process. Where tested, stable modules of code exist that can be immediately dropped into software, significant design, development, and testing effort can be eliminated and invested elsewhere. The aim of reusability is to develop software that is intended to be reused, rather than software that is reused solely through ad hoc or serendipitous implementation. Software reusability typically incurs a higher cost up front to design, develop, test, and validate. While this additional effort may not benefit the initial implementation of software, it represents an investment that pays dividends through future software reuse. In many cases, however, reusability principles—such as modular software design—can immediately benefit functional and performance objectives in the initial software use.

This chapter introduces software reuse, including two knowledge management artifacts—the reuse library and reuse catalog—that can benefit a team or organization by documenting and organizing reusable code and reuse implementations. It further explores the requirements of reusability and demonstrates through successive SAS examples how flexibility, modularity, readability, testability, and stability can individually and collectively improve software reusability. Moreover, where code modules are centrally located but universally implemented, the chapter demonstrates how software maintainability is benefited when a single version of software can be maintained while implemented across multiple software products.

REUSE

Software reuse occurs constantly and oftentimes unconsciously in software development. Whole products and snippets of code, as well as knowledge of their underlying concepts, design, techniques, and logic, are regularly incorporated into software as it's developed. When external sources of code such as Base SAS documentation, white papers, or other technical sources are all considered, it becomes virtually impossible to develop software without some reuse component. While code reuse alone is often touted as an objective of software development, the net value of code reuse will depend on the extent to which code can be efficiently and effectively incorporated.

Software reuse in fact can be a painful, inefficient, and ungratifying experience when the quality of software or its documentation is lacking, or when reuse is implemented without planning. By incorporating reusability principles into software, developers incur a greater up-front cost due to increased focus on dimensions of quality, but can be rewarded with painless future reuse of software modules. For example, by encouraging software testability during development, SAS practitioners can more effortlessly and effectively reuse code because it will have been more thoroughly tested. Reusability principles often benefit not only the software for which the code was originally written, but also future software that can reuse the modules. To be clear, however, a significant portion of software is functionally specific and not adaptable for software reuse.

Software use is an inherent prerequisite of software reuse. Thus, before software can be reused, it should have already passed through the software development life cycle (SDLC) design, development, testing, and validation phases. When software is reused, it can bypass much of the SDLC, making second and subsequent implementations more efficient than the original. A second prerequisite of software reuse is an awareness of the initial use, which can include an understanding of software objectives, strengths, weaknesses, vulnerabilities, demonstrated quality, testing results, and performance metrics. Armed with this information, developers can intelligently compare the benefits and risks of software reuse with the costs of new design and development.

From the User Perspective

From the user's perspective, software reuse describes the adoption or purchase of whole software products, such as open-source, commercial-off-the-shelf (COTS), or government-off-the-shelf (GOTS) software. Some software products are domain-specific, and users understand their limited scope and use. For example, a customer buys architectural software so he can remodel his kitchen but, once the remodel is complete, may believe he has no future use for the software. However, when he whimsically decides to dig a 10,000-gallon pond, he realizes that the software also has landscape design functionality and can be reused. Thus, even in niche markets, the extent to which software can flexibly meet diverse needs of users will increase its reuse potential.

With the exception of open-source products that end-users intend to modify, most users purchase software "as is" with the understanding that its function is controlled solely by software developers. Periodic software maintenance—released to users through patches or updates—can further facilitate software reuse by extending the software lifespan. Moreover, software reuse is encouraged where regular upgrades expand software functionality, responding to shifting customer needs and competing for market share against other new and expanding software products.

From the Developer Perspective

From the developer's perspective, software reuse describes the incorporation of existing modules of code into software as it's being developed. Rather than accepting or rejecting software as a whole, developers are able to realize and benefit from its composite nature, retaining and reusing useful elements and discarding the chaff. The extent to which existing software can be incorporated into later software will be determined by its functionality in addition to its flexibility, generalizability, reliability, and other performance attributes. Furthermore, software documentation, including demonstration of thorough and successful software testing, can instill confidence in developers that software will be of sufficient quality for its intended reuse.

The user and developer perspectives of reuse differ substantially. Users view architectural software products in terms of their ultimate functionality—the ability to create architectural designs and, in the case of some products, to additionally design superfluous water features. Developers instead view software functionality as a set of discrete components. For example, the architectural software may include a list of the nine most recently used floorplan files and display these in a drop-down menu. A developer designing unrelated music editing software might be able to reuse this recent file functionality to recreate an identical menu item in the music software. Thus, while end users might perceive no similarities between the overall functionality of these wildly divergent software applications, a developer can view disparate software products and functionally decompose them into the component level to extract functionality and code that can be reused in dissimilar software products. Functional decomposition is further discussed in the "Functional Decomposition" section in Chapter 14, "Modularity."

Reuse of External Code

Code can be acquired from external sources, such as textbooks, technical white papers, software documentation, blogs, or user forums. If you're lucky, the developers had reuse in mind when they designed and developed the software and instilled reusability principles. Perhaps they included documentation that outlines software functionality as well as vulnerabilities, enabling subsequent developers to better understand the intent and quality of the software and thus make a more informed decision about whether and how to reuse it. Software modularity also supports reuse because subsequent developers are able to pick and choose functionally discrete components without having to separate them like conjoined twins.

Knowledge of reusability principles benefits SAS practitioners as they incorporate external code because they can more accurately assess the level of effort required to modify, integrate, test, and implement the code. Imagine the experienced general contractor that you've contacted to give you an estimate for a bathroom remodel. With ease, he should be able to establish which components can remain, which can be modified, and which must be replaced. It might be possible to keep the antique claw-footed tub in place and remodel around it, but in other instances, this could prove to be so awkward and inefficient that it would be less expensive to jackhammer it, remove the pieces, and install a replacement.

Thus, while developers can't improve the quality or reusability of software that has already been developed externally, their knowledge of reusability principles can benefit code incorporation and reuse. As skilled artisans, savvy SAS practitioners should be able to review external code and more accurately decide whether and how to best reuse software. Moreover, to those who have suffered through the incorporation of software that lacked reusability principles, the distinction and benefits of reusable software will be evident and valued.

Reuse of Internal Code

In each of the previous reuse examples, software developers evaluated and reused code but were powerless to influence its reusability because the software had already been produced by some third party. However, as SAS practitioners develop new software, incorporation of reusability principles encourages reuse and can benefit not only current but also future software products.

The primary benefits of reuse of internal code are the same as external code—a reduction in design, development, and testing effort, since the code has already passed through the SDLC. But one of the primary benefits of internal code reuse is the ability to plan and anticipate the reuse. When a SAS practitioner searches online for a solution to a complex REPORT procedure that he is trying to create but doesn't want to engineer from scratch, he might find the solution spread across multiple white papers. He ingests the knowledge and techniques and, in an ad hoc fashion, formulates and empirically tests the software until the requirements are met.

This ad hoc reuse method contrasts sharply with intentional code reuse, in which anticipated future use of code may be planned even before the initial development or use has occurred. For example, in a planning and design meeting, a team might discuss the development of a specific data cleaning algorithm for use in an extract-transform-load (ETL) infrastructure. One SAS practitioner might realize that he could also utilize the cleaning module in a separate SAS analytic program being developed, and thus request that the module be developed to be as dynamic as possible. While this extra dynamism might not directly benefit the ETL software for which it was originally intended, it could dramatically increase the efficiency with which the later analytic software is developed.

In other cases, software modules are developed and reused within the same software, thus enabling reuse to immediately benefit the original software product for which the module was developed. This occurs most commonly with low-level functionality, such as basic I/O functions or basic data operations. For example, if during software planning it's decided that SAS software will need to be robust to missing or locked data sets, a reusable macro could be developed to confirm data set existence and availability. Especially by functionally decomposing software objectives into discrete, modular chunks, SAS practitioners will be better poised to identify software functionality that can benefit from reuse (from existing modules) as well as to implement reusability principles within software to support eventual reuse.

Reuse Artifacts

While this text focuses on software quality rather than the quality of the software development environment or specific software development methodologies, some development artifacts and best practices are so beneficial that they beg inclusion. Software reuse libraries and reuse catalogs provide a proven method to organize, discover, and archive software modules within an organization, facilitating security, reusability, and overall maintainability of code. For example, without some knowledge management method to track in which software products and in what ways modules of code have been reused, it becomes impossible to maintain the intent and integrity of code modules as their function may morph or expand over time.

Reuse Library

A *reuse library* is "a classified collection of assets that allows searching, browsing, and extracting."[2] An *asset* can represent software itself, accompanying documentation or, more abstractly, even software frameworks or templates. The library is not classified in the national security sense (requiring a polygraph to peruse it), but rather in the sense that it is structured and organized—think Dewey Decimal, not Treadstone or Blackbriar. Thus, a reuse library will include instances of modular, readable, tested, stable software that are sufficiently organized and documented to ensure rapid discovery and comprehension. Additional accompanying documentation such as test plans, test cases, or test results may also be included.

Myriad knowledge management software applications exist which organize software in reuse libraries. They range from exceedingly complex to mid-range solutions that are more generic, such as SharePoint, to well-organized folder structures saved on a server. In a SAS environment, a reuse library could be as simple as a segregated portion of the directory structure that ensures sufficient security and stability of its contents, so long as it yields search and retrieval capabilities. If SAS practitioners don't have at least read-only access permissions to the reuse library or can't locate software efficiently within it, the utility of the library is eliminated.

In some cases, production code could be run directly from a reuse library, but in a development environment that espouses software modularity, most modules of code won't be intended to be run in isolation, but rather called by parent processes. Thus, you wouldn't expect to find an ETL system codified within the reuse library because it would have been developed for a specific purpose. However, that ETL software might utilize multiple SAS macros that are contained within and called from the reuse library. For example, a macro that converts a comma-delimited text file into a SAS format would have value outside of the ETL software, so it would ideally have been dynamically created sufficiently to be inducted into the reuse library. Once the macro module would have been completed, tested, stabilized, and saved as a distinct SAS program file, it could have been included in the reuse library so it could benefit future software.

The following code, reprised from the "SAS Formatting" section in Chapter 17, "Stability," creates SAS character formats from comma-delimited text files:

```
*<DESC> creates a SAS format that categorizes or bins character data;
* must include at least two comma-delimited items;
* the first of which is the new format and the second (or additional);
* which will be converted to the first value;
%macro ingest_format(fil= /* logical location of format file */,
   form= /* SAS character format being created */);
filename zzzzzzzz "&fil";
data ingest_format (keep=FMTNAME START LABEL);
   length FMTNAME $32 START $50 LABEL $50;
   fmtname="&form";
   f=fopen("zzzzzzzz");
   rc1=fsep(f,",");
   do while(fread(f)=0);
      rc3=fget(f,label);
      label=strip(label);
      do while(fget(f,start)=0);
         start=strip(start);
         output;
         end;
      end;
run;
proc format library=work cntlin=ingest_format;
run;
%mend;
```

The extent to which duplication of development can be avoided is paramount because it reduces developer effort. First, it reduces the initial development process because code only needs to be written and tested once, and second, it substantially improves maintainability of code because only one copy of code needs to be modified, retested, and reintegrated into software when maintenance is required. In an end-user development environment that is not implementing separate knowledge management software to organize a formal reuse library, teams can facilitate code organization and search capabilities by including code descriptions in module headers. Where software header formats are standardized and parsed by automated processes, software reusability will be substantially improved.

For example, if comments are prefaced with *<DESC> or other symbolism, rather than using only a nondescript single asterisk, a code parser can quickly ingest and parse all software within a reuse library and automatically create sufficient documentation to allow rapid search, discovery, and comprehension of all modules. In this way, developers are able to differentiate comments that should be included in the reuse library to facilitate search and retrieval from those that should be viewed only from within the software itself. This type of automated comment parsing is demonstrated in the "Automated Comment Extraction" section in Chapter 15, "Readability."

A reuse library can include not only code but also templates such as common software requirements that are regularly implemented within code. For example, each time a data set is referenced in Base SAS, multiple assumptions must be met, such as the existence of the data set, accessibility of the data set, and valid structure of the data set. Threats exist when these assumptions fail to be met; a list of common threats to the DATA step is enumerated in the "Specific Threats" section in Chapter 6, "Robustness." By identifying and preserving these threats as threat templates within a reuse library, SAS practitioners can reuse these templates to identify threats in future software products.

Less common residents of reuse libraries include test data sets used in test plans to demonstrate that software meets functional and performance requirements. By formalizing common test data sets, test plans can reference these test data sets, thus ensuring that they adequately cover necessary use cases for data, and that developers do not waste time recreating test data sets. Moreover, if vulnerabilities are later discovered in test data sets, such as the identification of software use cases that were not adequately represented, all software that relied on those data sets for testing can be retested to ensure they still meet test requirements when the additional test data are utilized.

Reuse Catalog

A *reuse catalog* is "a set of descriptions of assets with a reference or pointer to where the assets are actually."[3] When the asset (typically a software module) is stored in a reuse library, the reuse catalog represents a brief description of the intent and content of the module and enumerates every instance in which the module is implemented within software. Reuse catalogs often contain basic information, including:

- Code module name
- Description
- Requirements (if module is coupled with other modules or data)
- Inputs (if coupled to parameters or data sets)
- Vulnerabilities

- Authors
- File location
- Creation date
- Modification date
- Test date
- Program file checksum
- List of all programs or modules that utilize the module

Software creation dates, modification dates, and test dates lend credibility to code modules by demonstrating adherence to the SDLC and testing methods. Combined with a description of the software and known vulnerabilities (if included), this information enables SAS practitioners to search and investigate existing code modules quickly to determine not only their relevance but also their relative quality and risk. The optional checksum utilizes a hash function to create a cryptographic representation of the SAS program file. In addition to inspecting modification and test dates, SAS practitioners can further compare checksum values between the software reuse catalog and other copies of the code that may exist, thus further validating software integrity. This quality control method is discussed in the "Checksums" section in Chapter 11, "Security."

One of the most important aspects of the reuse catalog is the comprehensive listing of all SAS software programs that implement each code module. This listing not only provides a metric by which primary software reuse can be calculated within a team or organization, but also ensures that when a module must be modified, all known uses are identified. Without this record, SAS practitioners might modify a core software module for one purpose and program, yet disrupt that module's prior function in unrelated software products. Thus, by tracking all uses (and reuses) of shared and centrally managed software modules, SAS practitioners can ensure that modifications maintain backward compatibility. And again, with thorough and standardized commenting within software modules, reuse libraries and catalogs can (and should) be generated automatically through programs that ingest and parse SAS program files.

REUSABILITY

Software reuse can make the software development process more efficient because redundant software design, development, and testing are eliminated. Reuse, however, too often occurs in an unplanned manner without central management. For example, chunks of software might be internally plundered from other existent software but, rather than maintaining a single copy of the software within a central repository, no record is maintained to link the two uses of the software. If improvements or other modifications are made to one module, they won't benefit its twin, and divergence creeps into the environment, which thwarts maintainability.

Because software reuse involves and can benefit multiple software products, reuse is more commonly implemented at the team or organizational levels. Thus, reuse artifacts such as the reuse library or reuse catalog may be adopted as a best practice within an organization to ensure that SAS practitioners both register reusable code modules and investigate the reuse library (to see if any existing modules would be beneficial) before beginning to develop new software.

Reusability principles further ensure that, to the extent possible, software modules can be efficiently reused within either the original program or future software. By infusing these principles into software design and development, and in coordination with reuse artifacts, SAS practitioners can author software that will be more likely to be reused without software modification. These principles incorporate other static performance attributes and are discussed in the following sections.

Flexibility

Reusable code must be flexible—rigid, hardcoded SAS software is more difficult if not impossible to reuse. Within Base SAS, the SAS macro language most commonly facilitates flexibility by enabling macro code that dynamically writes Base SAS code. Code that writes code for you—that sounds like an idea that those lazy, lava-fearing tourists could get behind!

Consider the scenario in which a developer needs to convert all character variables in a specific data set to uppercase. The Sample data set is created in the following code, which includes character variables Char1 and Char2, after which a hardcoded DATA step manually converts the variables to uppercase:

```
data sample;
   length char1 $20 char2 $20 num1 8 num2 8;
   char1="I love SAS";
   char2="SAS loves me";
run;

data uppersample;
   set sample;
   char1=upcase(char1);
   char2=upcase(char2);
run;
```

The code is straightforward, sufficient and, at least on this scale, efficient to write. But, as the number of variables increases, developers might want to implement a more scalable solution that dynamically converts all variables to uppercase. A dynamic solution, moreover, would be valuable if the variable names were subject to change. And, central to reusability, a dynamic solution would enable future software products to utilize the module irrespective of the number or names of variables or the name of the data set.

The %GOBIG macro dynamically determines all variable types and converts character variables to uppercase.

```
%macro gobig(dsnin= /* old data set in LIB.DSN or DSN format */,
   dsnout= /* updated data set in LIB.DSN or DSN format */);
%local dsid;
%local vars;
%local vartype;
%local varlist;
%local i;
%let varlist=;
%let dsid=%sysfunc(open(&dsnin,i));
%let vars=%sysfunc(attrn(&dsid, nvars));
%do i=1 %to &vars;
   %let vartype=%sysfunc(vartype(&dsid,&i));
   %if &vartype=C %then %let varlist=&varlist
      %sysfunc(varname(&dsid,&i));
   %end;
%let close=%sysfunc(close(&dsid));
%put VARLIST: &varlist;
data &dsnout;
   set &dsnin;
   %let i=1;
   %do %while(%length(%scan(&varlist,&i,,S))>1);
      %scan(&varlist,&i,,S)=upcase(%scan(&varlist,&i,,S));
      %let i=%eval(&i+1);
      %end;
run;
%mend;

%gobig(dsnin=sample, dsnout=uppersample);
```

The %GOBIG macro can be run on any data set, irrespective of the quantity and names of variables. This represents a much more flexible solution, but it still can be improved by infusing additional reusability principles, as demonstrated in the following sections.

Modularity

The smaller the module, the more likely it is to be reused. One of the central principles of modular software design is discrete functionality: the objective that a module should do one and only one thing. This principle indirectly limits the size of the modules, because *one thing* can only occupy so much space. Moreover, discrete functionality maximizes module generalizability because a smaller component can be flexibly placed into more software.

The %GOBIG macro is somewhat modular because it can be removed from SAS code, saved to a SAS program file, and included via the SAS Autocall Macro Facility or an %INCLUDE statement. However, its functionally is not discrete because %GOBIG opens an I/O stream to the data set, creates a list of character variables, closes the stream, and finally performs the DATA step. This functionality could be further decomposed into a separate module that only creates a list of all character variables within a data set. This module could subsequently be used to populate the variables in a data set DROP or KEEP statement dynamically.

To improve modularity by making %GOBIG more functionally discrete, the following code uncouples the function that identifies all character variables from the function that converts them to uppercase. Moreover, flexibility can also be improved by enabling &VARLIST to be created with all numeric variables, all character variables, or all variables of any type:

```
* creates a space-delimited macro variable VARLIST in data set DSN;
%macro findvars(dsn= /* data set in LIB.DSN or DSN format */,
   type=ALL /* ALL, CHAR, or NUM to retrieve those variables */);
%local dsid;
%local vars;
%local vartype;
%global varlist;
%let varlist=;
%local i;
%let dsid=%sysfunc(open(&dsn,i));
%let vars=%sysfunc(attrn(&dsid, nvars));
```

```
%do i=1 %to &vars;
   %let vartype=%sysfunc(vartype(&dsid,&i));
   %if %upcase(&type)=ALL or (&vartype=N and %upcase(&type)=NUM) or
         (&vartype=C and %upcase(&type)=CHAR) %then %do;
      %let varlist=&varlist %sysfunc(varname(&dsid,&i));
      %end;
   %end;
%let close=%sysfunc(close(&dsid));
%mend;

* dynamically changes all character variables in a data set
   to upper case;
%macro gobig(dsnin= /* old data set in LIB.DSN or DSN format */,
   dsnout= /* updated data set in LIB.DSN or DSN format */);
%local i;
%findvars(dsn=&dsnin, type=CHAR);
data &dsnout;
   set &dsnin;
   %let i=1;
   %do %while(%length(%scan(&varlist,&i,,S))>1);
      %scan(&varlist,&i,,S)=upcase(%scan(&varlist,&i,,S));
      %let i=%eval(&i+1);
      %end;
run;
%mend;

%gobig(dsnin=sample, dsnout=uppersample);
```

The %GOBIG macro invocation presumably occurs within some larger body of code—a parent process that calls %GOBIG as its child. Thus, an added benefit of the increased modularity is backward compatibility, since the macro invocation is identical to that in the previous "Flexibility" section. The functionality of the %FINDVARS macro has also been improved and now reads the TYPE parameter to determine whether all character, all numeric, or all variables of any type will be saved to the &VARLIST macro variable.

The second principle of modular design specifies that software modules should be loosely coupled. The %GOBIG macro, however, includes a DATA step which restricts the ways in which the module can be implemented. For example, consider that the initial objective of the software was not only to convert character variables to uppercase but also to perform other functions within a DATA step. The following DATA step simulates additional functionality by arbitrarily assigning NUM1 the value of 99:

```
data uppersample;
   set sample;
   char1=upcase(char1);
   char2=upcase(char2);
   num1=99;
run;
```

Because the current %GOBIG module includes a DATA step, the num1=99 statement must be included in a separate DATA step, which inefficiently requires a gratuitous DATA step given the current macro configuration. In other words, any macro that includes a DATA step cannot be run from within another DATA step. The following code demonstrates the lack of loose coupling and the gratuitous DATA step:

```
%gobig(dsnin=sample, dsnout=uppersample);

data uppersample;
   set uppersample;
   num=99;
run;
```

A more efficient and less coupled solution would enable %GOBIG to function anywhere, removing its dependency on the DATA step. This is demonstrated in the following %GOBIG macro, which now dynamically creates UPCASE statements rather than performing a DATA step.

```
* dynamically changes all character variables in a data set
   to upper case;
%macro gobig(dsn= /* old data set in LIB.DSN or DSN format */);
%local i;
%findvars(dsn=&dsn, type=CHAR);
%let i=1;
%do %while(%length(%scan(&varlist,&i,,S))>1);
   %scan(&varlist,&i,,S)=upcase(%scan(&varlist,&i,,S));;
   %let i=%eval(&i+1);
   %end;
%mend;

data sample;
   length char1 $20 char2 $20 num1 8 num2 8;
```

```
   char1="I love SAS";
   char2="SAS loves me";
run;

data uppersample;
   set sample;
   %gobig(dsn=sample);
   num1=99;
run;
```

This revised code illustrates a much more flexible solution that supports reusability principles. Because of their versatility, both the %GOBIG and %FINDVARS macros could be saved as SAS program files and included within reuse libraries for subsequent use in software products. In a production environment, exception handling would likely be included to support more robust execution, as demonstrated in the "Exception Handling Framework" section in Chapter 6, "Robustness."

Readability

Software readability improves reusability because software that is easily understood is more likely to be implemented. Where SAS practitioners are confused about the intent, logic, or vulnerabilities of syntax, they will be less likely to reuse it and more likely to find or develop another solution. As code is made more modular, software inherently becomes smaller and more comprehensible. However, as modules are updated, simulated by the successive modifications made to the %GOBIG macro in the "Modularity" section, software comments should also be updated. Note that the last two instances of the %GOBIG macro contained identical headers, yet the first instance performed a DATA step while the second only generated dynamic code—two entirely different outcomes unfortunately having identical comments:

```
* dynamically changes all character variables in a data set to upper
  case;
```

Readability should be improved with additional comments that clarify the way in which the two UPCASE functions are applied: uncoupled or coupled to a DATA step. Furthermore, as development and testing reveal vulnerabilities in software modules, these should be articulated in the code as well so that those seeking to reuse it can gain insight into its anticipated quality. Readability can also be improved through automated parsers that ingest standardized comments (such as program headers) within software, as described earlier in the "Reuse Library" section and as demonstrated in the "Automated Comment Extraction" section in Chapter 15, "Readability." When in doubt, it is always better to include no software comments than to include outmoded or incorrect comments that decrease software readability by confusing stakeholders.

Testability

Software testing is intended to demonstrate that software meets established customer needs and requirements. Moreover, it can expose defects and vulnerabilities, allowing developers to decide whether to eliminate or mitigate risk of software failure. Software that is thoroughly tested and that has been documented through a formalized test plan is more reusable because SAS practitioners who seek to use the software will understand the degree of quality to which it was written as well as threats to its functionality or performance.

This is not to say that all or any vulnerabilities must be resolved, only that their identification furthers software reuse. For example, within the %FINDVARS macro in the "Modularity" section, the OPEN statement will fail if the data set is exclusively locked by another user or SAS session. To eliminate the risk of failure due to a locked data set, the lock status would need to be tested, passed through a return code to the %GOBIG macro, and in turn passed to the parent DATA step, thus requiring an extensive exception handling framework to detect and eliminate this single threat.

If the original requirements that led developers to build the %FINDVARS module greatly prioritized function over performance, thus demanding neither reliability nor robustness, this additional exception handling would not be warranted because it would provide no added business value. For example, if the code were intended to be run by end-user developers who could quickly remedy a common failure (such as a locked data set), then this increased robustness would be wasted. Regardless of the degree of *performance* required by software, however, positive and negative test cases could still be created to validate module *functionality*.

For example, because the %FINDVARS macro is now functionally discrete, a unit test can validate the TYPE parameter when set to CHAR. The following positive unit test validates partial functionality of %FINDVARS:

```
* TEST FINDVARS: test CHAR=TYPE one variable;
data test;
   length char1 $10;
   char1='taco';
```

```
run;
%findvars(dsn=test, type=char);
%put FINDVARS: char1 --> &varlist;
```

The output demonstrates that the variable CHAR1 was discovered as expected and, even if the macro were later refactored to improve performance or repurposed to provide additional functionality, the same test case could be reused to show consistent, correct functionality. In more robust macros that include exception handling—for example, quality controls that validate macro parameters—negative unit tests should also be incorporated. A negative test might invoke the TYPE parameter with the invalid value "other" to evoke a return code demonstrating the failure. Unit and other tests are discussed throughout Chapter 16, "Testability."

Stability

I once had a team of developers waiting for a SAS module I was completing so that it could be included in their respective programs. We were collectively developing a comprehensive ETL infrastructure to process vast amounts of data, and my module was a self-contained quality control that would facilitate higher data quality. Because of the diversity of data being processed, the module was being built to be used by multiple programs. With software reuse imminent, I was aware of and focused on reusability principles, including software testing and testability.

I finished the code and did some preliminary testing, but in the interest of expediting the delivery of the code to coworkers, I checked it into our central repository before all testing was complete. Thereafter, I continued to test the module and of course was forced to make a substantive change—the generation of an additional return code. When I replaced the preliminary code with the thoroughly tested code, my team wasn't pleased because my subtle modification spurred subsequent modifications that each developer had to make in his respective software.

Code stability is essential where software reuse is likely or intended. Whenever possible, modifications to code that has been reused should ensure backward compatibility so that existing parent processes that rely on a software module will not have to be modified when a child process requires maintenance. Without code stability, modifications will undermine software reuse and reusability in software—SAS practitioners will be leery of reusing centralized code modules for justifiable fear of their instability. Moreover, if reuse is not supported within an environment, reusability principles may add little value to software.

Stability is increased when formal SDLC procedures are in place. In my premature release of code, all software testing was ad hoc, as no formalized test plan or test cases existed. Had a formalized test plan been established, accepted by stakeholders during software planning, and included in requirements specifications, I would have had a concrete enumeration of boxes to check before even considering peer review or release of the code into our central repository. Without these quality controls in place, however, it's much easier for even well-meaning SAS practitioners to circumvent best practices and hastily deliver a solution that may lack sufficient quality and require later (or even imminent) modification that thwarts stability.

FROM REUSABILITY TO EXTENSIBILITY

Software reuse is often epitomized as whole software modules that can be plucked effortlessly from a reuse library or other repository and dropped into (or referenced by) software with no modification to the code. In fact, where the %INCLUDE statement or SAS Autocall Macro Facility are utilized to reference SAS program files or macros, this reuse does only require a pointer to the software module and demonstrates nearly effortless reuse. This archetypal reuse also provides security because the module cannot be accidentally modified when it's maintained centrally in a secure location, thus the integrity provided by testing is conferred upon the reuse in later software.

In other cases, however, *reuse* depicts pillaging and plundering, as existing code is picked apart and only partially absorbed into new software. While this still constitutes code reuse, it more commonly is distinguished as software *repurposing* where functionality is modified, advanced, or extended. For example, consider an initial %FINDVARS macro intended solely to retrieve the list of variables in a data set:

```
%macro findvars(dsn= /* data set in LIB.DSN or DSN format */);
%local dsid;
%local vars;
%local vartype;
%global varlist;
%let varlist=;
%local i;
%let dsid=%sysfunc(open(&dsn,i));
```

```
%let vars=%sysfunc(attrn(&dsid, nvars));
%do i=1 %to &vars;
   %let vartype=%sysfunc(vartype(&dsid,&i));
   %let varlist=&varlist %sysfunc(varname(&dsid,&i));
   %end;
%let close=%sysfunc(close(&dsid));
%mend;
```

When executed, the %FINDVARS macro displays all variables in a parameterized data set.

```
data mydata;
   length char1 $10 char2 $10 num1 8;
run;

%findvars(dsn=mydata);
%put VARLIST: &varlist;
VARLIST: char1 char2 num1
```

However, if a subsequent software product requires similar functionality but also needs to differentiate between character and numeric variable types, a solution could be developed that repurposes the module while allowing backward compatibility to existing uses. The repurposed %FINDVARS module is demonstrated in the "Modularity" section earlier in the chapter. Because the default value for the TYPE parameter is ALL, the updated module can be invoked with or without the TYPE parameter. Repurposing software while ensuring backward compatibility facilitates maintainability, because only one version of the software needs to be maintained.

The ability to repurpose software for added functionality describes software extensibility, a component of software reuse often described as a separate performance characteristic. Because of the modular, structured, dynamic nature of the original %FINDVARS macro in this section, it was easily transformed with a couple of changes to deliver the additional functionality. In the next sections, extensibility is described further, including methods that can facilitate divergent reuse of software.

Defining Extensibility

Extensibility is defined by the ISO as a synonym of *extendibility*, "the ease with which a system or component can be modified to increase its storage or functional capacity."[4] Only the second use of extensibility is discussed, demonstrating the repurposing of software to extend functionality. Thus, extensibility builds upon an existing code base to meet new or shifting needs and requirements. In some cases, additional functionality can be incorporated into existing software when backward compatibility can be achieved. In other cases, extensibility represents not only reuse of some existent code but also a further divergence from that code to produce additional functionality.

Extensibility can be viewed as software dynamism with an eye to the future. Code reusability principles aim to make software flexible so it can deliver the same functionality in multiple situations or environments and, relative to data analytic environments, to diverse and possibly unpredictable data injects. Code extension, rather, delivers new or variable functionality to meet new objectives or functional requirements. Thus, extensibility principles aim to create software nimble enough to be repurposed through minimal or sometimes no change to the existing code base.

Facilitating Extensibility

Extensibility doesn't require you to predict how software will be repurposed in the future, but imagination does benefit this endeavor. Since extensibility is often difficult to pin down, an analogy captures the principle motivation. A friend of mine built a three-story townhome a couple of years ago, and it was exciting to watch him work through the design process as each week he made choices about flooring, lighting, cabinetry, and the general design of the house. In the basement, the builders gave him the choice of installing an optional bathroom, an optional kitchenette, an optional master suite, or a default "recreation room" that was an undefined rectangular space.

He chose the rec room because it was the most extensible option. Despite its simplicity, the rec room had been designed so that homeowners could easily customize or upgrade the room in the future. In one closet, supply and drainage plumbing had been run where a toilet and shower could eventually be installed. On a separate wall, 30-amp electric service had been installed, just in case a washer/dryer unit was relocated from the first floor to the basement. And, in one corner, gas, plumbing, and electrical connections could someday service a kitchenette.

In fact, without substantial construction, a bathroom, bedroom, and kitchen could be added so that the entire basement could be inhabited or rented as a separate dwelling. The builders had designed and built with extensibility in mind, employing a little extra cost and labor up front so that the homeowners could more easily modify and repurpose the space in the future. Rather than having to tear up floors to run PVC or walls to run gas and electric lines, tremendous future functionality—including functionality imagined yet not selected or prescribed—could be added with minimal effort.

In this scenario, the homebuilders had an advantage because they not only knew what common upgrades homeowners were likely to request but also had already drawn up architectural designs for several options. In software development, SAS practitioners might not know the specific scope or extent of all future functionality or performance of software, but typically have the experience to be able to anticipate ways in which software might be repurposed or extended in the future.

For example, if a rudimentary ETL process is designed but does not include a quality control module to clean ingested data, this represents a likely candidate for future *functional* extension if the process becomes more critical over time. Similarly, if a rudimentary child process doesn't initially require robustness, the creation of a placeholder return code—one that initializes but doesn't assign or validate the global macro variable—could facilitate future *performance* extension. In either case, the subtle hint of possible future functionality or performance provides some basic substrate upon which later improvements can be constructed.

WHAT'S NEXT?

Identification, valuation, and implementation of software quality.

Quality Identification. Multiple dimensions of software quality, representing both dynamic and static performance characteristics, have been described and demonstrated. An awareness and understanding of these respective dimensions is the critical first step toward advancing software quality; the next step is identification. Understanding the software quality landscape and nomenclature facilitates the identification and differentiation of software performance, providing a structured quality model—a lens through which to assess and measure software quality characteristics. With this knowledge—including critical yet complex pairings such as reliability and robustness, speed and efficiency, and modular and monolithic design—stakeholders can identify the degree to which software performance characteristics have been included or excluded, not only in software but also in software requirements.

Quality Valuation. The value of software quality is never assessed in a vacuum, but rather against competing priorities of project scope (including functionality), schedule, and cost. Especially in data analytic development environments, software quality must also be weighed against data quality and data product quality. Whereas the identification of software performance leans toward objectivity, asking the question *Is this software reliable?*, the valuation of software performance asks the more subjective (and substantive) question: *How reliable should this software be made?* Valuation forces stakeholders to acknowledge the benefits of software performance inclusion and the risks of software performance exclusion, and to prescribe realistic technical requirements that specify what functional and performance objectives software should achieve.

Quality Implementation. You're aware of software quality. You can identify dimensions of software quality. You've prioritized software quality—functionality and performance—into software requirements. Implementation of quality is the final step—that is, software development that achieves and measurably demonstrates functional and performance objectives. While the focus of this text has been on demonstrating a software product quality model to identify and advance these objectives, brief forays have introduced software development best practices (such as Agile development methodologies, risk management, and the SDLC) and software development artifacts (such as risk registers, failure logs, reuse libraries, reuse catalogs, test plans, and test data) that can further facilitate achieving software quality objectives. Software quality can and should be allowed to flourish in any environment.

NOTES

1. ISO/IEC 25010:2011. *Systems and software engineering — Systems and software Quality Requirements and Evaluation (SQuaRE)—System and software quality models*. Geneva, Switzerland: International Organization for Standardization and Institute of Electrical and Electronics Engineers.
2. IEEE Std 1517-2010. *IEEE standard for information technology—System and software life cycle processes—Reuse processes*. Geneva, Switzerland: Institute of Electrical and Electronics Engineers.
3. Id.
4. ISO/IEC/IEEE 24765:2010. *Systems and software engineering—Vocabulary*. Geneva, Switzerland: International Organization for Standardization, International Electrotechnical Commission, and Institute of Electrical and Electronics Engineers.

Index